T0271958

ROUTLEDGE LIBRARY EDITIONS:
ROMANTICISM

Volume 20

SCIENCE IN THE ROMANTIC ERA

SCIENCE IN THE ROMANTIC ERA

DAVID M. KNIGHT

Routledge
Taylor & Francis Group

LONDON AND NEW YORK

First published in 1998 by Ashgate Publishing Limited

This edition first published in 2016
by Routledge
2 Park Square, Milton Park, Abingdon, Oxon OX14 4RN

and by Routledge
711 Third Avenue, New York, NY 10017

Routledge is an imprint of the Taylor & Francis Group, an informa business

British Library Cataloguing in Publication Data
A catalogue record for this book is available from the British Library

ISBN: 978-1-138-64537-0 (Set)
ISBN: 978-1-315-62815-8 (Set) (ebk)
ISBN: 978-1-138-64444-1 (Volume 20) (hbk)
ISBN: 978-1-315-62882-0 (Volume 20) (ebk)

Publisher's Note
The publisher has gone to great lengths to ensure the quality of this reprint but points out that some imperfections in the original copies may be apparent.

Disclaimer
The publisher has made every effort to trace copyright holders and would welcome correspondence from those they have been unable to trace.

David M. Knight

Science in the
Romantic Era

Ashgate
VARIORUM

Aldershot · Brookfield USA · Singapore · Sydney

This edition copyright © 1998 by David M. Knight

Published in the Variorum Collected Studies Series by

Ashgate Publishing Limited
Gower House, Croft Road,
Aldershot, Hampshire GU11 3HR
Great Britain

Ashgate Publishing Company
Old Post Road,
Brookfield, Vermont 05036–9704
USA

ISBN 0–86078–693–5

British Library CIP Data
Knight, David M.
    Science in the Romantic Era. (Variorum Collected Studies Series CS615).
    1. Science – History – 19th Century. I. Title
    509'.034

US Library of Congress CIP Data
Knight, David M.
    Science in the Romantic Era / David M. Knight.
          p.   cm. – (Variorum Collected Studies Series: CS615)
    1. Science – History – 18th Century. 2. Science – History – 19th Century.
    3. Science – History – Sources. I. Title. II. Series: Variorum Collected
    Studies series: CS615.
    Q125.K569 1998   98–14851
    509'.034–dc21   CIP

The paper used in this publication meets the minimum requirements of the
    American National Standard for Information Sciences – Permanence of Paper for
    Printed Library Materials, ANSI Z39.48–1984. ∞ ™

Printed by Galliard (Printers) Ltd, Great Yarmouth, Norfolk, Great Britain

VARIORUM COLLECTED STUDIES SERIES CS615

CONTENTS

This volume contains xii + 352 pages

FOREWORD

It is a well-known fact that, like claret, historians of science improve and mellow with the passage of time. Returning to papers written in the vigour of youth is nevertheless a curious and chastening experience. S.T.Coleridge was advised at school to go through whatever he had written, and when he came upon anything he thought especially fine, to strike it out. His teacher was clearly right, and sometimes a fine phrase insinuated into an old paper may here trip up the reader, as it did this one. Looking back over more than thirty years makes one smile indulgently at brash remarks, or sometimes wince at overconfidence; and then sometimes feel surprise at the perceptiveness and sharp wittedness of the younger me. It was said of Joseph Priestley and of William Paley that they never wrote an obscure sentence: how splendid it would be to be confident of doing the same.

Pierre Duhem said that there are two kinds of mind: the broad and shallow, and the deep and penetrating. There is no doubt that mine is in the former category. Some of these papers were written because after reading or hearing something, I had a great urge to say my piece; but many of them were commissioned. What a joy it is to be asked to do something, if one is basically rather indolent: pulling things together; responding to a deadline; negotiating (like Christopher Wren over St Paul's) so that the editor, organizer or publisher has to accept one's strengths and limitations; getting it reasonably right on the night.

My first degree was in chemistry, where I retained my amateur status by never wearing a white coat; and after writing a thesis under Alistair Crombie, I moved into a post in history of science in Durham University's department of philosophy, and wide ranging teaching. Teaching and research go happily together in a two-way interaction; and I never had to fix too firmly upon a period or a science. Like a cuckoo, I laid my eggs in other birds' nests. It has been of great value to me to play a part in the British Society for the History of Science, the Society for the History of Alchemy and Chemistry, the Society for the History of Natural History, the Royal Institution Centre for the History of Science and Technology, the British Association, the European Science Foundation programme on the Evolution of Chemistry, and other such things which have provided an enormous stimulus, and also yielded many friends. Over the years, a magpie as well as a cuckoo, I have picked up a great deal from conversations, lectures and papers; and I hope that I have paid my intellectual debts in footnotes, if not always in letters.

As an editor, I sought to persuade authors that they must engage with each other: a paper without references to very recently published work lacks context; historians of science show that they are professionals this way, as opposed to amateurs who cite only primary sources. And in writing, I have tried to keep up to date, and fashion conscious. Those of us who have broad, shallow, Baconian minds try to connect; and surprise our more literary and philosophical, or sociological, colleagues by not being especially interested in what in these disciplines is called theory. Science certainly consists of facts ordered by theory: but such theory makes possible definite explanations and predictions. For the historian, it is particulars which are interesting; and, as Goethe wished scientists would, one looks for the Ur-phenomenon, the shining instance, the representative case.

What brought me into the history of science was an interest in what made people become scientists: after all, I was a lapsed chemist; and doing science is not an obvious way of spending a life. Indeed, it may mean retaining a juvenile curiosity at an age when most people have gone on to adult activities, seeking sex, power and money. I focused upon Humphry Davy because Crombie, and then Sir Harold Hartley, advised me to do so: and then wrote a thesis on atomic theory in the nineteenth century because I had been surprised to find how poorly it had been received among chemists (including Davy) when John Dalton proposed it. For Davy, chemistry was a vehicle for spectacular social mobility: a poor boy from Cornwall became President of the Royal Society and famous throughout Europe, just as his contemporary Thomas Lawrence became President of the Royal Academy – Lawrence's swagger portrait of Davy is particularly splendid. Davy cared about words, knew Coleridge, William Wordsworth, Robert Southey and Walter Scott: there were no two cultures in his day, which makes the Romantic period especially rewarding to the broad-minded historian.

Davy was also one of the first professional men of science; until he married a wealthy widow and could take early retirement, he lived by his science at a time when others had another profession (often in medicine, the army or navy, or the churches) to keep their heads above water. And then his safety lamp of 1815 for miners was one of the very first examples of 'applied science' where an eminent Londoner devised in the laboratory an instrument which worked down the mine far away in the north. This brought immense prestige to science; earlier industrial improvements (and many later ones, down to our own day) were the fruit of organized common sense rather than following on from recent scientific discoveries.

Thirty years ago, it was the 'scientific revolution' of the seventeenth century which attracted most attention; and because Bacon, Descartes, Galileo and Newton were all deeply involved in philosophical questions historians and philosophers of science had a lot to say to each other: indeed, they felt they were pursuing one subject, by complementary methods. As the new generation, we enfants terribles, began to look hard at the 'chemical revolution' of 1789 and the early nineteenth century which ushered in the age of science, social history rather than philosophy looked more promising. The scientists (and that new word was beginning to be used) of the nineteenth century were the heirs not only of the Enlightenment but also of the Romantic movement: they did not simply believe the world was an enormous clock, and all of us little clocks; they sought unity and aesthetic satisfaction in science, as well as hoping to improve the world; they saw themselves as natural philosophers with a message for the world.

In these papers, reprinted just as they appeared even when for clarity a few have been reset, we begin with reflections on the history of science, its place in the intellectual world, and its public. On the whole, like the centipede, I prefer to get on with the job rather than reflect too much about how it is possible. Examples are usually better than precepts, but sometimes it is good to think about what we are doing, or have been doing. Then come some writings about Romantic science, seen as characteristic of a time rather than of a cast of mind; and trying to focus also on the threat which the French Revolution posed to religion and social order: for behind it science seemed to lurk, in the writings of Diderot and Voltaire. Some delighted in science for just this reason; but for those like Davy moving into, and funded by, high society, it had to be made respectable again, a suitable activity for patriotic Britons.

It is an easy move into looking at institutions and careers, in Britain about 1800: and here the dominant figure is Sir Joseph Banks, President of the Royal Society from 1778 to 1820 – Davy, his successor, was born a few days after Banks took office. Banks saw the scientific establishment through the difficult years of the Revolutionary and Napoleonic wars, which made peace seem almost unnatural for a whole generation. He ruled as an autocrat, or perhaps a philosopher-king, over his learned empire, and was tireless in promoting his vision of science, based upon his own work on James Cook's first voyage, where he had botanized at what was duly christened Botany Bay. But he had little sympathy for the abstruse and recondite reasonings of physical scientists; and was thus delighted by Davy's safety lamp – its political importance he at once saw, writing a magnificent letter to Davy about it.

The language of chemistry with Lavoisier moved from metaphor towards algebra: so children learning science are taught to write with the authority of the passive voice and the abstract noun, and when they come to university we have to teach them to write plain English over again. The role of pictures, symbols and diagrams in natural history and in chemistry is a fascinating study. They make up a kind of visual language, in which it is possible to make complex theoretical points, express hierarchies and taxonomies, and even tell jokes. One picture may be worth a thousand words; certainly the compressed language of modern sciences depends upon symbols and diagrams to convey a world view, and perhaps to mystify outsiders. Blinding or dazzling with science does happen.

Language is the distinctive human attribute, and especially in the days before specialization – the Romantic period – men of science had to express clearly what they had done in order for anybody to understand them. In universities, students with examinations to pass had often to submit to tedious lectures going steadily through a syllabus. But elsewhere, in Davy and Faraday's Royal Institution, in Literary and Philosophical Societies, and in Academies, professing was a performance art. Attracting audiences whose subscriptions would pay for the research laboratory was essential. Popular lecturing or writing is nowadays rather despised within the scientific community; it was not so in the nineteenth century, where the reputation of John Tyndall and T.H. Huxley for example was enhanced by it, as Justus Liebig's was by his popular and accessible writings.

We conclude with some papers taking forward some Romantic themes into the latter years of the nineteenth century; concerned especially with science and world views, still in days before there was anything like 'two cultures'. Indeed, if there are separate cultures in modern academe, it seems very odd to suppose that there are merely two: there are well-policed frontiers between some sciences, and also within humanities and social sciences. Much science was conveyed to Victorians in the Reviews, coming out quarterly and then, as the pace of life picked up, monthly; and William Crookes launched what was in effect a quarterly review of science, from which we can learn a good deal. He was also a spiritualist; and psychical research, the attempt to take the methods of science into quite another domain, is a fascinating topic for the historian of nineteenth-century science. It attracted many eminent people, and should not be written off as pseudo-science but investigated for the light which these dark seances cast upon late Victorian thinking. And we end with A.J.Balfour, suggesting that all science (indeed all thought) rests upon belief: an idea that became timely in the revolution in physics that we associate with Planck and Einstein.

These papers represent a selection, some easy and some hard to find elsewhere. I hope that they pick up a fair range of my interests, and that they will arouse those of readers. They show I hope development, and some changes of mind; and perhaps may help to indicate intellectual fashions too. I hope that I have become more sensitive to contexts, better-read, more able to pull things together, and more open and entertaining with the passage of years; but maybe I was wiser when younger. If they stimulate anyone to fresh engagement with the history of science, then I shall be delighted.

Durham 1997   DAVID KNIGHT

ACKNOWLEDGEMENTS

Grateful acknowledgement is made to the following persons, editors, publishers, institutions and journals for their kind permission to reprint the articles included in this volume: *Zeitschrift für allgemeine Wissenschafts-theorie*, Wiesbaden (I); *British Journal for the History of Science*, London (II, XXIII); *Studies in Romanticism*, The Graduate School and the Trustees of Boston University (III), Cambridge University Press (IV, XXIII); Professor W.H. Brock on behalf of *Ambix* (V, XIX); *The History of Science*, Cambridge (VI); *Durham University Journal* (VII, XXV, XXVI); *Actes du Colloque Gay-Lussac* and the École Politechnique, Paris (VIII); Professor Robert Halleux on behalf of the International Union of History and Philosophy of Science (IX); *Annals of Science*, London (X, XIII); Kluwer Academic Publishing, Dodrecht (XI); the Royal Botanic Gardens Kew (XII); *Acta Universitatis Upsaliensis* and the Uppsala University Library (XIV); Martinus Nyhoff Publishers, Dodrecht (XV); *Archives of Natural History* and the Society for the History of Natural History (XVI); J. G. Cotta'sche Buchhandlung Nachfolger GmbH, Stuttgart (XVII); Biblioteca di Nuncius Studi e Testi and Leo Olschki (XVIII, XXIV); Dr Antonella Grandolini on behalf of Accademia Nazionale Delle Scienze Dei XL, Rome (XX).

I would like to thank Mrs Kathleen Nattrass, who retyped the papers onto a disk for reprinting; and Dr David Mossley who reset papers from disks and prepared camera-ready copy, and who was also responsible for the index.

I

The History of Science in Britain: A Personal View

Summary

Historians of science in Britain lack a firm institutional base. They are to be found scattered around in various departments in universities, polytechnics and museums. Their history over the last thirty-five years can be seen as a series of flirtations with those in more-established disciplines. Beginning with scientists, they then turned to philosophers, moving on to historians and then to sociologists: from each of these affairs something was learned, and the current interest determined which aspects of the history of science were seen as most interesting. At first it was settling who really discovered what; then an interest in concepts, methods and case-studies; then understanding the broader historical context of science; and after that seeing science in its social context, with special emphasis on institutions and professionalization. Where we shall go next is unclear: these vagaries may be no more than examples of intellectual fashion, but we may hope that they represent a zig-zag route towards deeper understanding.

Whereas Church History has an established institutional base usually in Theology Departments of British Universities, the History of Science remains marginal. Scientists seem less happy to take seriously the development of doctrine than are churchmen; though there are some who are interested in seeing how the spirit of truth has worked itself out through history. Among scientists, it is generally the development of theories now important which is attractive; they want to see how they have got to where they are now.[1] On this side of the German Ocean, this vision of history as more-or-less rapid progress towards the present is called 'whig history': after those like Macaulay in the last century who saw the reformed and liberal state of their day as the end to which the whole creation has been moving. Such ideas can make the more remote past seem very obscure, peopled with heroes and villains of superhuman proportions; while for contemporary history it can yield an anecdotal approach which is closer to popular science than to true history of science, seen as a critical discipline.

Scientists then perhaps tend to look into history for what is like the present, like geologists in the tradition of Charles Lyell and the so-called uniformitarians; while historians of science are more like Darwinians, seeing development over time leading to real differences. This means that relations between the groups are uneasy. Many scientists find history irrelevant to their activities, and fear that students will be muddled by being introduced to obsolete

---

[1] A recent textbook in this genre is J. Marks, *Science and the making of the Modern World*, London, 1983; a distinguished monograph might be C. A. Russell, *The History of Valency*, Leicester, 1971.

science; they would rather have them spend time on a microcomputer. Historians of science, on the other hand, find themselves tempted by a Kuhnian notion of science: in which it becomes a matter of 'paradigms' arrived at more-or-less intuitively by occasional men of genius, and taught dogmatically by lesser folk who practice 'normal science' within which they solve puzzles.[2] To look from one paradigm into another requires imagination and perhaps a steady nerve; it is easier to suppose that one is right and one's predecessors were ignorant, even if this may seem a little defensive. Historians of science also grow tired of meeting the assumption that they must be interested in how things turned out: as though the student of the Austro-Hungarian Empire should necessarily be concerned about local government in present-day Budapest.

Just as in the corrosive climate of the eighteenth century the history of the Church came to seem less then always edifying, so in the later twentieth century the reputation of science has become ambiguous. To T. H. Huxley a hundred years ago, science was unlike religion because it had never done anybody any harm. Whiggish historians saw progress in science that they failed to find in philosophy; the triumphs of the wave theory of light, and its extensions by Maxwell, Hertz and Röntgen seemed for example shining instances of real movement towards truth. The historian of science of the nineteenth century could feel that he was portraying an heroic enterprise, a search for truth above the petty nationalisms and cruelties of the ordinary world. Such episodes as the award to Sir Humphry Davy, by the Parisian Institut, of a prize for his electrochemical work in the middle of the Napoleonic Wars, were lovingly dwelt upon.

This story probably shows how unimportant science was in the early years of the nineteenth century: for with the World Wars of our century came the expulsion of enemy aliens from scientific societies and academies, the rise in secret government-supported research, and the development of chemical and nuclear weapons. The nineteenth century had produced terrible pollution, but science seemed a force that could cure it rather than its cause. The historian of science is dealing with something undoubtedly important and interesting, but not with something which transcends its cultural background like a good deed in a naughty world. Except perhaps in Faculties of Science and in the government Department of Education and Science, one meets alarm and unease about science, and there is no reason why its historian should also be its apologist.

For these reasons, historians of science are not often found in Britain in science departments in Universities and Polytechnics; and scientists there who are interested in the history of their field can feel lonely, and get the message that antiquarian researches should only be pursued after retirement from active science. Nevertheless, most historians of science have begun as scientists, at least to first degree level: and this arises from certain features of the cultural

[2] T. S. Kuhn, *The Structure of Scientific Revolutions*, 2nd ed., Chicago, 1970. On science and culture, see M. Pollock (ed.) *Common denominators in Art and Science*, Aberdeen, 1983.

and educational system in Britain. During the nineteenth century, *par excellence* the scientific age in which the authority of religion and literature gave place to that of science, the word 'science' came to be used in a narrower and narrower sense. In the first half of the century it was set against merely practical or narrowly professional knowledge, or rule of thumb.[3] Thus it was possible to speak of medical science, of the scientific parts of theology, and of cudgelling scientifically performed. Chemistry could be applied to the 'arts' – that is, in practical matters – but it was a science insofar as it was 'philosophical': that is, having a structure, a method, a collection of facts, and some theory which fitted them into a framework. The word 'science' could therefore include almost any academic discipline.

Whereas in Scotland there were universities which taught a fair range of academic disciplines, at Oxford and Cambridge only classics and mathematics were systematically taught. There were Professors to lecture to any enthusiasts on Chemistry, Botany, Modern History or Modern Languages, but they cut little ice in the university and attending their courses did not count towards a degree. With the founding of the British Association for the Advancement of Science in 1831, and the coining of the word 'scientist' at the third meeting, two years later, the word 'science' came to stand for those activities carried on at the BAAS. These included Geology, and Statistics if mathematical rather than political; but although in 1800 many Fellows of the Royal Society had also been Members of the Society of Antiquaries, by 1900 historians and archaelogists were not deemed scientists. To this day, geographers and psychologists aspire (generally without much hope) to scientific status; but the lines drawn in the last century seem to hold.

This means that while the historian of science does not feel at home among the scientists, there is also no very obvious place for him among intellectual historians. The rigid separation of science from other activities has meant that the history of literature, religion, philosophy, and art has been studied separately from that of science; which may be illuminating when looking at the history of physics in positivistic periods but is generally very unhelpful. 'Science' used to mean something like 'Wissenschaft', for which there is now no obvious translation: and although the useful phrase 'climate of opinion' was invented by Joseph Glanvill, a Fellow of the Royal Society contemporary with Boyle, Geisteswissenschaft has no English equivalent. Though certain Polytechnics do now lay on interdisciplinary courses on History of Ideas which may include history of science, there are no Departmens of Intellectual History in which an historian of science should find a place.

Indeed the historian of science over the last twenty years begins to look a little like the 'wallflower' at the Ball, making eyes at various handsome – or even uprepossessing – suitors, but all in vain. Scientists failed to take much interest even in technical and internal histories. Worse, the teaching of science in schools and universities began to be increasingly separated from its history.

---

[3] For terms and concepts in the history of science, see W. Bynum et al, ed., *Dictionary of the History of Science*, London, 1981; though one looks in vain for 'pollution', 'poison' or 'weapon'.

346

Chemistry in particular had been taught historically, and also to some extent heuristically: the student was encouraged to put himself in the position of Boyle, Priestley, Lavoisier, Dalton, Wöhler, Fischer and others, often attaching great names to reactions or techniques. This was hardly critical history, for those whose labours had not turned out positively, however eminent they were in their own day, were omitted; but it did give some idea of a discipline developing over time. From the 1960s new syllabuses in which matters were arranged logically rather then following the accidents of history became the vogue. It is ironical that since then chemistry has become the least popular of sciences among students in Britain: its presentation as a logical but dull and inhuman activity may have had something to do with this.

In the last ten years there has been more historical activity within the Royal Society of Chemistry, which now has a Historical Group arranging lectures at the Annual Congresses and also organising other meetings. But the numbers involved are small, and the prospects of much support for history of science from scientists seem poor: for them it remains a marginal activity. The next group with which historians of science therefore tried to ally themselves were the philosophers, who had since the time of Descartes (if not since Antiquity) been dazzled with the possibilities of scientific explanation. The ideas of the Vienna Circle had by the 1930s reached Britain, and the principle of verification – 'if a proposition cannot be empirically verified, then it is meaningless' – had been used to demolish the Hegelians and Idealists. Philosophers however tend to be like scientists in that they quarry the past, looking for nuggets to prise out of their matrix rather then seriously working out how things were.

Out of the alliance with philosophers came the idea that the history and the philosophy of science were really one subject. Unless anchored in history, with genuine case-studies, philosophy of science would be abstract and remote, a kind of intellectual game; while without some philosophy of science, the history of science would become mere erudition: the two joined together would form a really exciting and worthwhile activity.[4] In various universities, including Cambridge, Leeds, and Durham, History and Philosophy of Science entered the syllabus and remains there; and many of those studying history of science willy-nilly study philosophy of science also.

Like scientists, philosophers are keenest on internal history. Dragging external factors – economic, social or institutional – into intellectual history seems reductive. What is really wanted is a history of scientific thought, and only the crucial experiments which forced choice between theories need be investigated in any detail: while philosophy of science had affinities with logic rather than with ethics, so that the stress was all upon method and explanation rather than upon the application of science. The slightly arid history which can

---

[4] See M. Hesse, *Forces and Fields*, London, 1961, for a good example of this genre; A. R. Hallam, *A Revolution in the Earth Sciences*, Oxford 1973, applies Kuhnian ideas to recent history. On ethics in science, see J. Ravetz, *Scientific Knowledge, and its Social Problems*, London, 1971.

go with this narrowing of focus had a long and respectable ancestry in Britain. William Whewell, an admirer of German philosophy and architecture, wrote his *Philosophy of the Inductive Sciences* in 1840 as a systematic follow-up to his *History of the Inductive Sciences* of 1837. Whewell was opposed to the prevailing Baconian ideal of science, and stressed the importance of having appropriate leading ideas rather than merely collecting facts and generalising from them; and his books were written to vindicate his view historically. In view of Whewell's importance in the BAAS and at Cambridge, where he was Master of Trinity College, his writings are of considerable interest[5].

Whewell was well described by a wit of his own day: 'Science was his forte, and omniscience his foible'. But the suspicion of omniscience must hang more heavily over J. T. Merz, whose *European Thought in the Nineteenth Century* (1904–12) remains a monument to his enormous erudition. Two volumes of this work are devoted to science, in the narrow sense, and two more to philosophical thought outside the sciences. Much that has been later written on the sciences of that period can almost be seen as footnotes to Merz: except that the book has such formidable footnotes already that there is no room for more. Particularly interesting are the chapters where Merz compares the way science was carried on in France, Germany and Britain, because here he described the different educational and institutional systems which lay behind the different achievements. Other chapters cut across the divisions in the sciences, being concerned for example with the 'vitalistic' and the 'statistical' views of nature.

In more recent years it was probably the writings of A. O. Lovejoy and of Alexandre Koyré which boosted the vision of the history of science as a part of general intellectual history. An important figure in this development in Britain was A. C. Crombie at Oxford: his writings on medieval and early modern science were influential,[6] and his pupils occupy various important posts in the history of science. The strength of the discipline in the early 1960s was shown by the launching in 1962 of two new journals, which joined *Annals of Science* started twenty-five years before when history of science was an appendage of science. The British Society for History of Science was founded in 1947, and produced for some years a *Bulletin*: this with the new confidence of 1962 was transformed into the *British Journal for the History of Science*, a general journal containing original papers and a gradually-increasing number of book-reviews and occasional essay-reviews. The other journal, founded by Crombie and Michael Hoskin from Cambridge, was *History of Science*, which published essay-reviews.

Those engaged in history and philosophy of science tried to write critical history, and soon found that this was less helpful than it might have been to philosophers looking for case-studies. Many scientists seem to have been like

---

[5] Whewell features in S. F. Cannon, *Science in Culture*, Folkestone, 1978, and in J. Morrell and A. Thackray, *Gentlemen of Science*, Oxford, 1981. For bibliographies generally, see S. A. Jayawardene, *Reference Books for the Historian of Science*, London, 1982, and my *Sources for the History of Science*, Cambridge, 1975. The *British Journal for the History of Science* (*BJHS*) carries articles from time to time on collections in various centres.

[6] A. C. Crombie, *Augustine to Galileo*, London, 1952.

348

Einstein, unscrupulous opportunists who failed to follow the canons of method laid down for them by eminent philosophers. Indeed it seems to be a measure of a great scientist, as of a great writer, that he extends our notion of the range and methods available. History is also full of accidents: neither Kepler's Laws nor the Laws of Thermodynamics were found in the right order, Dalton failed to come to grips with the researches of Gay-Lussac, and Mendeleev predicted various non-existent elements as well as Scandium and Germanium. Historians found themselves getting fascinated by those who had supported the wrong side – they were secptical about atoms, or rejected evolution – and failed to show clearly how the rational activity (which 'science' must be!) goes on from strength to strength. Rather than give up belief in the rationality of science, and recognise it as human, all too human, philosophers resorted to 'rational reconstructions'[7]: these meant the creation of orderly if mythical pasts, bearing some relation to what the historians found but reconstructed to bring out what was deemed to have 'really' happened beneath the personalities, and the accidents of time and place. Historians are mostly disposed to place these efforts somewhere on that famous continuum: – lies, damned lies, and statistics.

If historical case-studies began to disappoint the philosophers because of their untidiness, they also proved less than attractive to some wanting more down-to-earth history. Something of my own in this genre was described as making science no more than an intellectual game, and as the sound of disembodied voices in a dusty library. Fully-satisfying history cannot be written on the assumption that external factors should not be invoked until internal ones have been exhausted; and in the later 1960s there was a debate between externalists and internalists which with hindsight seems just to represent two untenable positions. In fact, good historians had invoked both – like Merz – but differed in the weight they assigned to them: and this weighting is something which cannot be done by general rule but depends upon judgements about the subject matter in particular cases. Nevertheless, emphasis has generally changed over the last decade or so, so that whereas most histories emphasised the internal and the philosophical, they now stress the external: the social, economic and institutional factors. Works like Peter Harman's brief tour-de-force on nineteenth-century physics[8] have a restricted and old-fashioned air because they are in the 'philosophical' tradition.

The romance with scientists had not worked out, and neither did that with philosophers; but Herbert Butterfield, a distinguished historian notable for his attacks upon whig history had interested himself in the 'Scientific Revolution' of the seventeenth century.[9] It seemed to him that this was an event which

[7] I. Lakatos, *Philosophical Papers*, Cambridge, 1978; and on case studies, see the papers by T. H. Brooke and by N. Fisher on Avogadro's hypothesis, in *History of Science* 19 (1981) 235–73, and 20 (1982) 77–102, 212–231.

[8] My book was *Atoms and Elements*, London, 1967; P. Harman, *Energy, Force, and Matter*, Cambridge, 1978. Cf. also my *Transcendental Part of Chemistry*, Folkestone, 1978.

[9] H. Butterfield, *The Origins of Modern Science*, London 1949; an earlier work by an eminent historian was G. Clark, *Science and Social Welfare in the Age of Newton*, Oxford, 1937.

exceeded in importance the Renaissance and Reformation, those traditional discontinuities marking the beginning of Modern History. It is curious that Butterfield's book is itself a whiggish treatise: those who cannot see progress in the political history of the twentieth century can sometimes see it in the science. C. P. Snow, famous for his idea of the 'two cultures', believed that a schoolboy of the later twentieth century would know more physics then Newton – which is true but only in a banal sense.

Butterfield's book turned the attention of professional historians to the history of science; and courses on the scientific revolution began to appear, probably as an optional 'special subject', in undergraduate courses in history. Sometimes, as at the Universities of Loughborough and Lancaster, historians of science have posts in the history department. Historians, now joining those trained in a science or in philosophy, looked at different aspects of science. Indeed in many ways the different perspectives described by those with different backgrounds and affiliations is a source of great strength in the history of science in Britain, preventing the formation of any party line, or indeed any agreed set of textbooks: the tradition is always that the student must wrestle from the start with original materials, even though many of them must be in translation.

In the older traditions, the published materials of science were normally to be mastered before much value was to be expected from manuscripts. After all, if science is public knowledge,[10] what was never published was never science: and philosophers mostly agreed that a logic of discovery was a chimaera, so that one should look at published theories and explanations, rather than at notebooks or letters describing individual and unique strivings. These were materials for the 'mere' biographer.[11] To the historian, on the other hand, seeking to place the man of science in his historical context, manuscript material has an enormous appeal. Instead of going through the mathematical or experimental work published in books or journals, the historian investigating a milieu will probably be more interested in his hero's attitude to religion, his search for patronage, his income, and the circles in which he moved.

Thus with an increasingly 'historical' commitment, historians of science turned away from describing 'hard' science and began describing science as one human activity among others. Science, instead of some royal road to truth,

[10] J. Ziman, *Public Knowledge*, Cambridge, 1967; the literature of science has attracted various authors: see J. Meadows, ed., *The Development of Science Publishing in Europe*, Amsterdam, 1980; W. Blunt, *The Art of Botanical Illustration*, London, 1950; my *Zoological Illustration*, Folkestone, 1977, and *Natural Science Books in English*, London, 1972, lists author-bibliographies which are a very valuable tool for the historian of science; e.g. G. Keynes, *A Bibliography of Dr Robert Hooke*, Oxford, 1960.

[11] Notable biographies include A. J. Meadows, *Science and Controversy . . . Norman Lockyer*, London, 1972; A. E. Gunther, *A Century of Zoology*, Folkestone, 1975 (J. E. Gray and A. Günther); M. Crosland, *Gay-Lussac*, Cambridge, 1978. British historians have interested themselves in French science particularly: e. g., M. Crosland, *The Society of Arcueil*, London, 1967; W. A. Smeaton, *Fourcroy*, London, 1962; R. Fox, *The Caloric Theory of Gases*, Oxford, 1971; J. G. Smith, *The Origins and Develop ment of the Heavy Chemical Industry in France*, Oxford, 1979.

350

became a hobby or a route to fame and fortune; for many of its practitioners, especially during the period of the 'scientific revolution', it was certainly not the most important part of their lives, but a diversion in a career in politics, medicine, or the church. There is enough truth in Snow's picture of two cultures to mean that those who have specialised in history are not likely to know much science, and vice-versa; but in the study of seventeenth-century natural philosophy this need not put historians at a disadvantage. Recent studies have brought out how different Newton and his contemporaries were from modern physicists,[12] and how like they were to their contemporaries with interests in the interpretation of apocalyptical prophecies, in Christology, and in alchemy.

Baconian science has been duly examined to see if its exponents were the *philosophes* of the English Revolution of the 1640s which culminated in the execution of King Charles I; and how close the connections were between Calvinism, capitalism, and science in seventeenth-century England.[13] Here, investigation of Roman Catholic and Arminian Anglican scientific circles indicate that there was no necessary link between science and Calvinism. The early Royal Society seems chiefly to have been composed of men not especially interested in dogma; and the Reformation may have been chiefly significant for science because it led to a married clergy, whose poor but well-educated sons played a disproportionate role in science.[14] For Hanoverian England, connections between Newtonian philosophy, liberal religion, freemasonry and enlightenment have been investigated, with results that are tantalisingly suggestive.[15]

What historians have brought to the history of science is an interest in the normal, rather than the exceptional. Scientists and philosophers were not interested in the ordinary man of science, but only in those who did something exceptional. The newer interests have led to the technique of prosopography, originally devised by Ancient Historians, being applied in the history of science: as far as possible, biographies of members of a circle or a society, whether eminent or obscure, are drawn up to determine what sort of people worked in science. The historian's attention thus passes from public know-

[12] The standard Newton biography comes from America: R. Westfall, *Never at Rest*, Cambridge, 1981. See also Newton's *Correspondence,* ed. H. W. Turnbull, J. F. Scott, A. R. Hall and L. Tilling, Cambridge, 1959–78, and *Mathematical Papers*, ed. D. T. Whiteside et al., Cambridge, 1967–81. On magic etc., F. Yates, *Giordano Bruno,* London, 1964; K. Thomas, *Religion and the Decline of Magic*, London, 1971, and *Man and the Natural World*, London, 1983.

[13] C. Hill, *Intellectual Origins of the English Revolution*, Oxford, 1975; C. Webster, *The Great Instauration*, London, 1975.

[14] M. Hunter, *Science and Society in Restoration England,* Cambridge, 1981 and *The Royal Society and its Fellows, 1600–1770*, Chalfont St. Giles, 1983; these have useful bibliographies. J. Henry, 'Atomism and Eschatology: Catholicism and Natural Philosophy in the Interregnum', *BJHS*, 15 (1982) 211–39. A. R. and M. B. Hall (ed.) *The Correspondence of Henry Oldenburg,* Madison, 1965.

[15] M. C. Jacob, *The Newtonians and the English Revolution, 1689–1720,* Hassocks, 1976, and *The Radical Enlightenment,* London, 1981, and see G. C. Gibbs' essay-review, *BJHS*, 17, (1984), 67–81.

9

ledge to the public records, to parish registers of baptisms marriages and funerals, to local newspapers, and to university records. Scientists are studied as any other group – politicians or pigeon-fanciers or journalists – might be.

What such studies indicate is how science has gradually shifted from being a kind of hobby into being a career or profession, some time during the nineteenth century.[16] This question has particularly fascinated historians of science because it parallels what has been happening to them. In the 1940s and early 50s there was a part-time MSc course at University College, London, introducing various people – mostly teachers of science – to the history of science; but it was only from the late 1950s that the idea of becoming a professional historian of science could plausibly be entertained by anybody. Appointments in museums, universities and in polytechnics, undergraduate and graduate courses, and Open University programmes have gradually turned the British Society for the History of Science into a body with a core of professionals who fill its posts of responsibility and plan its meetings, surrounded by an apple of amateurs. While some of us might prefer the flavour of the apple to that of the core, and enjoy nineteenth-century science because it is pre-professional, there can be little doubt that studies of professionalisation and of scientific societies and communities are 'relevant' to historians of science in present-day Britain.

This interest was not shared by many historians; and it is also true that historians of science have not gone in for detailed historical comparisons between science and the established professions of the Church, the Law, and the armed Services in nineteenth-century Britain. While historians followed Butterfield in emphasising the Scientific Revolution, they have perhaps not always found our prosopographical studies compulsive; and have not been convinced that they ought to investigate provincial 'literary and philosophical' societies rather than high politics or social history. So historians of science have had to seek a new suitor, and have made eyes at sociologists. In the years around 1968 sociologists enjoyed high prestige, unlike historians of science: but the sociological model of scientists as marginal men seeking to make their way in a world dominated by the rich, the powerful, and the classically-educated was appealing, especially perhaps to those who themselves felt marginal.[17]

Marginal men were unable to enter the world of high culture because Horatian tags and aesthetic judgements only come easily to those who have been learning Latin and looking at works of art since childhood. Science however was open to manufacturers, sectarian ministers and surgeons, who formed the élite in the industrial cities of the early nineteenth century. Once science had helped to give them status as leading citizens, and the Reform Bill

[16] Crosland, Gay-Lussac, note 11 above; C. A. Russell, N. G. Coley, and G. K. Russell, *Chemists by Profession*, Milton Keynes, 1977.

[17] I. Inkster and J. Morrell, *Metropolis and Province*, London, 1983; social anthropology is used as a key by C. B. Wilde, 'Matter and Spirit as Natural Symbols in 18th-century British Natural Philosophy', *BJHS*, 15 (1982) 99–131, and by Thomas, *Decline*, note 12 above. I. Inkster, 'Science and the Mechanics' Institutes', *Annals of Science*, 32 (1975), 451–474.

of 1832 had given them votes, they then used it to close the door behind them: science then formed an instrument for social control. The working class became those who relied upon rule of thumb, and in 'Mechanics Institutes' could be shown by their betters how difficult real science was. Dazzled by science lecturers – and as Liebig remarked, public lectures in Britain were often very well done – and persuaded that iron laws of nature governed the social as well as the material world, workers became docile believers in self-help, and the Chartist agitation of 1848 fizzled out. This at least is the story told by sociologically-minded historians, whose sympathies are with the radicals, with the obscure whose scientific attainments are small (those like Davy and Faraday, who made their way by science, being vaguely seen as having sold out); which means that their version of history is even further from that of the scientists or philosophers.

The social history of science has been used with great effect to cast light on the origin and early years of the British Association for the Advancement of Science, where a Cambridge network took control of what had been a provincially based body.[18] It can also help to illuminate the emergence of new sciences in the nineteenth century: a specific society, a journal, and university posts being as much a sign of the appearance of, for example, physical chemistry, as any change in intellectual habits or in apparatus. But there is no doubt that this kind of approach can be reductive; it can neglect the intellectual excitement which is what makes science attractive to its most able practitioners, who have not chosen it simply as a job. When we look at empire-building, at manipulation of committees, black-balling of candidates, appointments of examiners, and so on we are only seeing a part of science: the same would be true of this kind of perspective on literature, painting or religion. We need to look at what makes science different from other activities as well as what makes it ordinary.

Perhaps therefore this prolonged and unsuccessful wooing has not all been a sad story. In British universities, historians of science are sometimes found in small departments of their own; sometimes linked to science departments; sometimes with philosophy, with history or with sociology. A diversity of perspectives is necessary to see science whole, for it is at once an intellectual, a social and a practical activity.[19] There are various specialised societies which

---

[18] Morrell and Thackray, see note 5 above; also there, Cannon, ch. 2 on the 'Cambridge Network'. R. Macleod and P. Collins, *The Parliament of Science*, London, 1981; a social explanation is also prominent in D. A. Mackenzie, *Statistics in Britain, 1865–1930*, Edinburgh, 1981, featuring Galton, on whom see R. E. Fancher, 'Francis Galton's African Ethnology', *BJHS*, 16 (1983) 67–79. C. A. Russell, *Science and Social Change*, London, 1983, produces evidence against the view that science functioned as a method of social control.

[19] This was the theme of my book, *The Nature of Science*, London, 1976. Different sciences have their different histories; compare Harman, note 8 above, with A. Desmond, *Archetypes and Ancestors*, London, 1982, and my *Ordering the World*, London, 1981. But the boundaries between the sciences have been differently drawn in different times and places. On science in various countries, see M. P. Crosland, ed., *The Emergence of Science in Western Europe*, London, 1975. Probably because French is always the first foreign language learned in Britain, German science has been less studied here than French, which is misleading for those working on the nineteenth

deal with specific histories, of chemistry, of natural history, of technology, and of medicine; there are also journals which just cover the history of chemistry, or of astronomy. All this may indicate a fragmentation, and indeed to some extent it does. It is not clear that the histories of different sciences are best separated: different periods seem a more sensible historical division. But in a world of cuts, being hydra-headed may be an advantage: certainly the History of Science remains an active and flourishing discipline.

Who knows who will be the next suitor to whom the historians of science will make their blushing advances? Perhaps it is the turn of theologians or psychologists, though the utilitarian impulse in today's educational funding might propel us towards the engineers. With each wooing comes a new perspective: it may involve a caricature, like Kuhn's normal science, or the sociological vision of science as primarily a method of social control. Nevertheless, as politicians know, a caricature is a vision of truth, distorted; and from these different visions we can build up a three-dimensional and lively picture of science. Perhaps we should change our metaphor. When Maxwell was appointed to his chair at Cambridge, the Cavendish Laboratory was not yet built for him; and he had to lecture in rooms generally used for different subjects. He described himself as a cuckoo, laying eggs in other birds' nests. Historians of science are perhaps a kind of academic cuckoo; if so, we can hope even in 1984 for a coming spring.

century: but histories of chemistry have been unable to omit Germans! See e. g. W. H. Brock, 'Liebigiana', *History of Science*, 19 (1981) 201–18; and Eric Forbes, who has worked on Mayer and other astronomers – *The Euler-Mayer Correspondence*, London, 1971.

II

Background and Foreground: Getting Things in Context

Historians generally grumble at the liberties taken with letters and papers by editors and biographers in the past, while reviewers may complain at the professorial pomposities which interfere with the reader's interaction with the text. Certainly, reading is not a mere matter of information retrieval or of source-mining, but a meeting of minds, and any over-zealous editing which makes this more difficult will have failed. Editors, whether of journals or of documents, are midwives of ideas—self-effacingly bringing an author's meaning and style into the world. What reviewers praise is the unobtrusive, and what they damn is 'a manner at once slapdash and intrusive', making allowances perhaps for an 'introduction which is as admirable as his footnotes are useless'.[1] When in the 1960s new technology brought us a flood of facsimile reprints of scientific works, some avoided these problems by appearing naked and unashamed:[2] but for a text on phrenology, or for Goethe's *Theory of Colours*, a fig leaf or two of commentary is really necessary to help the innocent reader to interact with the book. Facsimiles of nineteenth-century editions of Wilkins' papers, of some Newton correspondence, or of Henry More's poetry are even more problematic; the reader should know that these editors' assumptions cannot be taken for granted, and that their introductions are themselves historical documents. The exact reproduction of misprints and misbindings (giving pages out of order and misnumbered) is of dubious assistance to the modern reader.

The real question is, who is this modern reader of works in the history of science and technology? Who do we think we are trying to help when we prepare texts for them? We may feel that we should edit and present old scientific material because, like Mount Everest, it is there, but this is perhaps not enough to justify a life spent in the history of science.[3] It is not immediately obvious why many people should be interested in past

1 The great example of bad documentary editing is Hawkesworth's account of Cook's voyage of 1773; see J.C. Beaglehole (ed.), *The Journals of Captain James Cook*, I, Cambridge, 1955, ccxlii ff. The remarks quoted come from recent issues of the *TLS*.

2 G. Combe, *The Constitution of Man*, (1847), Farnborough, 1970; *Goethe's Theory of Colours*, (1840), London, 1967, (with new index); J. Wilkins, *Mathematical and Philosophical Works*, (1802), London, 1970; J. Edelston (ed.), *Correspondence of Sir Isaac Newton and Professor Cotes*, (1850), London, 1969; H. More, *The Complete Poems*, (ed. A. B. Grosart) (1878), Hildesheim, 1969; R. Hooke, *Posthumous Works*, (1705), London, 1971, is an uncorrected reprint, misdescribed as a second edition, where the confusions of the page numbering in the Royal Institution copy reproduced are understated on the verso of the title page.

3 See my 'History of science in Britain: a personal view', *Zeitschrift für allgemeine Wissenschaftstheorie*, (1984), **15**, pp. 343–353.

This article is reproduced with the permission of the Council of the British Society for the History of Science. It was first published in *BJHS* (1987) **20** 3–12.

science, for the sciences are notoriously a sphere of built-in obsolescence. The test of a really original piece of science, or of technology, is that it makes what went before seem hopelessly old-fashioned. To us, this means that old science cries out for careful editing; to those working in science it may mean that its history is bunk.

On the other hand, science and technology generally seem, at least to practitioners, spheres where real progress is going on: which should mean that their history is a more interesting process than the sound and fury, signifying nothing, which political history seems to be. Those whose working lives have been spent in science, medicine or technology may often be interested in making sense of them through historical study. They may even want to write on the history of science, and indeed memoirs and obituaries are an important genre in our discipline. Scientific societies are increasingly, in Britain at least, forming historical sections or groups, which hold meetings separately or at major conferences and which publish newsletters. There is here clearly a constituency which professional historians of science have not fully succeeded in claiming. Our sceptical and ironical erudition passes them by; while we may lose their insights into the everyday and humdrum activities in and around the laboratory and the seminar room, which make up the reality of the scientific life.

J. T. Merz, working a century ago on his *European Thought in the Nineteenth Century,* reckoned that somebody whose memory carried them back to the middle of the century, and whose education embodied the ideas of a generation before that time, could claim to have some personal knowledge of the whole century.[1] That indeed seems to be about the limit of first-hand knowledge; I got to know an eminent man who had begun his scientific career in the last century, and who had himself known one of the delegates at the Karlsruhe Conference of 1860: this took him back to the mid-nineteenth century. It is this span of nearly 100 years which is likely to be of most interest to those whose own lives have been spent in science and who in understanding their time will be understanding themselves. Merz added that this recent history had the great advantage that the hidden life of an age was still accessible; we could see for our own epoch how the co-operation of the many was what made the few, whose names become famous, succeed. He believed that to do justice to the unknown of remoter periods was hardly possible; since his century was the period in which the exponential growth in scientists took off, and the number of lesser thinkers was thus much greater, this was particularly important. The trouble with recent history of science, however, is that it can shade into 'general science', becoming a mere historical introduction to be associated with popularizing. There is no need for the historian of science to be its apologist in this way.

Even recent scientific texts can be problematical for the reader familiar with modern science, because words change meaning, and the reputations of people and institutions (even centres of excellence) rise and fall rapidly. Social and political factors of sixty years ago are not obvious to us all without tactful reminders, and inflation—something Merz did not have to put up with—makes prices and salaries hard to evaluate. But the real problem is that one reads back into the science of a generation or two ago the concerns of the present. Such recent science has a specious modernity, which offsets the advantages

4 J.T. Merz, *A History of European Thought in the Nineteenth Century,* London, 1904–1912 Vol. I, pp. 12, 8. My friend was Sir Harold Hartley.

to the historian which Merz perceived: the science of Paracelsus' day needs plenty of glossing, but the reader is less tempted to read modern conceptions into it than into that of Rutherford's time.

The scientist, active or retired, is likely to be impatient with what seem to be distractions or digressions on the progress towards truth, or anyway towards the present. And when reading a text of past science, he may well be most interested in what 'really happened', if some experiment or observation is in question, or how it can be set out in modern notation if it is a description in natural history or a piece of mathematics. This essentially lets us in for an exercise in translation.[5] To write a modern chemical reaction for a nineteenth-century chemical test, or a genetic analysis of some work on selective breeding, is a dangerous business; well done, it may illuminate the text, but it can just be confusing. Like a translation, it is no substitute for the real thing.

An editor may be tempted to improve on his text in this way: thus, the Harvard reprint of the first edition of *The Origin of Species*, done in facsimile,[6] had the original index omitted and, in a fit of misplaced editorial industry, replaced by one of which Ernst Mayr wrote:

> I have attempted to include primarily items of interest to the evolutionist and to students of the history of ideas. I call special attention to Darwin's views on competition and its role in evolution and as the cause of extinction, and his discussions of isolation, dispersal, population size, reproduction, isolating mechanisms, behavior, and the meaning of taxonomic characters.

The reader thus finds in the index such items as 'cladogenesis', 'genotypes' and 'polyphenism', themes which may or may not have been in Darwin's mind, and many references to 'evolution', which is a word that does not appear in the text. Editing (perhaps) helpful to the working neo-Darwinian here seems intrusive to anyone more sensitive to the context within which the book was published; there may be occasions in which it is impossible to please all the people all the time.

The Origin of Species now seems to be a scientific classic; it is so described in a new book by Derek Gjertsen,[7] which sets out to be a study of twelve enduring scientific texts. His is a curious list, in which Darwin's book is the last; it includes Ptolemy's *Almagest,* which undoubtedly had a long run but could hardly be described as an enduring classic by anybody but the most enthusiastic historian of science—indeed, Seth Ward in the middle of the seventeenth century found it already slightly embarrassing to be doing so. The idea that there are classics of science in the same way as there are classics of literature, works which the world should deeply note and long remember, is an odd one.

---

5 A.I. Sabra, *Theories of Light from Descartes to Newton,* 2nd edn., Cambridge, 1981, has some rather dangerous 'translations' on pp. 97, 228, 252, 281, where modern views and notations are used to express old ideas. For translation in the ordinary sense, see the two translations of Hegel's *Philosophy of Nature* which appeared in 1970: by A.V. Miller (Oxford), and by M.J. Petry (London). Petry's is fully annotated, in three volumes; without such editorial guidance it is almost impossible to make sense of the text; merely putting it into English is not enough.

6 C. Darwin, *On the Origin of Species,* (1859), Cambridge, Mass., 1964, p. 497. For Darwin's vocabulary, see P. Barrett, D.J. Weinshank and T.T. Gottleber, *A Concordance to Darwin's Origin of Species,* 1st edn., Ithaca, 1981.

7 D. Gjertsen, *The Classics of Science,* New York, 1984; T.S. Kuhn, 'The function of dogma in scientific research', in A.C. Crombie (ed.), *Scientific Change,* London, 1964, pp. 347–369. G.J. Toomer's translation of Ptolemy's *Almagest,* London, 1984, appeared too late to be noted by Gjertsen.

6

Gjertsen's Kuhnian collection also includes Copernicus' *de Revolutionibus* and Dalton's *New System*, for which the most that can really be claimed is that they started off something important. The idea that anybody but a research student should read either of these from cover to cover is grotesque; they are not analogues of works of Shakespeare or of Mark Twain.

Classics of science, and the *Origin* is easily the most accessible on Gjertsen's list, seem to need editors in a way that other classics do not; and the editor will have to do his best, if he is thinking in this way to show how his text is relevant today—a difficult task indeed if it is the *Almagest*. The same problems face those writing about scientists for the series of 'Past Masters' or 'Modern Masters', where the underlying assumption is that the hero is not just to be placed in context but is to be brought up to date.[8] Here the scholar has to become the apologist, or even worse, finds himself ticking off his hero for not getting his science a bit closer to what is the received opinion in the late twentieth century. How fortunate are historians of Byzantium or of the Maya, who at least do not have this problem, while we seem to get few advantages from chronicling the expensive and progressive world of science. But because there are scientists, and because some of them are interested in their past, we must take due note of their requirements when editing or commenting upon texts in our field. They represent one part of our constituency, and in some cases the most important part, and we must therefore show some concern with what was 'really' happening, set out in today's terms, and with the consequence of it for the science of today. We may also find ourselves let in for some of that ranking so beloved of democratic America.

Mayr worked not only for 'evolutionists' but also for students of the history of ideas, and in the evolution of our discipline the emphasis duly moved that way. Indeed, our own histories must often have recapitulated this development, for many historians have begun as scientists and gone into intellectual history, and several of us indeed teach in philosophy departments. Philosophers, even if they are philosopher–scientists like John Herschel, like case-studies which make some interesting point about explanation or method. Paradigm shifts, crucial or falsifying experiments, and fruitful or declining research programmes are the kind of things that those who come to the history of science through philosophy seek. In editing texts for such as these, one should no doubt signpost ambiguous words like 'prove'—which may be used to mean 'test'—and perhaps call attention to the nature of the reasoning going on in a document or letter.

Certainly one of the striking things in the sciences seems to be that the criteria of good science change with time. Just as a great poet or novelist will expand the limits of the medium, or a great painter will find new things to do with paint, so a scientist may expand a generations's view of what can be done in science.[9] To approach the history of science as a matter of an expanding range of methods applied to an expanding range of problems can be fascinating, and the Kuhnian distinction of revolutionary and normal science, concerned respectively with questions difficult to ask and with questions difficult to answer, can be illuminating. The trouble with the 'History and Philosophy of Science' seems to be that it leaves out 'the hidden life of the unknown' to which Merz referred, and that it emphasizes the differences between science and other activities: there is a tension

8 J. Howard, *Darwin*, Oxford, 1982; W. George, *Darwin*, 1982; in her *Lamarck*, Oxford, 1984, L. Jordanova aims simply to set him in context, and the result is a *tour de force*.

9 D.L. Hull, *Darwin and his Critics*, Chicago, 1973.

between viewing science as cultural behaviour, and as *the* road to knowledge.[10] Elegant but schematic, even 'rationally reconstructed', case histories can be curiously bloodless. Even the life of the mind fails to fit the tidy categories of philosophers.

Both scientists and philosophers tend towards internal history, seeing the science as the foreground and the other parts of the life and times of the scientist or the institution as the background. This may sometimes be the way participants saw it too, but science is an activity pursued by humans, who do need to eat, to occupy some social position, to love, to suffer pains and to enjoy pleasures—their life, liberty and pursuit of happiness are not entirely devoted to the discovery of laws of nature. We shall find that even the most eminent of scientists will often have other concerns in the foreground for all or most of their time.

Science is not an activity that needs no explanation.[11] At different times and places it has flourished or languished, and emphasis on utility or on intellectual interest has also varied. Pure science may be seen as a curious form of self-indulgence, not to be overdone; as the highest aim of the human mind; or valuable only as a guide to practice. Once historians decided that their affinities were with history, that if their field was not history it was nothing, they were forced to bring context into focus, if not into the foreground. There are even those who believe that matters as recondite as the development of statistical theory in Britain in the years around 1900—the work of Galton and Pearson—only make sense when explained in terms of external factors: in this case the rise of a professional middle class, and the associated ideas of eugenics.[12] The job is easier when the rise of organic chemistry in the later nineteenth century is connected with synthetic dyes.

Historians of science have perhaps had their life made more complicated by the demise of the Whig view of history, or of the more general theory of progress—or even the theory of decline. That is, the historians of our day are the heirs not of the philosophical historians of the eighteenth century—Vico, Voltaire, Gibbon and Herder—but of the antiquaries. It is not the broad view, good style and wide reading which is much admired; rather it is hard work among the archives of some institution or scientist, eminent or obscure. The avoidance of broad synthetic views, of wide-angle lenses in favour of close-ups, makes the move from history of science written for scientists or philosophers to that written for historians more difficult. It is a real journey from one world to another: the eagle's to the mole's.

If science is public knowledge, then generally speaking what was not published was not science; and if there is no such thing as a logic of discovery, but only perhaps of confirmation or falsification;[13] then the manuscripts of scientists might be only of very moderate interest. Even published work of the past cries out for editorial intervention, as

10 See my *The Nature of Science*, London, 1976 [actually 1977], for science as an intellectual, social and practical activity; for a new view of the interaction of science and Christianity, see C.A. Russell, *Crosscurrents*, Leicester, 1985.

11 I. Inkster and J. Morrell (ed.), *Metropolis and Province*, London, 1983; C.A. Russell, *Science and Social Change*, London, 1983.

12 D.A. Mackenzie, *Statistics in Britain, 1865–1930*, Edinburgh, 1981; W.J. Hornix at Nijmegen University is working on chemistry and the dye industry.

13 M.J. Petry, *Hegel's Philosophy of Nature*, London, 1970, I, p. 62 (see note 5).

8

Jacob's work on the writings of Stubbe indicates:[14] texts can be oblique or deceitful; they may not be what they seem, whether they are apparently vindicating a king or describing an experiment, and if they are across a 'paradigm', as with an alchemical text, their meaning and significance may be hard to grasp.

Yet as we all know, the manuscripts of scientists—even in tantalizingly edited form, as in Victorian *Lives and Letters*—are fascinating documents which can be used to illuminate the workings of past science. As the formality of scientific prose increased from Priestley's[15] to Richards' (taking two eminent American chemists a century apart), so the value of informal letters or notebooks increased—and along with them, bills for apparatus or for services, contracts with publishers or for consultancy, drafts of examination papers, and so on. Such material allows us to see the science of the past in perspective, without committing ourselves to the tunnel vision involved in an exclusively internal or external methodology. We want to know how people lived, and not merely how they thought or worked; and the historian can make us see the value of particularity.

The trouble is that moving into history has not created a space within which historians of science find it easy to work, and where they find an appreciative audience. This is partly due to the antiquarian flavour of historical research. Work on provincial literary and philosophical societies in Britain, or indeed on science in mid nineteenth-century America, does not necessarily rivet the attention of the learned world. County histories were chiefly aimed at those who lived there and wanted to learn about their ancestors or predecessors; the same is true of college histories, aimed at alumni. To get beyond a similar, and rather small, body of readers remains a problem: a concern with the prominent, and with how things turned out, is what publishers at any rate often seem to suppose is required if the writings in the history of science are to sell—or are even to be noticed, which matters to the most ivory-towered of us.

To give general interest to studies which are narrowly focused on a person or institution, it may be possible to use them to test a theory about social mobility or marginality; and after a rather fruitless pursuit of historians, it was to sociologists that those interested in the history of science turned their attention.[16] Just as natural historians turned from the search for 'non-descripts' towards describing the whole fauna and flora of a place, so historians of science should perhaps cease to be dazzled by the great and look for the typical. They might even find themselves becoming as critical of the great in science as of any other 'establishment'. Thus, the historian of the Royal Institution in London[17] (not exactly a typical place) may see Davy and Faraday as having

14 J.R. Jacob, *Henry Stubbe, Radical Protestantism, and the Early Englightenment*, Cambridge, 1983; for an edition of alchemical emblems, see M. Maier, *Atlanta Fugiens*, (ed. H.M.E. de Jong), Leiden, 1969; a lightly-edited reprint of an astrological text is W. Lilly, *Christian Astrology*, (1647), reprint London, 1985, with afterword by P. Curry and modern astrological perspectives by G. Comelius.

15 Recent papers on Priestley are interred in: Royal Society of Chemistry, *Oxygen and the Conversion of Future Feedstocks*, London, 1984, pp. 305–471.

16 J. Secord, 'Natural history in depth', *Social Studies of Science*, (1985), 15, pp. 181–200.

17 M. Berman, *Social Change and Scientific Organization: The Royal Institution, 1799–1844*, London, 1978; S. Forgan (ed.), *Science and the Sons of Genius*, London, 1980, is the proceedings of a conference on Davy—Dr Forgan is currently working on the architecture of scientific institutions: see her paper 'Context, image and function', *BJHS*, (1986), 19, pp. 89–113.

sold out to the landed interest, rather than as having produced triumphantly successful scientific research and teaching. Those who remained poor and in garrets might thus be congratulated upon their avoidance of the seductions of capitalism, rather than commiserated with, even though this is probably not the way they would have seen their own lives.

An interpretation thus at odds with contemporary opinion may be justifiable, but such anachronism is dangerous. In this case it came out of the close study of the Managers' Minutes, rather than of the scientific papers; and the documents which we select, like the phenomena a scientist deems relevant, will affect the conclusions at which we arrive. History of science which ignores the public knowledge,[18] the published papers and the course of research, is like a college history written entirely from the point of view of its treasurer: it makes its subject look like any other, which is only half the story. Nevertheless, if we are setting out to tell something like the whole story, we must include this half. The historian of science, or the editor of a collection of documents or even of a single document, must be aware of the sociological as well as the historical, philosophical and scientific aspects of the story, and must have in mind readers in these various categories.

It is perhaps not only from these diverse reactions that the history of science may be approached; all roads, almost, lead to science. Science must be expressed in language— sometimes in the international languages of mathematics or of drawing—and it has strong connections with literature. The occasional masterpiece—Goethe's *Elective Affinities* or Mary Shelley's *Frankenstein*—depends upon the science of its day for the springs of its action, while many novels, plays and poems use language resonant with science.[19] Metaphors in descriptive passages, and the terms used in remarks displaying the wisdom or ignorance of characters in fiction, may now require editorial intervention if they are to be understood, while correspondingly scientific writing may echo literature of a different kind. Even to know that Lord Kelvin's yacht was called *Lalla Rookh* after an oriental poem by Tom Moore may tell us something about nineteenth-century engineering! The two cultures, scientific and literary, which C. P. Snow described (and bridged in his own career) were not a feature of the last century.

Works of literature can help us see how words were used and how certain imagery was powerful at a certain time in the past: Coleridge going to Davy's lectures to improve his stock of metaphors may have been unusual, and indeed current science usually sounds comic rather than profound when put into poetry, but he was not unique, and the

18 J. Ziman, *Public Knowledge*, Cambridge, 1967. On the publication of science, see A.J. Meadows (ed.), *Development of Science Publishing in Europe*, Amsterdam, 1980; W.H. Brock and A.J. Meadows, *The Lamp of Learning: Taylor and Francis and the Development of Science Publishing*, London, 1984; M.H. Black, Cambridge University Press, 1584–1984; R.E. Kohler, *From Medical Chemistry to Biochemistry*, Cambridge, 1982, is a strongly institutional history. On sources, see my *Sources for the History of Science*, Cambridge and Ithaca, 1975; readers of Japanese should use the version ed. and tr. H. Kasiwagi, Tokyo, 1984. *Collections* of documents are regularly published by the Hakluyt Society: e.g. W. Kaye Lamb (ed.), *George Vancouver's Voyage of Discovery, 1791–1795*, London, 1984, 4 vols. Many such voyages were scientific; similar methods might be applied to other episodes or institutions in the history of science: J. Morrell and A. Thackray (ed.), *Gentlemen of Science: Early Correspondence of the British Association for the Advancement of Science*, London, 1984, prints only letters (without 'tops and tails', see note 20).

19 See my 'Chemistry and poetic imagery', *Chemistry in Britain*, (1983), **19**, pp. 578–582.

10

edition of his *Notebooks* shows the close relationship of science and imaginative literature. Davy's rhetoric at the Royal Institution shows his connections with non-scientific writing, and we can also learn something from such simple things as the beginnings and ends of scientists' letters.[20] These may often be conventional: but Davy's generally were not, and we can trace, for example, through the tops and tails of his letters, his relationship with Faraday as he progressed from amanuensis/valet to what in modern terms might be 'research student', to 'research assistant', and on to colleague. They reached the stage of being nearly as cordial as those addressed to Davy's own brother, but then came one of those monumental rows so characteristic of nineteenth-century science and a new formality.

Mathematics is a language which has changed with time, and old mathematics may cry out for translation, or it may be admired by those with an eye for elegance in perhaps a rougher tongue than ours. But mathematics is a language in which most of us are only semi-literate, whereas illustration in natural history and in technology is a visual language which almost everybody seems to admire.[21] Elderly illustrations seem often to pass that test of time which distinguishes the classic, or the great work of art. This is odd because scientific illustrations all had their context in the science of their day, and were generally done in conjunction with text so that their context is easier to fix than that of other works in the visual arts. We might expect, therefore, that like scientific texts, scientific illustrations would have rapidly dated.

We need perhaps here to distinguish the diagram, where the theory-loading may be extreme—as in discussions of mechanics—from other illustrations. The gulf is not very wide, for pictures of chemical apparatus, showing retort stands and often an elegantly-cuffed hand, gradually give way over the nineteenth century to conventional diagrams. Similarly, it may be difficult to place engineering drawings,[22] some of which indeed were done to decorate the walls of boardrooms, while most were done to be worked from and are more diagrammatic. Most of these will require essentially an engineering commentary, drawing attention to the conventions in which colour indicates materials, and pointing out the characteristic or novel features of the ship or locomotive depicted. But in some cases the elegance of the draughtmanship and the placing of the illustration on the page will elevate it into being a work of art rather than an object of curiosity. This judgement may let one in for some art–historical comments.

Technological illustration may also shade into topographical art, where a railway or a canal is depicted; here clearly artistic conventions are as important as technical ones,

20 See my 'Davy and Faraday: fathers and sons', in D. Gooding and F. James (eds), *Faraday Rediscovered*, London, 1985, pp. 33–49. G. Hartcup and T.E. Allibone, *Cockcroft and the Atom*, Bristol, 1984, p. 142, annotate a letter of 1946 beginning 'My dear Cockcroft' to point out 'that at that time the use of surnames alone was the usual informal manner of address between colleagues'; in assessing tones of letters this kind of thing is important.

21 M. Pollock (ed.), *Common Denominators in Art and Science*, Aberdeen, 1983; M. Rudwick, 'A visual language for geology', *History of Science*, (1976), **14**, pp. 149–195; papers, including mine, in A. Ellenius (ed.), *The Natural Sciences and the Arts*, Uppsala, 1985.

22 K. Baynes and F. Pugh, *The Art of the Engineer*, London, 1981; my *The Age of Science*, Oxford, 1986, chapter 7.

and it may well be that the artist has been so careful to get the background right that he has not understood the machinery in the foreground, or has made the whole scene too picturesque.[23] In zoological and botanical art there may be the same tension between decorativeness and utility.[24] The early bestiaries had essentially symbolic pictures of animals; we still expect lions and eagles to look kingly, or maybe presidential, but we demand of natural history pictures that they should enable us to identify the species, that is, they must show diagnostic features: and where illustrations have the advantage over photographs is that they need not include individual peculiarities, and may emphasize important characters. The extreme here might be the modern field guide, to help bird-watchers identify what they have only perhaps glimpsed.

The more diagrammatic and theory-laden a picture is, the more it is a part of the science of its day, but even the most decorative studies of plants and animals show their time and place. Birds in seventeenth- and eighteenth-century plates look stiff because they were painted from dead specimens; if they were exotic, then probably from skins shipped back and stuffed to resemble what the artist imagined the bird must have looked like when alive. Later artists did their best to paint from living creatures, perhaps in an aviary or zoo, but the effect of the great natural history pictures of the last century is still the odd one of seeing how an owl might look if one could get much nearer than in fact is possible.

Audubon's birds, and some of Joseph Wolf's plates, have a romantic air appropriate to their time, and plates of fish drawn against a background of the river where they may be caught should be judged not as ecological but as ingeniously getting tonal character. Animals and plants were drawn under European patronage by American Indians, East Indians and Chinese artists; the placing of the subject and the technique makes such pictures subtly different from native European ones. The medium in which they were printed—and only printed pictures are really public knowledge—is also important; William Swainson[25] did the first autolithographed series of natural history plates, but like many pioneers he failed to realize the implications of the new medium. He used short lines so that his prints look very like engravings, and on his pattern plates, done to show the colourist what to do, he even referred to his lithographs as 'engravings'. Later artists like Edward Lear saw how lithography allowed the artist to use long flowing lines. Commentary upon illustrations must draw upon artistic conventions and techniques. Because illustrated works often appeared in parts, not in the final systematic order, all kinds of bibliographical problems can arise.[26]

The modern reader of works in the history of science and technology is then a very diverse creature; indeed, it is our weakness and our strength that we do not have a single

23 D.J. Carr (ed.), *Sydney Parkinson: Artist of Cook's Endeavour Voyage*, London, 1983; especially chapter 10 (by J.R.H. Spencer).

24 See W. Blunt, *The Art of Botanical Illustration*, London, 1950, and my *Zoological Illustration*, Folkestone, 1972.

25 See my papers, 'William Swainson: naturalist, artist and illustrator' (the Ramsbottom Lecture for 1984), forthcoming in *Archives of Natural History*; and 'William Swainson: types, circles and affinity', in J.D. North and J.J. Roche (ed.), *The Light of Nature*, Dordrecht, 1985, pp. 83–94.

26 G.D.R. Bridson, 'The treatment of plates in bibliographical description', *Journal of the Society for the Bibliography of Natural History*, (1976),7, pp. 469–488.

22

12

group of readers, and that we are to be found scattered in various corners of academe and outside it. Editors of documents do not have to wrestle with exotic questions, such as: 'if there was a Church Scientific, who was its Constantine—or its Luther?', or with exactly how the term 'professional' ought to be used of nineteenth-century scientists; but they may have to gloss these terms when they find them in T. H. Huxley or in Faraday. Whereas the author of a book can have some audience in mind, and indeed must do so if the book is to have any coherence, and anyone submitting a paper to a learned journal can similarly make some assumptions about likely readers, the editor of letters or papers does not know for sure which of the many groups forming our public may use his work. The difficulties of avoiding professorial pomposity or intrusiveness are correspondingly greater, for little can be taken for granted: while some things will be familiar to any particular reader, even editors cannot be expected to know everything, and historians of science know at least that they have much to be thankful for.

III

The Scientist as Sage

Oh, most magnificent and noble Nature!
Have I not worshipped thee with such a love
As never mortal man before displayed?
Adored thee in thy majesty of visible creation,
And searched into thy hidden and mysterious ways
As Poet, as Philosopher, as Sage?[1]

A GREEK of the late fourth century would have been able to describe a man as a mathematician or a botanist, but would not have been able to translate the English word "scientist" except by saying "philosopher."[2] An Englishman in the early years of the nineteenth century would have had the same problem, for the word "scientist" was coined, by analogy with "artist," at a meeting of the British Association for the Advancement of Science in the 1830s.[3] Before that time, men of science were described as philosophers; natural philosophers, chemical philosophers, and other such persons composed the scientific community. Faraday, for example, continued to the end of his life to consider himself a natural philosopher, to dislike the narrower modern term. The distinction is not purely verbal: men of science in England in the first half of the nineteenth century were expected not to be mere specialists, but to see their subject in a general light, as part of a world view; and to be able to communicate their insights to anybody with a general education.

This cosmological interest showed itself in various ways.[4] Numerous works were produced on natural theology, a subject which had been of great general interest in the eighteenth century when much

1. *Fragmentary Remains of Sir Humphry Davy, Bart.*, ed. John Davy (London, 1858), p. 14.

2. Moses I. Finley, *The Ancient Greeks* (London, 1963), p. 117.

3. "On the Connexion of the Physical Sciences, by Mrs. Somerville," *The Quarterly Review*, LI (1834), 59.

4. See *Science Before Darwin*, ed. Howard Mumford Jones and I. Bernard Cohen (London, 1963).

66

popular science appeared in that form, and which reached its final flowering in the works of Paley, and in the Bridgewater Treatises (which were published in the 1830s and went through numerous editions). But by 1800 this rationalist tradition was being undermined from both sides. The sciences needed the hypothesis of God less and less; and thinkers under German influence, but following the lead of Hume's *Dialogues Concerning Natural Religion*, began to criticize a scheme in which God Almighty was congratulated by his arrogant and almost omniscient creatures on the sensible way He had created a mollusc or a man. More advanced thinkers among the natural philosophers turned therefore towards Idealism, and towards the German manifestation of this spirit which was called *naturphilosophie*. Professor Pearce Williams, in his recent monumental biography of Faraday, has underlined the importance of this tradition; and in this paper I propose to examine two works by predecessors of Faraday in the field of electrical science, Humphry Davy's *Consolations in Travel* and Hans Christian Oersted's *The Soul in Nature*.[5]

These works were intended to be popular, but they are not "popular science" as that term is usually understood. It was not the aim of the authors to make recent discoveries better known, but to disseminate a view of the nature of scientific investigation and scientific knowledge. Both of them, it seems fair to say, considered that experimental science was but one road to knowledge; intuition or pure reason were others. Both were greatly interested in poetry, and Davy wrote a good deal of verse;[6] neither could be described as a detached observer of nature, but rather as a wrestler with her, fascinated always and sometimes close to despair in the effort to find the great general law which would unify all the sciences into a harmonious whole as a true reflection of nature herself. Any scrutiny of these works must undermine the notion that the early nineteenth century was a time when scientists followed a "Baconian" program and eschewed theory; if there was ever a period in England when this was true of chemists, it was only in the period of consolidation in the generation after Davy, Wollaston, and Dalton.

5. Sir Humphry Davy, Bart., *Consolations in Travel, or the Last Days of a Philosopher* (London, 1830); Hans Christian Oersted, *The Soul in Nature*, tr. Leonora and Joanna B. Horner (London, 1852), as reprinted in facsimile (London, 1966)—hereafter cited as Oersted. This latter work, a collection of essays, contains a short biography of Oersted.

6. See J. Z. Fullmer, "The Poetry of Sir Humphry Davy, "*Chymia*, VI (1960), 109.

Davy's *Consolations* and the English translation of Oersted's book both appeared after the deaths of their authors, but it would be wrong to think of them as senilia. The *Consolations* and *Salmonia*, a work describing Davy's favorite sport, were written "during the time of a partial recovery from a long and dangerous illness," and shortly before disease struck its final blow.[7] Both of them sold better than his epoch-making contributions to chemistry and went through several editions, whereas, rather surprisingly, a second edition of his *Elements of Chemistry* was never required. *Salmonia* and *Consolations* both take the form of dialogues, and the latter, in particular, illustrates Davy's capacity for writing expository prose of a high order. There are clear echoes in the book of the lectures which Davy gave earlier in his life to large and fashionable audiences at the Royal Institution in London,[8] lectures which Coleridge said he attended to improve his stock of metaphors.

Cuvier wrote in his *Eloge* of Davy that "once escaped from the laboratory, he had resumed the tranquil reveries and sublime thoughts which had formed the delight of his youth; it [the *Consolations*] was in some measure the work of a dying Plato."[9] The wild theorizing of Davy's youthful paper on the chemical role of light, and his published resolve to avoid the repetition of such speculative excesses are well known. In his notebook he wrote, in chastened vein: "I began the pursuit of chemistry by speculations and theories: more mature reflection convinced me of my errors, of the limitation of our powers, the danger of false generalizations, and of the difficulty of forming true ones."[10] But we should be careful not to take too seriously such remarks or the disclaimers added to some of the papers written in his scientifically creative period (the first fifteen years or so of the century), to the effect that his aim was not to advance hypotheses but simply to announce the discovery of new facts. On such occasions he implied that hypotheses were but heuristic or mnemonic devices,

7. *Consolations*, "Preface."

8. See G. A. Foote, "Humphry Davy and His Audience at the Royal Institution," *Isis*, XLIII (1952), 6–12; and Roger Sharrock, "The Chemist and the Poet," *Notes and Records of the Royal Society*, XVII (1962), 57–76.

9. *Mémoires de l'Académie Royale des Sciences*, XII (1833), xxxv.

10. John Davy, *Memoirs of the Life of Sir Humphry Davy, Bart.* (London, 1836), I, 80. See also *The Collected Works of Sir Humphry Davy, Bart.*, ed. John Davy (London, 1839–40), II, 3–120; II, 126; III, 2; and Colin A. Russell, "The Electrochemical Theory of Sir Humphry Davy," *Annals of Science*, XIX (1963), 255–257.

useful for the making of discoveries, but not fundamental; and his rejection of the atomic theory ostensibly in favor of the wholly empirical doctrine of equivalents is well known.[11]

But it would be a mistake to suppose that Cuvier's picture of Davy as a wild youth who curbed his speculative intellect in his maturity, and let it go again towards the end, is complete. Davy's contemporaries were struck by the boldness of his imagination throughout his life; and those who attended his lectures were enthralled by the cosmic conclusions which he drew from his magnificent experimental research. The comparison between Wollaston and Davy, the one scrupulous and cautious, and the other scintillating and intuitive, was commonplace. Dumas wrote of Davy's electro-chemical theory: "When we know that there was poetry in his brilliant imagination, and how he had made for himself a system of Nature which he thought could embrace everything; when we know that he had studied the alchemists; when we know of his pantheistic ideas, we can understand with what ardent curiosity he had to follow an idea which seemed to him so vast, and with what troubled respect he had to try its power."[12] Davy is a difficult man to see whole, but we can say that throughout his life he seems to have clung to the belief that it is impossible with an inadequate world view to reason perfectly on particular topics. John Davy ended the biography of his brother with a tribute in verse by a Mr. W. Sotheby, F.R.S.:

> Thou! from whose lip the word that freely flowed
> With all a poet's inspiration glowed,
> Lamented friend, farewell! Thou liest at rest,
> A world of wonders buried in thy breast!
> High aims were thine,—all nature to explore,
> Make each new truth developed gender more,
> And upward traced through universal laws
> Ascend in spirit to the Eternal Cause.
> Such was thy ardent hope, thy view sublime.
> But ah! cut off in manhood's daring prime,
> Thou liest where genius leans upon thy tomb,
> And, half eclipsed, mourns thy untimely doom.[13]

11. See my article, "The Atomic Theory and the Elements," *Studies in Romanticism*, v (1966), 185–207.
12. J. B. Dumas, *Leçons sur la Philosophie Chimique* (Paris, 1837), pp. 406–407.
13. *Memoirs*, II, 412.

Berzelius' famous remark,[14] that Davy's work was "a brilliant fragment," and that if only he had been more thoroughly trained and
written more, he would have advanced the course of chemistry by a
century, is another way of making the same point. In fact the *Royal
Society Catalogue* lists a total of eighty-six papers by Davy; but
whereas Berzelius, the great systematizer, would have considered
works like the *Consolations* of much less value than a few more solid
chemical papers, Davy clearly felt that the propagation of his metaphysical notions was of extreme importance. Shortly before beginning the *Consolations*, he told his friend Thomas Poole: "I do not
wish to live, as far as I am personally concerned; but I have views
which I could develop, if it please God to save my life, which would
be useful to science and to mankind."[15]

In his early lectures a strong technological interest is apparent, in
accordance with the aims of the Royal Institution; thus, in January
1802 he declared that: "we do not look to distant ages, or amuse
ourselves with brilliant, though delusive dreams concerning the infinite improveability of man, the annihilation of labour, disease, and
even death. But we reason by analogy from simple facts. We consider only a state of human progression arising out of its present
condition. We look for a time that we may reasonably expect, for a
bright day of which we already behold the dawn."[16] Many of Davy's
early researches were on technological topics; his *Agricultural Chemistry* became a classic; and he is probably best remembered for his
safety lamp for coal miners. In the *Consolations* he returned to the
discussion of technical progress, the dialogue form providing for the
introduction of a leaden echo of religious orthodoxy to this youthful
enthusiasm and belief in progress.

The first dialogue in the book is in fact primarily concerned with
the question of progress, and the relationship between technological
innovation and improvements in the quality of civilization. After the
characters have been introduced, Philalethes—who represents Davy
—is abandoned, at his own request, by his friends in the Colosseum
in Rome, and sees a vision in which a Genius, appearing only as a
voice, shows him scenes from the past and from other planets in our

14. Ernst von Meyer, *A History of Chemistry*, tr. G. McGowan (London, 1891), p.
186 n.
15. *Memoirs*, II, 403.
16. *Works*, II, 323.

70

system. These glimpses of the whole sweep of human history from the primitive beginnings show steady and rapid technical progress, and lead to remarks like: "these arts will never be lost; another generation will see them more perfect; the houses, in a century more, will be larger and more convenient; the flocks of cattle more numerous; the cornfields more extensive; the morasses will be drained, the number of fruit-trees increased."[17] These improvements are due to "a few superior minds."

Even in the collapse of the Roman Empire technical skills had not been irretrievably lost: " 'See,' said the Genius, 'the melancholy termination of a power believed by its founders invincible, and intended to be eternal. But you will find, though the glory and greatness belonging to its military genius have passed away, yet those belonging to the arts and institutions, by which it adorned and dignified life, will again arise in another state of society.' "[18] And now society had taken its "modern and permanent aspect" on account of the invention of the printing press: "by the invention of Faust the productions of genius are, as it were, made imperishable, capable of indefinite multiplication, and rendered an unalienable heritage of the human mind. By this art, apparently so humble, the progress of society is secured, and man is spared the humiliation of witnessing again scenes like those which followed the destruction of the Roman empire."[19]

Another discovery of enormous importance was gunpowder, by which "civilized man is rendered . . . infinitely superior to the savage," and civilized nations secured "from ever again being overrun by the inroads of millions of barbarians."[20] By its aid, the white races were gradually and inexorably displacing the red Indians and the Negroes. Numerous other discoveries in physics and chemistry had made great differences to the life of everyone; the steam engine, for example, placed in man's hands a power which seemed almost unlimited. Empires come and go, monarchs and governments change their minds, but "To whatever part of the vision of modern times you cast your eyes you will find marks of superiority and improvement, and I wish to impress upon you the conviction, that the results

17. *Consolations*, p. 21.
18. *Consolations*, p. 27.
19. *Consolations*, p. 29.
20. *Consolations*, pp. 29–30.

of intellectual labour, or of scientific genius, are permanent and incapable of being lost."[21] History has been written, according to the Genius, too much in terms of the doings of great men, whereas in fact governments depend upon public opinion and the spirit of the age.[22] The really great improvers of mankind have usually been despised or neglected; and they can only have continued their labors because of "the pure and abstract pleasure resulting from the exertion of intellectual superiority and the discovery of truth and the bestowing of benefits and blessings upon society."[23] This picture of the wretched inventor struggling in his garret comes oddly from Davy, who had advanced from humble origins to an exalted position in society through his scientific abilities.

Although this passage resembles some of the writings of contemporary Americans eulogizing technology and inventors,[24] Davy's attitude toward science was never that it was of value only insofar as it led to technical advances. In a lecture of 1803 he had set such usefulness and the satisfying of intellectual curiosity side by side, adding that discoveries with no obvious utility might, like newly discovered countries, be found valuable in the future. But, he went on: "independently of all these considerations, all truths in nature, all inventions by which they can be developed, are worthy of our study, for their own sake, rather than with any idea of profit or interest. . . . [The] noblest faculties are reason and the love of knowledge. . . . Man is formed for pure enjoyments; his duties are high, his destination is lofty; and he must, then, be most accused of ignorance and folly when he grovels in the dust, having wings which can carry him to the skies."[25]

A quarter of a century later this conception reappeared in the *Consolations* when the Genius showed Philalethes other worlds, the abode of beings more exalted than we. Those writing on the strange and fascinating topic of a plurality of worlds must emphasize either the uniformity of nature, or her endless diversity;[26] and Davy, unlike

21. *Consolations*, pp. 33–34.

22. The taste for heroes soon recovered from attacks of this kind; see Walter E. Houghton, *The Victorian Frame of Mind* (New Haven, 1957), Ch. XII.

23. *Consolations*, p. 35.

24. See Perry Miller, *The Life of the Mind in America* (London, 1966), pp. 288 f.; for the situation in England, see Houghton, Ch. II.

25. *Works*, VIII, 164, 166.

26. See my article, "Celestial Worlds Discover'd," *The Durham University Journal*, LVIII (1965), 23–29.

most scientists, chose the latter alternative. Strange creatures of beautiful colors flew around Saturn in his vision; and the Genius sensibly remarked that: " ' you want analogies and all the elements of knowledge to comprehend the scene before you.' "[27] These beings, whose portrayal narrowly avoids being rather ridiculous, possess more exact scientific knowledge than we, and "their highest pleasures depend upon intellectual pursuits, so you may conclude that those modes of life bear the strictest analogy to that which on the earth you would call exalted virtue." They have no wars, their only ambitions are intellectualism and "a love of glory of the purest kind." All intelligent beings in the universe must possess an organ for receiving light— throughout his career Davy insisted on the importance of light in the world—but otherwise they are utterly different from one another. "Spiritual natures" pass in their progress from planet to planet, and from system to system: "The universe is every where full of life, but the modes of this life are infinitely diversified, and yet every form of it must be enjoyed and known by every spiritual nature before the consummation of all things." This will clearly take some time.

Even the comets, which Derham had a century earlier[28] "imagined to be destined rather for a place of torment, than for any other sort of living," had for Davy their populations, figures with an "awful and unnatural" resemblance to humans. These beings, "so grand, so glorious, with functions to you incomprehensible" once belonged, according to the Genius, to our earth; now all their dust had been left behind, and they carried with them only their intellectual power. Their earthly life was as unknown to them as our foetal existence is to us. Only the love of knowledge is carried from one state to another, and this "in its ultimate and most perfect development [is] the love of infinite wisdom and unbounded power, or the love of God." If this on earth has "been misapplied and assumed the forms of vague curiosity, restless ambition, vain glory, pride or oppression, the being is degraded, it sinks in the scale of existence and still belongs to the earth or an inferior system, till its errors are corrected by painful discipline."[29] These remarks remind us of the eighteenth-century cos-

27. *Consolations*, p. 48; he then paraphrased Pope on flies and microscopic eyes (*An Essay on Man*, I, 193-194). The other quotations in this paragraph are from *Consolations*, pp. 51, 54—such grouping of citations is frequent practice below.

28. William Derham, *Astro-Theology* . . . , 10th ed. (London, 1767), p. 208.

29. *Consolations*, pp. 55, 57.

mologist and astronomer Thomas Wright, to whom the heavenly
bodies were "the manifest Mansions of Rewards and Punishments."[30]

These notions seem closer to Buddhism than to the Christian orthodoxy of the 1820s, and in the next dialogue Philalethes discusses
the vision with his friends: Ambrosio, a liberal Roman Catholic, and
Onuphrio, a sceptical patrician. The former was shocked at the
"scepticism" revealed in the vision, and thought that in particular,
"your view of the early state of man, after his first creation, is not
only incompatible with revelation but likewise with reason and
every thing that we know respecting the history or traditions of the
early nations of antiquity."[31] Man's physical endowments are so
small, and his infancy so protracted and defenseless, that unless the
race had been especially protected at first it could never have survived. The Garden of Eden is a much more likely beginning than a
state of barbarism; and present-day savages have declined from civilization, rather than civilized nations have progressed from a state
of savagery.

The discussion of the vision terminates in Philalethes' acceptance
of Ambrosio's general points, and in his stressing that man is not
simply a reasoning animal, but needs faith to submit himself to the
inscrutable will of God. "We may compare," he said, "the destiny of
man in this respect to that of a migratory bird; if a slow flying bird,
as a landrail in the Orkneys in autumn, had reason and could use it
as to the probability of his finding his way over deserts, across seas
and of securing his food in passing to a warm climate 3000 miles off,
he would undoubtedly starve in Europe; under the direction of his
instinct he securely arrives there in good condition."[32] In other
dialogues in the book we find the same distrust of the unaided reason.
Davy, the friend of the young Coleridge and Southey, was passionate
and imaginative, and owed to these qualities his scientific creativity
and success. We may rejoice that he had never been forced to subject
himself to a prolonged and formal training in the sciences, which
would surely have bored him. His temperament was almost the
antithesis of that of Lavoisier, an incarnation of rationality who made
no discoveries himself but reasoned with precision on the experi-

30. Thomas Wright, *An Original Theory . . . of the Universe . . .* (London, 1750),
p. 35.
   31. *Consolations*, p. 73.
   32. *Consolations*, p. 101.

ments of others. Davy made numerous discoveries, but shared with some of the Romantic poets a short creative life. His most important work was done in his twenties and early thirties, and when he died prematurely at fifty he had done no scientific work of the first rank for more than a decade.

He was very conscious of a falling off and in *Salmonia* wrote:

Ah! could I recover anything like that freshness of mind which I possessed at twenty-five, and which, like the dew of the dawning morning, covered all objects and nourished all things that grew, and in which they were more beautiful than in mid sunshine, what would I not give? All that I have gained in an active and not unprofitable life. How well I remember that delightful season, when, full of power, I sought for power in others; and power was sympathy, and sympathy power; when the dead and the unknown, the great of other ages, and of distant places, were made by the force of the imagination my companions and friends; when every voice seemed one of praise and love; when every flower had the bloom and odour of the rose; and every spray or plant seemed either the poet's laurel or the civic oak, which appeared to offer themselves as wreaths to adorn my throbbing brow.[33]

This melancholy appears in the *Consolations* too when he describes how after his mother's death in 1826 the love of natural scenery was the only feeling he retained.[34] But we should note that in 1827 he wrote to Lady Davy: "You talk of honours; I ought to have been made a Privy Counsellor and a Lord of Trade, as my predecessor was."[35]

In the later dialogues of the *Consolations* we find the same themes: the intellectual satisfaction, and utility, to be derived from natural knowledge; the ultimate incomprehensibility of God, and also of life; and the notion that, whatever the orthodox might say, there is in nature not merely change but progress. In the third dialogue there appears a new character, the Unknown, a profound chemical philosopher whose biography, like that of Philalethes, overlaps with Davy's. This third dialogue is mostly geological. Davy claimed to have given the first public geology lectures in London and had made some field trips, so he was an authority on what was rapidly becoming the most popular of the sciences. The Huttonian, or uniformitarian view, he rejected, in the person of the Unknown, because it

33. *Memoirs*, II, 304.
34. *Consolations*, pp. 169–170.
35. Letter to Lady Davy, August 1, 1827, at the Royal Institution. I should like to express my gratitude to the Managers of the Royal Institution, in London, for permission to quote from Davy MSS in their possession.

33

seemed to rule out progress: "It is supposed that there are always the same types both of living and dead matter. . . . Now to support this view, not only the remains of living beings, which at present people the globe, might be expected to be found in the oldest secondary strata; but even those of the arts of man, the most powerful and populous of its inhabitants, which is well known not to be the case. On the contrary, each stratum of the secondary rocks contains remains of peculiar and mostly now unknown species of vegetables and animals."[36] A refined Huttonian system (that of Playfair and Hall) might, he admitted, save the phenomena; but it was necessary to admit that the present order of things was recent, and that earlier orders had been destroyed. But the absurd, vague, atheistical, false, and feeble views of the evolutionists who supposed that "living nature has undergone gradual changes by the effects of its irritabilities and appetencies; that the fish has in millions of generations ripened into the quadruped, and the quadruped into the man," must be rejected.[37] Such "sophisms" were on the level of talking about plastic powers, or giants, or suggesting that fossils had been created in the rocks to confound our speculations.

In general, Davy tended to oppose "systems," whether the antiphlogistic system in chemistry or wide-ranging theories in geology or biology. And in the latter science he was particularly scornful of mechanistic hypotheses. In the fourth dialogue—"The Proteus, or immortality"—he appears as a vitalist. Philalethes was on a fishing holiday in Austria with a physician friend Eubathes, when his boat was swept over a cataract and capsized; he was rescued by the Unknown, who used his fishing tackle for the purpose, and they continued their conversation. The Unknown affirmed that no mechanistic hypotheses had been able to elucidate the functions of life. Even the source of animal heat was still unknown, although "half a century ago the chemists thought they had proved it was owing to a sort of combustion of the carbon of the blood."[38] The Unknown's idea of

36. *Consolations*, pp. 143–144. Lyell's *Principles of Geology*, which made everybody uniformitarian, began to appear in the same year. Loren Eiseley (*Darwin's Century* [London, 1959], Ch. IV) stresses the difficulty of reconciling progress and uniformity, a task Darwin finally achieved.

37. *Consolations*, p. 150. Darwin's remarks about Lamarck were very similar—see Francis Darwin, *The Life and Letters of Charles Darwin*, 3d ed. (London, 1887), II, 23, 29.

38. *Consolations*, p. 194. The history of theories of the vital heat up to Lavoisier's

what happened in respiration was that oxygen contained an ethereal component capable of assuming the form of heat and light; "and that, in the course of the circulation, its ethereal part and its ponderable part undergo changes which belong to laws that cannot be considered as chemical,—the ethereal part probably producing animal heat and other effects, and the ponderable part contributing to form carbonic acid and other products."[39] Remarks like these are at first sight surprising from a chemist; but it is a fact that some of the most strenuous opponents of a wholly mechanistic biology have been physical scientists. That oxygen gas was a compound containing light had been the thesis of Davy's earliest paper (in 1799), which had been censured as excessively speculative; and it is interesting to see the speculation reappearing in his latest work.

The Unknown believed that the laws of chemistry were invalid in physiology: "I will not allow any facts or laws from the action of dead matter to apply to living structures." The blood is a living fluid, and it does not burn; the gastric juices act chemically to dissolve food, but they do not affect living matter. "There can be," he affirmed, "no doubt that all the powers and agencies of matter are employed in the purposes of organization, but the phenomena of organization can no more be referred to chemistry than those of chemistry to mechanics." Speculations about the role of electricity in animals were to be compared to the hypotheses of "some of the more superficial followers of the Newtonian philosophy, who explained the properties of animated nature by mechanical powers."[40] These men had been followed, on the invention of "pneumatic chemistry," by those— equally foolish—who imagined living creatures to be laboratories.

Physiology has always borrowed fashionable models from physics and chemistry in the attempt to provide explanations; some have seen progress in this, but Davy was prepared to admit only change. He argued elsewhere for the importance of analogy in scientific reasoning; but in mechanistic accounts of physiology "the analogy is too remote and incorrect; the sources of life cannot be grasped by such machinery; to look for them in the powers of electro-chemistry is

---

broadly satisfactory solution of the problem is given in Everett Mendelsohn, *Heat and Life* (Cambridge, Mass., 1964). The rise of vitalism is discussed in William Whewell, *The Philosophy of Inductive Sciences*, 2d ed. (London, 1847), I, Bk. IX.

39. *Consolations*, p. 196.
40. *Consolations*, pp. 198, 200, 201.

35

seeking the living among the dead;—that which touches, will not be felt, that which sees will not be visible, that which commands sensations will not be their subject."[41]

It was not merely that mechanistic hypotheses had hitherto been crude in conception and lacking in detailed explanatory and predictive power; rather, living organisms were in principle inaccessible to the objective procedures of physical science, and no models derived from inanimate nature could be of value in biology. This vitalistic view was not uncommon among chemists at the time,[42] though the Unknown's is perhaps a rather extreme statement of it; Wöhler's "synthesis" of urea in 1828 had very little effect on the climate of opinion. It was not until after the work of Berthelot in the second half of the century that belief in the chemical basis of life became general among chemists. For Davy, to say that God might have bestowed a power of thinking on matter was as absurd as saying that He might have made a house its own tenant. Life ceases if the heart stops, and vision cannot happen without an eye; but the living principle is not in the heart, nor is the percipient principle in the eye or brain. All these are but instruments of a power which has nothing in common with them; in our terms, they are in different categories.

We must, said the Unknown, admit our ignorance of "the *divine fire* which is the cause of the mechanism of organized structures. Profoundly ignorant on this subject, all that we can do is to give a history of our own minds. The external world or matter is to us in fact nothing but a heap or cluster of sensations, and in looking back to the memory of our own being, we find one principle which may be called the *monad*, or *self*, constantly present, intimately associated with a particular class of sensations which we call our own body or organs."[43] This passage recalls the effusion of the youthful Davy under the influence of nitrous oxide: "Nothing exists but thoughts! —the universe is composed of impressions, ideas, pleasures and pains!"[44] In the *Consolations*, the Unknown continues with the argu-

41. *Consolations*, p. 202. For Davy on analogy, see e.g., *Works*, VIII, 167. Thomas Jefferson explicitly compared fashions in mechanistic biological theories with those in dress; see E. J. Martin, *Thomas Jefferson, Scientist* (New York, 1952), p. 39.

42. See e.g., Justus Liebig, *Animal Chemistry*, ed. William Gregory (Cambridge, Mass., 1842), pp. 1 f., 185 f. William Prout, *Chemistry Meteorology and the Function of Digestion considered with reference to Natural Theology* (London, 1834), p. 431, tried to give a chemical account of the vital force.

43. *Consolations*, p. 211.

44. *Works*, III, 290. Davy's subjective accounts of anaesthesia are among the best

78

ment; the history of the intellect is one of change according to law, but only useful changes are remembered. Thus we forget what happened to us in earliest infancy, though many of the habits then acquired are retained through life. This chain of thought leads to speculations, similar to those in the first dialogue, on the immortality of the soul, which only recalls what is useful to it from its previous existence. And again the inscrutability of God is stressed; nothing could be more absurd or impious than "to *reason* respecting the decrees of eternal justice."[45]

The fifth dialogue brings us down to earth again, and its title indicates the subject, "The Chemical Philosopher." It is a discussion of the status and value of chemistry, and of the qualities required of one who would become a chemical philosopher. The first paragraphs, in which the Unknown gives a resumé of his life history, are clearly autobiographical. Once he had acquired an independent fortune, said the Unknown, he really became a philosopher, and travelled with the object of instructing himself and benefiting mankind: "My life has not been unlike that of the ancient Greek sages. I have added some little to the quantity of human knowledge, and I have endeavoured to add something to the quantity of human happiness."[46] Davy was very conscious that an advantage of a scientific career was that it provided a route by which to rise in society; and further that it was open to the talents, demanding neither the ruthless ambition of the eagle, as the law did, nor the "creeping powers of the reptile," necessary for success in politics. Men of fortune and rank should apply themselves to natural philosophy, getting permanent distinction—"independent of vulgar taste or caprice"—thereby, and benefiting their fellow-men. Above all, science should be pursued with true dignity rather than for profit; as Davy had said in 1807, men of science had made discoveries by examining "with reverence and awe the substantial majesty of nature."[47]

The Unknown's friends suggest that while mathematical sciences undoubtedly offer noble objects of contemplation, chemistry is a humbler business, the province of the apothecary and the cook, and

---

ever written; for discussion, see F. F. Cartwright, *The English Pioneers of Anaesthesia* (Bristol, 1952).
45. *Consolations*, p. 217.
46. *Consolations*, p. 224.
47. *Works*, VIII, 179.

unworthy to be called philosophy. The Unknown begins the defense of his science by showing the many useful and refined arts which flow from its cultivation. The origin, progress and improvement of society is founded in mechanical and chemical inventions, the latter including tanning, dyeing, and bleaching, metallurgy, the making of glass and porcelain, and even artists' colors. Individual processes have indeed often been discovered inadvertently: "But it has been by scientific processes and experiments that these accidental results have been rendered really applicable to the purposes of common life. Besides, it requires a certain degree of knowledge and scientific combination to understand and seize upon the facts which have originated in accident."[48] Technological discoveries, he went on to claim, are responsible not only for material progress, but ultimately for moral improvement.

This did not exhaust the reasons for studying chemistry, which like the other sciences, while in its applications it transformed the earth, in sublime speculations it reached to the heavens. It is a pure delight to know the laws governing the phenomena happening around us; moreover, the true chemical philosopher sees good in everything: "Whilst chemical pursuits exalt the understanding, they do not depress the imagination or weaken genuine feeling; whilst, they give the mind habits of accuracy, by obliging it to attend to facts, they likewise extend its analogies; and, though conversant with the minute forms of things, they have for their ultimate end the great and magnificent objects of nature." Chemists, in short, have as much opportunity for metaphysics and for cosmological speculations as any other kind of scientist. Every discovery, continues the Unknown, opens a new field for fresh investigations, and shows the imperfections of our theories; chemical inquiries are therefore "wonderfully suited to the progressive nature of the human intellect, which by its increasing efforts to acquire a higher kind of wisdom, and a state in which truth is fully and brightly revealed, seems as it were to demonstrate its birthright to immortality."[49] This passage, linking Davy's opposition to "systems" (since theories should continually be falsified and emended) with his speculations on the soul, is also an answer to the mathematical physicists who believed that

48. *Consolations*, pp. 233–234.
49. *Consolations*, pp. 245, 246.

chemistry had not yet become a real science.[50] According to Davy's argument, the most splendid period of the history of a science is its tumultuous youth, before very powerful theories which last for many generations have been developed, and when all is in flux. Most of his contemporaries expected to derive quasi-aesthetic pleasure not from the contemplation of this kind of situation, but from that in Newtonian mechanics, a science in which it seemed that the fundamental laws were already known, and that the edifice must stand unshaken and essentially unchanged throughout all ages.

The next section is of interest as revealing what Davy believed to be the qualities essential to a chemical philosopher. What is surprising is that although he (like Dalton and Faraday) had had little formal schooling, he recommends a formidable program of education before one should begin the pursuit of so lofty a science as chemistry. The would-be chemist must have a good grounding in mathematics and physics; preferably he should know German and Italian in addition to Latin, Greek, and French. As far as natural history and literature are concerned, it will be enough if he has had a liberal education such as that provided by a university: "indeed a young man who has performed the ordinary course of college studies, which are supposed fitted for common life and for refined society, has all the preliminary knowledge necessary to commence the study of chemistry." This same advice was given in 1840 by Sir Benjamin Brodie, a physiologist who later became President of the Royal Society.[51]

The chemist, the Unknown continued, requires also a good eye and a steady hand; but these qualities would probably be lost in the dangers of the laboratory, and therefore assistants who know nothing of what one had in mind on commencing an experiment, should be employed. This observation on the humble role which assistants should play came from Davy's experience; in his laboratory notebook at the Royal Institution we find the instruction: "No experi-

50. Whewell wrote that sciences were most popular when they were in an early form, when all was in flux, and when amateurs could expect to make contributions; but he preferred the more mathematical maturity of a science (*History of the Inductive Sciences* [London, 1837], III, 19). On the demand for a mathematical chemistry, see my paper in *Studies in Romanticism*, v (1966), 186–187; and the exchange between Davy and Laplace in the *Memoirs*, I, 470.

51. *Consolations*, p. 250. For Brodie's advice, see G. Kitson Clark, *The Making of Victorian England* (London, 1962), p. 264.

ments are to be made without the consent and approbation of the Professor of Chemistry and the attempt at original experiments unless preceded by knowledge merely interferes with the process of discovery. There is a sufficient number of new and interesting objects which a modest student would wish to pursue and in which the path is marked and distinct."[52]

The sixth and last dialogue—one on atomic theory was incomplete and not published until later—concerns time. Here Davy sought to present a dynamic view of a changing world, in which apparent rest is not more than an equilibrium of forces, and in which decay is revealed to be the necessary counterpart of birth: "the principle of change is a principle of life; without decay, there can be no reproduction; and every thing belonging to the earth, whether in its primitive state, or modified by human hands, is submitted to certain and immutable laws of destruction, as permanent and universal as those which produce the planetary motions."[53] Under the force of gravity everything is slowly levelled. The other great agent of decay, chemical change, is also a form of attraction; and these two together would, if unopposed, "produce rest, a sort of eternal sleep in nature." In heat there is an antagonist and vivifying power, but changes of temperature are as destructive of all material forms as chemical agencies.

Most of the dialogue is taken up with this discussion of decay, but the feeling of mild melancholy which this might produce is dispelled in the final paragraph, where Philalethes reiterates that decay is a necessary complement to rebirth and progress:

Time is almost a human word and change entirely a human idea; in the system of nature we should rather say progress than change. The sun appears to sink in the ocean in darkness, but it rises in another hemisphere; the ruins of a city fall, but they are often used to form more magnificent structures as at Rome; but, even when they are destroyed, so as to produce only dust, nature asserts her empire over them, and the vegetable world rises in constant youth, and, in a period of annual successions, by the labours of man providing food, vitality and beauty upon

52. On the use of ignorant assistants, see Claude Bernard, *An Introduction to the Study of Experimental Medicine*, tr. Henry Copley Greene (New York, 1957), p. 23, where he describes the work of a blind scientist and his serving man. Bernard's whole discussion of experimental method is of considerable interest. Davy's orders to his assistants are in his laboratory notebook at the Royal Institution, under August 30, 1810.

53. *Consolations*, pp. 259, 260. Alternations of decay and renovation and heat as the renovating power are to be found in John Playfair, *Illustrations of the Huttonian Theory of the Earth* (Edinburgh, 1802), pp. 497 *et passim*.

the wrecks of monuments which were once raised for purposes of glory, but which are now applied to objects of utility.[54]

With this the book ends. Although Davy was presumably to some extent influenced by German thought, first perhaps through Coleridge and later through his travels and wide acquaintance, he liked to think of himself as in the tradition of Bacon, Locke, and Hartley. His belief, that in a world of flux only laws are permanent, was to be shared by those who throughout the nineteenth century appealed for a dynamic chemistry, based on forces, rather than the static eighteenth-century model, based on weights, which Lavoisier had bequeathed to the science. Many of the same issues are raised in Oersted's book and in Davy's: the ennobling nature of scientific research, the feeling for flux and process, and the question of the plurality of worlds. But Oersted's treatment is throughout more a priori in character, a difference shown in Davy's remark: "*Oersted* is chiefly distinguished by his discovery of electro-magnetism. He was a man of simple manners, or no pretensions, and not of extensive resources; but ingenious, and a little of a German metaphysician."[55] Oersted's acquaintances in Germany included Schelling, the Schlegel brothers, Fichte, and Schleiermacher.

The last essay in Oersted's book is in some ways the most interesting; it concerns the spirit and study of universal natural philosophy, that bold enterprise in which we aim to comprehend the infinite multitude of objects that there are in the world. The undertaking is not hopeless, because a "deeply penetrating search into nature shows us an admirable unity in all this infinite variety." In the study of biology our attention is directed to homologies and types, and by analogical reasoning we can even reconstruct creatures long extinct. In astronomy, the laws which govern the revolutions of the Earth, are also applicable to stellar motions: "These examples show us

54. *Consolations*, p. 281.

55. *Memoirs*, II, 216. Oersted's most important scientific mentor seems to have been Ritter, of whom Davy in 1810 said: "And though Ritter, in some of his conclusions, seems to have followed the impulses of a strong imagination rather than the results of observation, yet the science is indebted to him for the invention of several happy combinations. His errors, as a theorist, seem to be derived from the peculiar literary taste of his country, where the metaphysical dogmas of Kant, which as far as I can learn are pseudo-platonism, are preferred before the doctrines of Bacon, Locke, and of Hartley—excellence and knowledge being rather sought for in the infant than in the adult state of his mind" (*Works*, VIII, 272). The list of Oersted's acquaintances, which also included Rumford, appears in *The Soul in Nature*, p. xi.

clearly what is accurately proved by philosophy, *that every well-conducted investigation of a limited object, discovers to us a part of the eternal laws of the Infinite Whole.*"[56]

The laws of Nature are, in Oersted's view, simply the laws of Reason; but our knowledge of them can never be complete, or "more than a faint image of the great whole; for our Reason, although originally related to the infinite, is limited by the finite, and can only imperfectly disengage itself from it. No mortal has been permitted entirely to penetrate and comprehend the whole." The scientist must be filled with "devout awe," and be ever mindful of the limits of his powers; then reason will not be his only guide, but by an intuitive process he will be able to arrive at an "anticipating consonance with Nature." If science be a mixture of reasonable and inspired jumping to conclusions, then naturally it would be wrong to make utility its aim; for such an activity needs no justification beyond itself: "It should be pursued on its own account, both as an expression of our inward life, and as an acknowledgment of that which is divine within us."[57] Because the world is a dominion of harmonious reason, science will also produce excellent results on a lower plane, that of utility. But it should be followed as a liberal study; indeed as the liberal study *par excellence*, perfecting the mental powers, and bringing the soul into inward peace and unison with the whole of Nature.

But this is not the whole story of the relationship between science and technology, for science teaches us to govern Nature so that we are no longer confined to a narrow sphere. We can roam to all regions of the globe, and extend by artificial means (such as the microscope) the limits of our senses. Science teaches us to play the wild powers of Nature against one another, to control the thunderbolt and the torrent, to force fire into our service, and to destroy disease. Barbarians, unguided by Reason, may be thought of almost as a "crude and hostile work of nature," but even against them science has provided protection in the weapons of modern war, which can only be made by nations which have attained some degree of development. The value of technology seems to have been, for Oersted, that it created the situation in which progress was possible:

56. Oersted, pp. 448–449. The italics are Oersted's.

57. Oersted, pp. 451, 452. For a study of this kind of science, see Rom Harré, *The Anticipation of Nature* (London, 1965).

84

"In short, science facilitates, enlarges, and in various ways secures our condition in life, and removes various obstacles, which prevent the free activity of the spiritual development of mankind."[58] This position is not altogether unlike that presented in Davy's vision in the Colosseum, but Oersted's view seems even further from crude utilitarianism.

Oersted's cosmological interest is apparent in his notion that science is essentially the reading of the book of nature. The "fundamental truth" on which all precepts for scientific method must be based is that "the whole of nature is the revelation of an infinite rational will, and that it is the task of science to recognize as much of it as is possible to finite powers."[59] In such a world, the natural laws must be in harmony with one another, and must be unchanging and invariable. Similarly, the fundamental forces of Nature—"original indications by which the creative power is recognized in external nature"—must be indestructible.

The laws governing the motions of the heavenly bodies had been shown by Newton to have a "natural necessity," proving them to be laws of Reason beyond what we could have devised: "divine dictates of reason, which, happily for us, we are able to comprehend."[60] Laws of Nature, Oersted remarks, frequently have been deduced a priori and confirmed afterwards; indeed in his view this hypothetico-deductive approach is that usually employed by scientific men. The world and the human mind were created according to the same laws: "If the laws of our reason did not exist in Nature, we should vainly attempt to force them upon her; if the laws of nature did not exist in our reason, we should not be able to comprehend them." But the speculative intellect must be controlled by knowledge of natural science; and it was want of this knowledge which had hitherto, and still, held back the best philosophers in their attempted syntheses. The intuitive and prophetic views of nature must be more exactly defined, and further developed, by means of the profounder knowledge derived from reflection. A vital task facing the nineteenth century must be the reconciliation of the worlds of imagination and reason. Positivism—"a tendency to reject all those immediate truths

58. Oersted, p. 454.
59. Oersted, p. 454. On the notion of the indestructibility of force, see Thomas Kuhn's article in *Critical Problems in the History of Science*, ed. Marshall Clagett (Madison, 1959), pp. 321–356.
60. Oersted, p. 80.

which do not rest upon the impressions of the senses, and to found its entire faith on these, and on the decision of the logical understanding"—is a state of degeneracy as bad as superstition.[61]

The role of experiment in Oersted's account of scientific method is somewhat limited, as one might expect; but he is careful to say that hypotheses and established truths must not be confused. Experience shades into observation, and observation into experiment; the difference being that we have little choice in what we experience, somewhat more in what we observe, and complete freedom in what experiments we elect to perform. The peculiarity of modern natural science is in fact this cultivation of Experience, an elaboration of the whole riches of external knowledge. The experimentalist requires dexterity and "spiritual and material endowments," but these would be useless without the guidance of a mind attuned to Nature. And "To have seen a great number of natural phenomena is not to have an insight into nature." Only he who knows what questions to lay before Nature can expect to choose the right experiments and make discoveries. Such a man must not be a narrow specialist; he "should constantly retain the whole in his view—for otherwise it is impossible to have a clear representation of the parts; on the other side, . . . he should regard nothing as beneath his attention, for it still belongs to the whole."[62] Reverence and awe are the qualities fundamentally necessary; the scientist possessing them will not find even chores tiresome. He will try to turn every experience and observation into an experiment, and thus be master of his investigations. But Natural Philosophy must not be a collection of tricks. It must have a firmer basis, which can be provided by examining it from a "still higher point of view."

Experiment is then viewed as a creative art, which can "transfer our souls into creative activity, so as by that means to produce a more harmonious, living, and powerful knowledge of the constant development of nature." The highest forms of experiment are purely mental; differential and integral calculus, for example, are essentially "thought experiments," and so is the definition of a circle in geometry. The creative, genetic approach inherent in this kind of "experiment" led to real knowledge of causes; for we come to know thereby the *origin* of every truth, and, according to Oersted's Kantian meta-

61. Oersted, pp. 11, 18, 43, 60.
62. Oersted, p. 457.

physics, we cannot but accept it: "The origin of its existence and of our certainty therefore coincide, so that if it is represented in this manner, it is already proved."[63] Mathematics is certain because we see how all its truths originate.

Mathematics is a vital tool in the progress of science, for each change has its magnitude; but one must not strive to impose upon Natural Philosophy the form of mathematics, which is not its natural form. Mathematicians are happy to save the appearances, to produce equations to which the phenomena conform; the natural philosopher must seek for real forces. Both must meet when a certain degree of perfection has been arrived at. Science has come close enough to mathematics, and the latter might perhaps with advantage try to approach closer to natural philosophy; an experimental mode of explanation in geometry might be able to subsist alongside the present form.

More strongly than Davy, Oersted urged that the world of appearances is a world of flux, and even followed Heraclitus in his example to illustrate this. The material of a waterfall is not the same on different occasions, yet it is the same waterfall: "Perhaps the invariable in this phenomenon might be superficially termed the thought of nature inherent to it." The laws of nature must be unchanging; but past epochs in the history of the globe might have been very different from ours, for if it be a principle of nature that "everything developes itself in the course of time, different conditions must necessarily succeed one another, or I would rather say, proceed out of one another."[64] Things are "incessantly passing from one condition to another"; their invariable components, or essences, are the natural laws according to which they are formed.[65] Knowledge of the essence of things is simply knowledge of the natural laws in action.

For Oersted, matter was space occupied by forces,[66] and apparent rest an equilibrium of forces rather than an inactive state of existence. Bodies at rest are continually resisting compressing forces; and those on the Earth receive their share of the forces which maintain our

63. Oersted, p. 460. These thought experiments differ somewhat from those for which Galileo and his predecessors are censured—experiments which they implied they had carried out in the laboratory, when in fact they had not left their armchairs.
64. Oersted, pp. 9–10.
65. Oersted, p. 450.
66. Oersted, p. 450. Cf. the theory of matter in Michael Faraday, *Experimental Researches in Electricity* (London, 1839–55), II, 284–293.

planet in its orbit. Everything is affected perpetually by external forces tending to produce internal changes; and an "incessant alternation is maintained between [each body] and the rest of the universe by means of heat, electricity, and magnetism. A constantly renewed giving and taking of influences is inseparable from material existence." Where things seem to be unchanging, this is because changes in them are slow: "inaction is only apparent, as the hour hands of a clock appear to be stationary, when we take a cursory glance at them. This is, however, but a feeble simile, when we speak of changes which are hardly perceptible in the course of many thousand years. Imagine a clock with a hand which took ten thousand years to pass over the space which the hour hand does in one hour, and the simile will be more expressive."[67] Perpetual change is possible because all bodies are composed of the relatively few chemical elements in different arrangements; but it would be wrong to expect to find passive, permanent, indecomposable bodies or elements, for "matter is not an inanimate existence, but an expression of activity."[68] This is the ultimate rationale for a "dynamic" chemistry.[69]

So far Oersted's conclusions have been not unlike Davy's, but in that purely a priori region of science dealing with the plurality of worlds he was in the main stream of thought, stressing the uniformity of Nature, where Davy had been impressed by her diversity. All existence, in Oersted's account, is a dominion of Reason; the same laws must apply throughout the cosmos. A being on another world "might differ from us in the form and mode of his perception; but so far as the harmonious laws of nature are rightly appreciated, his understanding faculty must agree with these laws, and consequently with our powers of thought." The inhabitants of, say, Jupiter, might be more advanced than we in certain respects, in their capacity for mental arithmetic perhaps; but their conclusions must tally with ours. The laws and effects of light must, again, be the same, for astromomy has established that light from celestial and terrestrial sources behaves

67. Oersted, pp. 8–9. Cf. Davy on the generation of gold in nature: "We, it is true, have never seen it composed, or decomposed; but our works are in moments,— those of nature in ages" (*Works*, VIII, 329).

68. Oersted, p. 88.

69. Oersted, p. 312. Cf. Samuel Taylor Coleridge, "Essays Introductory, XIII," *The Friend* (London, 1818), I, 163n.; and the remarks of Carey Foster and Edmund Mills in the Chemical Society debates on atomic theory in the sixties (William H. Brock and David M. Knight, "The Atomic Debates," *Isis*, LVI [1965], 5–25; and *The Atomic Debates*, ed. W. H. Brock [Leicester, in press]).

88

in precisely the same manner, and that therefore: "there are no limits in the immeasurable range of the universe, beyond which the laws required by our reason would be invalid." Inhabitants of other worlds must be mortal like us, and must receive sense impressions as we do; they must be rational, and feel pleasure and displeasure. With the increase of knowledge here on Earth and on other planets, we may hope that our posterities will know more about each other: "there is an arrangement in finite existence by means of which one part of the universe may comprehend the other by its mental faculties; that, consequently, each essential portion of the universe may recognize the whole; even that every one may possess a knowledge of the Knowledge, the Faith, and the Recognition of a God in other worlds."[70] A splendid community of reason would be thus established.

Compared to Oersted, Davy seems to have been less enamored of reason, and he therefore advocated a less a priori science.[71] He was also less trustful of the notion of "development," which may be accounted for by the greater religious and scientific orthodoxy prevalent in England at this period. More interesting perhaps than the differences are the similarities in their outlooks. Both attached great importance to the moral qualities of the scientist; he must not be bullying or arrogant in his dealings with Nature, but must approach her in a respectful and almost religious spirit, with due humility. Both men were indubitably great scientists—Davy the more important—and both were remote from the inductivist tradition, and also rather cool about the conversion of natural philosophy into applied mathematics. Metaphysics seemed highly important to both of them; and both would have set out on experimental inquiries with fairly clear ideas of what conclusions their investigations would yield. The ideas expressed in these essays found echoes down the century in a tradition, the importance of which is now being realized by historians; for chemists throughout most of the nineteenth century a dynamic, harmonious, and all-embracing ideal was available as an alternative to the somewhat humdrum main stream of chemical thought.

70. Oersted, pp. 95, 100, 132.
71. On Davy and reason, see *Memoirs*, I, 29; and for an interesting discussion of his attitude towards development and other biological topics, see Philip C. Ritterbush, *Overtures to Biology* (New Haven, 1964), pp. 180–186.

IV

Romanticism and the sciences

'Romanticism' refers to a period and not just to a state of mind. While therefore one can classify the men of science of any period into the Romantic and the classical, we shall be concerned only with those active in approximately the life-span of Humphry Davy, 1778–1829. We all have an idea of what a Romantic author, painter or musician was like, but the various groups which they formed or which were identified by friends or critics were shifting. Consistency was not seen as a virtue, and personal relations were important; so one cannot define members of the Romantic Movement as one might members of a political party.[1] There was no membership card; and as the movement affected different arts in different countries at different times we cannot expect to find it easy to generalize, or to say who exactly was or was not influenced by Romanticism: almost everybody was, in some degree.

The sciences also lacked sharp and natural frontiers. Polymaths were not uncommon: John Dalton not only published an atomic theory but also worked on meteorology and colour-blindness, and Thomas Young proposed a wave theory of light, identified astigmatism in the human eye, published tables of chemical affinities, did fundamental work in mechanics and began the decipherment of the Rosetta stone. Neither of these were what one would think of as a Romantic man of science; taking a wide range was an acceptable and expected thing to do. Specialization was indeed beginning, but not yet necessary for success in the sciences.

Again, some of the most popular sciences of the day have not survived, such as mesmerism and phrenology. Then as now what looks most showy may collapse soonest, like William Beckford's gothick tower at Fonthill Abbey; but it is important that we do not convict of credulity or absurdity those who adhered to what seemed to their contemporaries exciting disciplines on the frontiers of knowledge. They did not have our benefit of hindsight; and to know in one's own time what is going to turn out well is a great gift. Excessive caution at any period may lead to one's missing out on important discoveries; playing safe is not necessarily the best strategy.

14

Around 1800 'science' was not opposed to 'arts'; there was nothing like the 'Two Cultures' of C. P. Snow's famous essay. Indeed the then current classification of subjects would have put engineering among the arts, a useful rather than a fine art, while almost all other subjects now taught in universities, such as chemistry, history and theology, would have been sciences. The real division was between the realm of science, governed by reason, and that of practice, or rule of thumb; and apostles of science hoped to replace habit by reason in the affairs of life. Some aspects of natural science, such as pharmacy, were seen as professional: to discuss them in public would have been impolite, 'talking shop'; but discoveries in the sciences were generally very appropriate for civilized conversation even when it included women.[2]

We can take as our text a poem by Davy, not published by him but in a notebook:[3]

> Oh, most magnificent and noble Nature!
> Have I not worshipped thee with such a love
> As never mortal man before displayed?
> Adored thee in thy majesty of visible creation,
> And searched into thy hidden and mysterious ways
> As Poet, as Philosopher, as Sage?

Here nature is not 'it' but by implication is 'she'; personified and active, 'natura naturans' rather than 'natura naturata', in process rather than complete: God is working his purpose out.[4] Davy did not see the universe as an enormous clock; and insofar as there is any Romantic natural science, it seems to involve rejecting mechanical metaphors in favour of organic ones. Under S. T. Coleridge's tutelage, the young Davy had had 'an experience of nature' outside in the sunshine; a mystical feeling of oneness in which it would have pained him to pull a leaf from a tree.

Davy shared with William Paley, the natural theologian, a love of fishing, unaffected by this experience;[5] and indeed records an anecdote about Paley being bullied by the Bishop of Durham to finish his great book which would put atheists to silence, and replying that he would get down to it again as soon as the fly-fishing season was over. Paley's *Natural Theology* of 1802 begins with the finding of a watch: the reader sees how all the components fit together, and cannot accept that the particles which compose the watch could have come together by chance. The rest of the book is taken up with the argument that the world is like a great watch, the work of a Designer. Paley's rhetoric is very skilful, but his argument is very old, going back to the great clocks of the Renaissance:[6] and to Davy and anybody associated with the Romantic movement, Paley's mechanical world-view (like his Utilitarian moral philosophy) was unwelcome.

Nature if she is magnificent and noble does not respond to the mere watch-repairer or anatomist, accustomed to taking to pieces things which do not work, and perhaps moved to praise God because he could not have done a better job himself. Instead, an attitude of admiration, love and worship is what Davy calls for: that is, a personal response. Coleridge[7] called his time 'the age of personality', referring to the large egos he found amongst his contemporaries; but, in other senses too, that is what it was. Davy was impressed by the alchemists' humble search for wisdom, with the belief that knowledge is only given to those who deserve it: natural science is a personal interaction with nature, not an autopsy.

This is expressed in Coleridge's 'Dejection Ode', in the lines:

> O Lady, we receive but what we give
> And in our life alone doth nature live,
> Ours is her wedding garment, ours her shroud.

Coleridge had got the idea from Plotinus that our minds are active, and that we project onto nature what we expect to find in her. The Kantian doctrine of categories imposed by us upon experience went well with such ideas; and indeed Kant's work was interpreted by early reviewers, including Davy's mentor Thomas Beddoes, as a revival of Neoplatonism. In the sciences, Alexander von Humboldt called his book about his travels in Central and South America a *Personal Narrative*; this classic work formed a model for Charles Darwin when writing up his diary kept on board HMS Beagle. Humboldt sought to balance the subjective and the objective; and his later *Aspects of Nature* was an attempt at word-pictures of exotic scenery, informed by botany, geology and zoology. Descriptive natural science in Humboldt was passionate, in line with Davy's poem; and particularly united the aesthetic and the scientific.

Davy's searching into the hidden and mysterious ways of nature may be prosaically rephrased as seeking a satisfactory explanation of phenomena. In 1830, John Herschel in his famous *Preliminary Discourse*[8] urged that explanation must be based upon a 'vera causa', that is analogy from known causes which produce similar effects elsewhere. Those rejecting a mechanical world-picture would find some analogies unpersuasive; in the natural sciences there is always the danger that one may seek to explain the known and familiar in terms of the unfamiliar or the incomprehensible, and here personal and cultural attitudes determine which is which. For Hegel, 'when philosophy paints its grey on grey, a form of life is over: the owl of Minerva flies only at the coming of the dusk', but this view of natural science as a post-mortem examination was not shared by all who admired organic analogies and sought what they called a dynamical science, based upon forces and the understanding of process.

Davy saw himself in his poem first as poet; and indeed both Coleridge and Robert Southey had been impressed with Davy in this role, and believed that

16

he might have been a great poet if he had not become a great chemist.⁹ The
poetry he has left behind does not give much support to this claim; but there
was one poet of the first rank who spent a good deal of time on investigations
in natural science, and that was Goethe. While he never thought of himself as a
Romantic (even in his period of 'Sturm und Drang'), loving Italy and the
classical tradition, his natural philosophy is resonant with Romanticism.

In natural history, in his studies of plants and skeletons, Goethe sought
unity of plan.¹⁰ Behind the variety of nature, he believed he could perceive the
Ur-plant: the simplest plant, upon which nature had, as it were, played a series
of variations. Flowers were thus to be understood as modified leaves. The
development of flowers was a kind of transcendental evolution; but this idea is
not like Darwin's, because Goethe was not seeking an explanation of how
plants came to be the way they are, but an understanding. To see the Ur-plant
in the rose or the daisy is akin to seeing the 'form' of something in Plato's or in
Bacon's philosophy: one is not in the business of genealogy, constructing
family trees, but searching for natural kinds. In the same vein, Goethe
discovered the inter-maxillary bone in the human skull. This was not like
finding a little bone which nobody had noticed before: rather it was a matter
of identifying a part of our skull with something much more prominent in
other animals. Again, it was a perception of underlying unity, based upon
careful looking.

This work created much less stir than Goethe's experiments in optics. In
Italy he had been impressed by the colours he saw in shadows; and he took up
some textbooks to see how light and colours were explained. He found there
ambiguous descriptions of Newton's 'crucial experiment' with prisms.¹¹
Newton had caused a narrow beam of light to fall upon a prism, and to be
refracted onto a screen some considerable distance away; he found that the
rays were not merely bent but that the image was elongated into a spectrum.
When he put a second prism near the screen in such a way that rays of only one
colour went through it, he found that they were refracted but not further
decomposed into colours. The second prism put near the first but the other
way up would recombine the rays so that a circular white image was formed
on the screen. Newton concluded that white light was therefore a mixture of
rays of the other colours.

Goethe misunderstood the descriptions he found, and looked through the
prism instead, when he found that he saw colours only when there was a sharp
boundary between black and white in his visual field. He believed that this
was a fundamental or primordial phenomenon, an *Urphänomen*; and
although Newtonians assured him that it could be explained in terms of
current physics, he believed that he had shown that Newton's work was not
based upon open-minded experiment but upon dogma: an axiomatic theory, a
mathematical abstraction based on the notion of rays, which had been

perfunctorily tested. The innocent eye recording phenomena was the proper foundation for natural science; in optics Goethe believed that Robert Boyle had worked this way, but that Newton had propelled the science in quite the wrong direction. This objection to overweening confidence and premature abstraction is quite different from Keats's feeling that Newton had destroyed all the associations of the rainbow, reducing the world to a prosy factuality. Goethe wanted a science of optics, but its aim would be to cast light upon what we see. Its basis was that colours were generated at boundaries of black and white, and he had experiments to show it.

Goethe went on to do various experiments in perception; but as Dennis Sepper argues in his *Goethe contra Newton: Polemics and the Project for a New Science of Color*,[12] the most interesting feature of his attack upon Newton was that he sought to undermine the whole notion of a 'crucial experiment'. For Bacon, the crux was the signpost at a crossroads, and a crucial experiment was one which indicated the most promising line of inquiry to follow. Newton, with the conviction that certainty could be found in geometrical demonstration, believed that a crucial experiment could be used to prove a theory: and this has become the modern use of the term. Given a theoretical proposition, some consequences could be deduced from it, preferably unexpected, surprising or almost paradoxical ones; and if these could be demonstrated experimentally, then the theory must be right. This process was illustrated in our century when Einstein's astonishing prediction that because space was curved, some stars actually behind the sun would be visible beside it in a solar eclipse, was verified in 1919, thereby falsifying Newton's own system of physics.

For Goethe, there were more or less plausible interpretations of experiments, but no experiment could entail a theoretical interpretation. Goethe hoped in optics, and in the sciences generally, for competing positions and critical debate, dreading dogma: this was something which in his own day he saw in chemistry, where all was in flux and where crowds came to hear public lectures. Goethe's ideal of natural science was thus anarchic; it was personal knowledge, not gained at second hand but based upon flashes of insight or disclosures, and tested by its comprehensiveness, unity and truth to experience. Those engaged in natural science should never take the groundwork upon authority: natural philosophy would then be a real education, an essential part of their individual *Bildung* (self-cultivation), ethics and politics. This is in some respects a curiously modern vision, and an attractive one in many ways, though hardly doing justice to the masses of information which constitute elementary natural science; but Goethe failed to persuade his contemporaries. This was largely because his tone was so polemical; feeling ignored, he assailed the Newtonians in the tones of a heresy-hunt or an election, and thus made sure that outside the narrow ranks of a few disciples

18

he made no converts. His insights into natural philosophy were lost; and though some of his psychological optics and his advice to artists on the use of colour were taken up in the nineteenth century, his work was generally seen as a dreadful warning of what can happen when able but untrained people venture into unfamiliar territory.

Goethe also used concepts from the natural sciences, especially chemistry, in writing literature. His *Faust* is full of alchemical ideas; but one could hardly say that these were widely current in the chemical community of 1800. They were from an occult tradition, in which texts have meanings on different levels, and are therefore open to both the ignorant and the initiated. Alchemy had earlier links with poetry, for instance in George Herbert, so there was nothing Romantic about using it. In his novel *Die Wahlverwandtschaften* (*Elective Affinities*), however, Goethe did something very different. The idea that metals and other substances had their particular and quantifiable elective affinities was a strong feature of the chemistry of the eighteenth century, and lasted into the nineteenth with Young and in elementary works, perhaps especially in Britain.

When a solution of what we call barium chloride is added to a solution of a sulphate, say sodium sulphate, then a 'double decomposition' happens: barium sulphate comes down as a white precipitate, and sodium chloride is left in solution. This is one of the standard tests for sulphates. Goethe picked up this idea in writing about a marriage and what happens when another man and woman are brought into the household. Goethe exploits the general interest in chemistry of the epoch, the time of A. L. Lavoisier, Joseph Priestley and C. W. Scheele, to set the scene with great skill, and uses chemical metaphors right through. But whereas in the test-tube there is no pain and the reaction goes speedily and quietly to completion, in life everything is very different.[13] The book is the history of a disaster, which makes the reader reflect how different people are from particles of matter. A book begun with metaphors from the inorganic world shows how inappropriate and reductive they are. This is indeed a Romantic message, conveyed with subtlety and intensity in a masterpiece of fiction. The cobbler should perhaps have stuck to his last: Goethe's novel is much more successful than his *Zur Farbenlehre* (*On the Doctrine of Colours*), but of course it raises questions of natural philosophy less directly.

Goethe may stand then for our poet; but Davy's next category was 'philosopher'. Around 1800, this term included the 'natural philosopher', and on board HMS Beagle Darwin was 'Philos' to his shipmates. Indeed the word by itself might almost mean what we call 'scientist', except that it lacked any professional connotations. But there were philosophers in our sense of the word who were involved in Romanticism and in natural science.[14] Friedrich Schelling in his *Naturphilosophie* presented a world of opposed polar forces, in which apparent rest was in truth only dynamic equilibrium, and solid bodies only in reality endured in the way waterfalls or columns of smoke do: his was a Heraclitean world where all is flux. His most eminent disciple among men of science was Hans Christian Oersted, who believed that electricity and magnetism, both polar powers, must be effects of one underlying polar force and be interconvertible: and proved it so with his discovery of electromagnetism. Oersted also wrote philosophical essays, later translated into English as *The Soul in Nature*, setting out his metaphysical beliefs; Davy described him on the strength of some of his earlier papers as a German metaphysician, an epithet which he did not mean to be polite. Another disciple of Schelling was J. W. Ritter, who discovered the ultra-violet rays in the spectrum. It is possible to talk of Romantic men of science in Germany as a genuine group.

Hegel was not strictly a Romantic, but one of his great interests was in classifying the sciences: as a part of his project for an encyclopaedia, and generally in his scheme of circles, levels and transitions between them.[15] He was strongly opposed, like Romantics, to reduction: he did not believe that the sciences of inorganic matter were at the top of the hierarchy, but at the bottom. One moved upwards through chemistry and electricity to the sciences of life and then on to psychology; and confusion must result when ideas appropriate to one level are unthinkingly applied on another. In Hegel's scheme, getting appropriate general ideas was as important as getting authenticated facts; but both were important. In both Schelling and Hegel, we find dialectic emphasized; truth is realized through resolution of the clash of apparent opposites. Similarly, chemical synthesis results from the fusion of opposites, as water is generated when oxygen and hydrogen are sparked together – this being then a rather new piece of chemistry. Men of science in Germany were aware of their use of and need for metaphysics,[16] while in Britain empiricism prevailed, at least in public.

The Romantic period was one of rising interest in physiology, the science of life: and in Britain the great man here had been John Hunter, who had raised surgery from a craft to a science. One of his well-known investigations concerned self-digestion. Our stomachs are made of flesh, and yet contain juices which are capable of digesting raw meat if we care to gulp it down. After death, it does sometimes happen that the fluids in the stomach do begin to attack it; and for Hunter and his disciples this showed that dead and living matter are subject to different laws. Only after death do the ordinary laws of inorganic matter supervene, and decomposition is the result.

Hunter was also impressed with the way we endure despite the flux of the material particles which compose our bodies; and in the Animal Chemistry Club in the first decade of the nineteenth century, which included Davy, William Allen the Baconian and Quaker pharmacist and Benjamin Brodie the

20

physician, these questions were discussed. One of Brodie's experiments concerned Lavoisier's idea that in respiration oxygen is absorbed and converted into carbonic acid, with production of heat. Brodie found that when oxygen is blown into the lungs of a corpse then carbonic acid is produced, but the body cools down quicker than if left alone: which was evidence for him that analogies between the living and the dead are to be suspected. Philippe Ariès notes how medical men at this period sought to make a sharper distinction between life and death.[17] Concern with the processes of life was very appropriate for Romantics; John Keats was a surgeon, and Mary Shelley's *Frankenstein*, with its portrait of the man of science as a sorcerer's apprentice, starts from the assumptions of a materialistic physiology.

Davy was probably the most important man of science in Britain who can be described as a Romantic. His abilities were first recognized by Gregory Watt, son of James Watt, who got him invited to Bristol as assistant to Thomas Beddoes at the Pneumatic Institute where 'factitious airs' (synthetic gases), recently discovered by Priestley, were to be administered to invalids, especially sufferers from tuberculosis. Here he met Coleridge, Southey and William Wordsworth, and might therefore be said to have become a Romantic, though this could never be more than one characteristic among others: it is more important in understanding Davy that he had always been a storyteller, and it was through rhetoric, in public lectures, that this small dark man from the Celtic fringe made himself the best-known natural philosopher of his day, and eventually President of the Royal Society.

Appointed first Lecturer and then Professor at the Royal Institution in London, Davy became the apostle of applied science, seeking to understand, and then perhaps to improve, the processes of tanning and of fertilizing crops in the years of Continental Blockade. From these rather unromantic and smelly bits of natural science, Davy returned in 1806 to work he had begun in 1800 in Bristol: the study of electrochemistry. Alessandro Volta's view had been that the mere contact of dissimilar metals would generate electricity, but this Davy, like most of his compatriots, could not believe. He saw electricity as the power responsible for chemical affinity, and therefore believed that a chemical reaction was necessary to generate electricity in the 'pile' of Volta. William Nicholson and Anthony Carlisle had in 1800 applied the terminals of a pile to water, and found that they got oxygen and hydrogen at the positive and negative poles respectively. Later investigations showed that the water around the positive pole became acidic, and that round the negative pole alkaline, so it seemed that some complex reactions must be going on. Davy refused to believe that electricity could be more than an agent of analysis, and in 1806 went on refining his experiments until they fitted his expectations – not the sort of thing Goethe would have approved of. For Davy, this meant using gold, silver and agate apparatus and then putting it in an atmosphere of hydrogen, so that dissolved nitrogen could not react with the nascent oxygen and hydrogen round the poles, forming nitrous acid and ammonia.

Having proved to his satisfaction that chemical affinity and electricity were manifestations of one power, Davy investigated the chemical effects of charging metals with electricity. He found that negatively charged metals are inert, and positively charged ones reactive, something he was later able to use in proposing cathodic protection for the copper bottoms of warships. He was particularly delighted with this finding, because it chimed in with his belief that chemical properties are not determined by the nature of components. For Lavoisier, oxygen was the generator of acids; acidity was a property one could expect if and only if oxygen were a component. Davy could not accept this; like his hero Newton, and in a reaction to the materialism of Priestley and the atheism of the French *philosophes* characteristic of his generation, he believed that matter was inert rather than active. God had chosen to impose powers upon brute matter, which otherwise could not think, live or even form crystals. Davy demonstrated at Bristol how different are laughing gas, air and nitric oxide, though all composed of oxygen and nitrogen; in London that the acid from sea salt contains no oxygen, but the substance which he called the element chlorine; and in Florence, after his marriage to a wealthy widow, that charcoal and diamond, which look so different, are chemically identical, both burning to give the same quantity of carbon dioxide. It was not the components, but the powers associated with them, which gave character to substances.

At the end of his life, Davy put together earlier thoughts into a series of dialogues about natural science and life, which was published posthumously as *Consolations in Travel*:[18] here Davy emerged as the sage, passing on experience to the younger generation. But although the book sold well, later editions having attractive engravings after drawings by Lady Murchison, it does not seem to have had much influence. It is mentioned as something someone was reading in a novel by Anne Brontë, and some of Davy's remarks on geology were taken up by Charles Lyell in his *Principles of Geology*, perhaps because Davy was safely dead and could not challenge Lyell's extreme uniformitarianism. Davy's research programme was developed by Faraday,[19] and his insight that chemical affinity was electrical is still a fundamental tenet of chemistry. This was a special case of the Romantic belief that all force was one, which led some men of science in the next generation towards the conception of conservation of energy, and the creation of classical physics.

And yet Davy's electrochemistry was very similar to the ideas developed independently by J. J. Berzelius in Sweden, who was not in any obvious way a

22

Romantic; and in the later development of conservation of energy some of those involved were not Romantic in any way, an example being James Joule.[20]

Romantic ideas about the unity of organisms went with a transcendental evolutionary scheme, which like the idea of the conservation of energy brought unity into the various branches of natural science which were otherwise separating into specialized disciplines; but Darwin's great synthesis in the *Origin of Species* was not rooted in Romanticism but in the very different tradition of Paley and Thomas Malthus.[21] The theory of the conservation of energy and evolutionary theory in the mid-nineteenth century developed in Germany and in Britain, where Romantic natural science had been strongest, and not in France; and this was a factor in the relative decline of French science in the course of the nineteenth century. We can no longer simply assent to Justus von Liebig's view that *Naturphilosophie* was the Black Death of the nineteenth century.

In scientific illustration, we also see the transition from the portrayal of stiffly mounted corpses on studio stumps to lively pictures of birds and animals in their habitats; but this development of visual language in natural science depended partly upon new methods of reproduction, wood-engraving and lithography, and partly upon the growth of zoos and the simplification of travel which made it almost as easy to get the artist to the creature as to get the corpse to the artist.[22] A pioneer like Thomas Bewick does not fit into our idea of the Romantic movement, though perhaps William Swainson might and Edward Lear[23] and J. J. Audubon in their different ways do so.

We are back where we started: Romanticism was not something to which one formally subscribed, and while it was not a polar opposite of natural science as some have supposed, one cannot outside Germany draw up a table of Romantic men of science and expect them to form a coherent group. Collective biographies, prosopographies, which can illuminate disciplines, societies and other institutions, do not seem very promising for so indefinite a group; but in this period of political and intellectual ferment, we shall find that some men of science and some writers and painters found that they had things to learn from one another. The biography of Davy or of Alexander von Humboldt must include references to Romanticism if it is to be adequate; and similarly anybody seeking to understand Goethe, Coleridge, J. M. W. Turner or William Martin will have to refer to some natural science. It is only because we are still dominated by the idea of the Two Cultures that we find this surprising; in the early nineteenth century natural science was a component of the one culture of western Europe. Romantics preferred some parts of it, the organic rather than the mechanical; but understanding nature was not something quite separate from other concerns, and was still open to those without formal training. In the Romantic period, natural science could still be fun.

NOTES

1 J. R. Watson (ed.), *An Infinite Complexity: Essays in Romanticism* (Edinburgh, 1983).

2 See my 'Accomplishment or Dogma: Chemistry in the Introductory Works of Jane Marcet and Samuel Parkes', *Ambix*, 33 (1986), 94–8.

3 J. Davy (ed.), *Fragmentary Remains, Literary and Scientific, of Sir Humphry Davy* (London, 1858), p. 14.

4 H. W. Piper, *The Active Universe* (London, 1962).

5 [H. Davy], *Salmonia*, in J. Davy (ed.), *The Collected Works of Sir Humphry Davy*, 9 vols. (London, 1839–40), IX, p. 11; D. Knight, 'Davy's *Salmonia*', in S. Forgan (ed.), *Science and the Sons of Genius: Studies on Humphry Davy* (London, 1980), pp. 201–30.

6 T. Cosslett, *Science and Religion in the Nineteenth Century* (Cambridge, 1984), pp. 25–45; C. M. Cipolla, *Clocks and Culture, 1300–1700* (new edn, New York, 1977); O. Meier, *Authority, Liberty and Automatic Machinery in Early Modern Europe* (Baltimore, 1986).

7 T. H. Levere, *Poetry Realized in Nature: Samuel Taylor Coleridge and Early Nineteenth-Century Science* (Cambridge, 1981).

8 J. F. W. Herschel, *A Preliminary Discourse on the Study of Natural Philosophy* (originally published 1830; with Introduction by A. Fine, Chicago, 1987), pp. 144ff.

9 T. H. Levere, 'Humphry Davy, "the Sons of Genius", and the Idea of Glory', in S. Forgan (ed.), *Science and the Sons of Genius*, pp. 33–58; J. Heath-Stubbs and P. Salman (eds.), *Poems of Science* (Harmondsworth, 1984).

10 See W. D. Wetzels, 'Art and Science: Organicism and Goethe's Classical Aesthetics', in F. Burwick (ed.), *Approaches to Organic Form* (Dordrecht, 1987). On forms in Bacon, see his *Advancement of Learning*, II, 5, and *Novum Organum*, esp. II, 11, on the form of heat: it is interesting that S. T. Coleridge, in his celebrated comparison of Bacon and Plato in *The Friend* (ed. B. E. Rooke, London, 1969), I, pp. 482ff., does not refer to this aspect of Bacon's philosophy.

11 See G. Cantor, *Optics After Newton: Theories of Light in Britain and Ireland, 1704–1840* (Manchester, 1983), esp. chs. 1 and 2.

12 Cambridge, 1988.

13 See my 'Chemistry and Poetic Imagery', *Chemistry in Britain*, 19 (1983), 578–82.

14 D. von Engelhardt, 'Romanticism in Germany', in R. Porter and M. Teich (eds.), *Romanticism in National Context* (Cambridge, 1988), pp. 109–33.

15 R.-P. Horstmann and M. Petry, *Hegels Philosophie der Natur* (Stuttgart, 1986).

16 D. von Engelhardt, 'Bibliographie der Sekundarliteratur zur romantischen Naturforschung und Medizin 1950–1975', in R. Brinkmann (ed.), *Romantik in Deutschland* (Stuttgart, 1978), pp. 307–30.

17 G. Averly, 'The "Social Chemists": English Chemical Societies', *Ambix*, 33 (1986), 99–128, esp. pp. 101f. P. Ariès, *The Hour of our Death* (trans. H. Weaver, Harmondsworth, 1981), ch. 9.

18 J. Z. Fullmer (ed.), *Sir Humphry Davy's Published Works* (Cambridge, Mass., 1969); *Consolations* is reprinted in Davy's *Collected Works*, IX, pp. 213–83.

24

19 D. Knight, 'Davy and Faraday: Fathers and Sons', in D. Gooding and F. A. J. L. James, *Faraday Rediscovered* (London, 1986), pp. 33–49.
20 See T. S. Kuhn's paper 'Energy Conservation as an Example of Simultaneous Discovery', reprinted in his *The Essential Tension* (Chicago, 1977).
21 J. C. Greene, *Science, Ideology, and World View* (Berkeley, Calif., 1981), esp. ch. 1.
22 W. Blunt, *The Ark in the Park: The Zoo in the Nineteenth Century* (London, 1976); D. Knight, *Zoological Illustration: An Essay towards a History of Printed Zoological Pictures* (Folkestone, 1977); A. Moyal, *A Bright and Savage Land: Scientists in Colonial Australia* (Sydney, 1986).
23 S. Hyman, *Edward Lear's Birds* (London, 1980); M. J. S. Rudwick, 'A Visual Language for Geology', *History of Science*, 14 (1976), 149–95; D. Knight, 'William Swainson: Naturalist, Author and Illustrator', *Archives of Natural History*, 13 (1986), 275–90.

V

STEPS TOWARDS A DYNAMICAL CHEMISTRY

"Science itself is becoming dynamical rather than mechanical; powers and agencies are discovered in nature itself, not less mysterious than those which miracle-workers spoke of. Man is able, through science, to exercise such powers as seem to attest the dominion of spirit over nature more completely than any signs they wrought".[2]

In the first years of the present century the leading opponent of atomism in the field of chemistry was Wilhelm Ostwald, and we know from Morris Travers' *Life of Ramsay* that Ostwald's views exerted a strong pull on chemists in England, though apparently not on physicists. In 1904 Ostwald delivered before the Chemical Society in London the Faraday Lecture, which he devoted to an explanation of how chemical changes and radioactive decay could be accounted for without introducing atoms. Shortly afterwards the researches of Einstein on the Brownian Movement forced him to recant, and to accept in some sense the truth of the theory that matter is particulate or granular. Ostwald was under the influence of Ernst Mach, who never accepted an atomic theory, and one might be tempted to treat him as a positivist, as the last great chemist seriously to challenge the use of unobservable entities in the science.[3]

But on closer examination the Faraday Lecture seems more interesting than that, because it can be read as the last attempt to achieve a chemical *Naturphilosophie*.[4] Ostwald's opposition to atomism would then be in a

---

[1] Based on a paper read in Oxford, at a seminar arranged by Mr. T. H. Levere under the aegis of Dr. A. C. Crombie, on 29 May, 1967.

[2] F. D. Maurice, *Theological Essays*, ed. E. F. Carpenter, London, 1957, p. 195. This edition reprints the text of the 2nd ed., of 1853.

[3] I have done so myself in *Atoms and Elements*, London, 1967, p. 143; on Ostwald's influence, see M. W. Travers, *A Life of Sir William Ramsay*, London, 1956, chap. XVII.

[4] For a negative view of *Naturphilosophie*, see H. G. Schenk, *The Mind of the European Romantics*, London, 1966, pp. 180–1. For a general account of philosophies of nature, see J. T. Merz, *A History of European Thought in the Nineteenth Century*, London, 1904–12, vol. III, chap. VI. On Ostwald, see E. P. Hillpern, "Some Personal Qualities of Wilhelm Ostwald", *Chymia*, II (1949), 57–64; and on the general topic of this paper, see E. Farber, "The Theory of the Elements and Nucleosynthesis in the Nineteenth Century", *Chymia*, IX (1964), 181–200; and V. M. Schelar, "Thermochemistry and the Third Law of Thermodynamics", *Chymia*, XI (1966), 99–124. C. C. Gillispie, *The Edge of Objectivity*, Princeton, 1960, chap. VI, describes the confrontation of the chemistry of Lavoisier and that of Lamarck; this drama was re-enacted at intervals throughout the succeeding century.

tradition which had consistently rejected mechanical accounts of phenomena in favour of dynamical ones, based on forces and not on brute matter. Indeed Ostwald, who was very interested in history (he published editions of the chemical classics) and would not have chosen names by accident, called his system a philosophy of nature. The lecture of 1904[5] begins with the remark that Faraday had consistently directed all his attention to the study of the transormations of force, or energy. He had in consequence been preoccupied with the question of the nature of the chemical elements. Faraday and J. B. Dumas, among others, had refused to believe that the number of irreducibly-different fundamental particles in the universe could be very large. The chemical elements could not be composed of true atoms, for this would offend against the simplicity and harmony which they knew to be characteristic of nature. They had produced various arguments in favour of the complexity of the elements, and Dumas' advocacy of the radical theory, and the analogy between radicals and elements, had led to the Periodic Table. Ostwald believed that every generation of chemists must formulate its views anew on this fundamental problem

Physicists generally held, he remarked, mechanical—or, if they were up to date, electrical—theories of the construction of the chemical atoms. But many chemists believed that hypotheses taken from another science prove ultimately insufficient, and that only chemical evidence should count in chemistry. On that basis—at first view surprising in the founder of physical chemistry—Ostwald set about giving an account of "chemical dynamics". From chemical dynamics, which included the study of reaction rates but meant in this case primarily the empirical Phase Rule, it was possible to deduce the laws of chemical composition. Therefore atomic hypotheses were not required. Ostwald's fundamental definition was that: "a substance or chemical individual is a body which can form hylotropic phases within a finite temperature and pressure range". A hylotropic phase is one of which the composition is not changed on change of phase; that is to say, it boils or freezes unchanged. Elements are then bodies which never form other than hylotropic phases; whereas compounds subjected to certain changes of conditions become non-hylotropic, and separate into components. From the requirement that hylotropic phases be formed, the law of constant composition follows; and according to Ostwald, though he did not elaborate, isomerism could also be explained.

Matter is not atomic but is simply a complex of energies which we find together in the same place. In explaining radioactivity, Ostwald supposed the elements to be regions of low energy. He took as his example a cave with

[5] W. Ostwald, "The Faraday Lecture", *Journal of the Chemical Society*, LXXXV (1904), 506–22.

stalactites on the roof; elements then correspond to drops of water on the ends of the stalactites, and to transform one into another energy is required to lift the drop up to the top of its stalactite so that it can flow down another. But the radium stalactite is so short—no more than a corrugation in fact— that a drop thereon may flow spontaneously onto a longer stalactite; and this is what happens in radioactive decay. Since helium, on the other hand, is stable and unreactive, its stalactite must be long. The only real difference between elements and compounds is that the available energy of the former is a minimum compared with that of all adjacent bodies.

There was very little in this lecture that had not previously been presented in English, sometimes by scientists of equal distinction, during the nineteenth century. Since in this kind of story it is almost impossible to falsify hypotheses about who influenced whom, we shall not be concerned in this essay with the question of Ostwald's sources. Such a study would demand a wide acquaintance with German chemical literature of the last century. Instead, we shall glance at some of the works in English which had appeared earlier which contained views rather similar to those of Ostwald, and which show the existence of a non-Daltonian tradition.

The first point is the opposition to atomism, and it is this which must have struck contemporaries most forcibly. From its first appearance Dalton's atomic theory—the first atomic theory to be of real significance in chemistry— had met with opposition.[6] Positivists like William Wollaston, Benjamin Brodie, and Mach refused to accept the existence of hypothetical entities having what seemed to them such a small cash value; mathematicians looked for laws like those of astronomy rather than for causal explanations; and Humphry Davy and other Romantics demanded an explanation more in accordance with the simplicity and harmony of nature. Some of Davy's utterances would seem to imply that he was a positivist: but his friendships; his opposition to systems and even to consistency; the fragmentary and original nature of his *opus*; his short creative life; the evidence of his more hypothetical papers, his lectures, his poetry, and his *Consolations in Travel*; and even his love for the falls of the Traun; all put him in the mainstream of Romanticism.[7] Ostwald comes much too late for it to be possible to say this of him; but nevertheless he seems to fit as easily into this tradition as into the positivistic one.

---

[6] W. H. Brock (ed.), *The Atomic Debates*, Leicester, 1967.

[7] See my paper on Davy and Oersted, "The Scientist as Sage", *Studies in Romanticism*, VI (1967), 65–88. The criteria for describing a man as a Romantic are derived from Schenk, *The Mind of the European Romantics*. It is possible to argue that in the laboratory Davy did not use his speculative abilities, but relied on soundly based analogies; see R. Siegfried, "Humphry Davy on the Nature of the Diamond", *Isis*, LVII (1966), 325–35.

62

Well before Dalton's theory appeared, Joseph Priestley had taken up a
dynamic corpuscularianism, following Roger Boscovich.[8] In this model, the
atom degenerates into a mere point and attention is directed to the forces
which emanate from it. Davy took up this idea, but his dialogue advocating
point atoms was unfinished at his death and was not included in his extraordinarily interesting *Consolations in Travel*, which Cuvier likened to the work
of a dying Plato.[9] The most famous popularization of the theory of point atoms,
as a minimum form of atomism, is Faraday's "Speculations" of 1844.[10] This
paper was referred to explicitly by Ostwald in 1904; and he remarked that[11]:
"Faraday ever held up the idea that we know matter only by its forces, and
that if we taken the forces away, there will remain no inert carrier, but really
nothing at all". Ostwald was prepared to be more radical than Faraday, who
retained point atoms, and he suggested that the energy distribution in matter
was probably uniform rather than grained.

An advantage of the atom of Boscovich was that it was not incompatible
with the belief that chemical synthesis was not a mere juxtaposition of particles
but a genuine amalgamation in which the constituents lost their separate
identities.[12] This was even simpler to account for if all forms of atomic theory
were rejected; and this is what many Romantics did, believing the atomic
theory to be a crude mechanical account put forward by physicists who were
trying to take over chemistry. Priestley had written[13]: "Hitherto philosophy
has been chiefly conversant about the more sensible properties of bodies;
electricity, together with chymistry, and the doctrine of light and colours,
seems to be giving us an inlet into their internal structure, on which all their
sensible properties depend. By pursuing this new light, therefore, the bounds
of natural science may possibly be extended, beyond what we can now form

[8] R. E. Schofield, "Joseph Priestley, Natural Philosopher", *Ambix*, XIV (1967), 1–15;
and R. E. Schofield (ed.), *A Scientific Autobiography of Joseph Priestley*, 1733–1804, Cambridge, Mass., 1966, pp. 194–7.
[9] G. Cuvier, *Mémoires de l'Academie Royale des Sciences*, XII (1833), xxxv; S. T.
Coleridge, *Philosophical Lectures*, ed. K. Coburn, London, 1949, pp. 33, 158.
[10] M. Faraday, *Experimental Researches in Electricity*, vol. II, London, 1844, pp. 284–93;
L. P. Williams, *Michael Faraday*, London, 1965, chap. 2.
[11] W. Ostwald, *J. Chem. Soc.*, LXXXV (1904), 519.
[12] See the article "Atom" in *The Scientific Papers of James Clerk Maxwell*, ed. W. D.
Niven, Cambridge, 1890, vol. II, pp. 448–9.
[13] J. Priestley, *The History and Present State of Electricity*, 3rd ed., London, 1775,
vol. I, p. xiv. *Cf.* F. Soddy on the study of radioactivity, *Science and Life*, London, 1920,
p. 103: "Prior knowledge of the atoms of matter has been superficial in the literal sense
—confined entirely to the outermost shell of the atom. We have now penetrated to the
interior . . .". The excitement of chemical speculations can be caught in the extract from
Coleridge's *Friend* reprinted in K. Coburn, *Inquiring Spirit*, London, 1951, pp. 248–9.

an idea of. New worlds may open to our view, and the glory of the great Sir Isaac Newton himself, and all his contemporaries, be eclipsed, by a new set of philosophers, in quite a new field of speculation". A dynamical chemistry, electricity (and hence polarity), and light, did indeed form the basis of the cosmologies of *Naturphilosophie*; and Newton was made to share his pedestal with Davy, the greatest of the Romantic scientists.

Davy advocated a dynamical chemistry from which solid atoms were excluded—this must have been the reason for his distaste for John Herapath's paper of 1821, which Davy found "speculative"—although molecules found a place in his system, as his unfinished Boscovichean dialogue shows.[14] He was therefore able to have both a kinetic theory of heat, and matter not composed of Daltonian atoms. And like Ostwald, but unlike for example, Lorenz Oken, Davy was able to maintain the distinction between speculation and observation. This was particularly clear in his work on the chemical elements; he used the criterion of Antoine Lavoisier, that elements resist decomposition, with more rigour than Lavoisier had; and yet he adhered firmly to a belief that most of the elements were complex. S. T. Coleridge pointed this out in a note on a work of Hans Christian Oersted's[15]: "It is of the highest importance in all departments of knowledge to keep the Speculative distinct from the Empirical. As long as they run parallel, they are of the greatest service to each other; they never meet but to cut and cross". Davy had properly proved chlorine to be a simple body on the one hand, and on the other speculated about one prime matter—what he called "the transcendental part of chemistry".

In *The Friend* Coleridge proposed an account of chemical change more radical and speculative than any to be found in Davy's works[16]: "EVERY POWER IN NATURE AND IN SPIRIT *must evolve an opposite, as the sole means and condition of its manifestation*: AND ALL OPPOSITION IS A TENDENCY TO RE-UNION.

[14] *The Collected Works of Sir Humphry Davy*, ed. J. Davy, vol. IX, London, 1840, pp. 383–8; see my paper "The Atomic Theory and the Chemical Elements", *Studies in Romanticism*, V (1966), 185–207. On Herapath, see G. R. Talbot and A. J. Pacey, "Some Early Kinetic Theories of Gases: Herapath and his Predecessors", *British Journal for the History of Science*, III (1966), 133–49. J. J. Thomson, *Recollections and Reflections*, London, 1936, p. 50, reports a conversation between Stokes and Lord Kelvin in which the latter declared that he did not believe in atoms, but only in molecules.

[15] K. Coburn, *Inquiring Spirit*, p. 249; a similar remark about the facts and theories of chemistry running side by side appears in the article by A. Kekulé, "On Some Points of Chemical Philosophy", *The Laboratory*, I (1867), 303–6; *The Collected Works of Sir Humphry Davy*, vol. VIII, p. 328.

[16] S. T. Coleridge, *The Friend*, London, 1818, vol. I, p. 155, footnote. *Cf.* J. B. Stallo's dismissal of the vortext atom, *The Theories and Concepts of Modern Physcis*, ed. P. S. Bridgman, Cambridge, Mass., 1960, p. 30.

184

This is the universal law of Polarity or essential Dualism, first promulgated by Heraclitus. . . . The Principle may be thus expressed. The *Identity* of Thesis and Antithesis is the substance of all *Being*; their *Opposition* the condition of all *Existence* or Being manifested; and every *Thing* or *Phænomenon* is the Exponent of a Synthesis as long as the opposite energies are retained in that Synthesis. Thus Water is neither Oxygen nor Hydrogen, nor yet is it a commixture of both; but the Synthesis or Indifference of the two; and as long as the copula endures, by which it becomes Water, or rather which alone *is* Water, it is not less a *simple* Body than either of its imaginary Elements, improperly called its Ingredients or Components. It is the object of the mechanical atomistic Philosophy to confound Synthesis with *synartesis*, or rather with mere juxtaposition of Corpuscles separated by invisible Interspaces. I find it difficult to determine, whether this theory contradicts the Reason or the Senses most; for it is alike inconceivable and unimaginable". Because, as we shall see, oxygen and hydrogen played an important part in the cosmologies of *Naturphilosophie*, the question of the composition of water was a particularly interesting one; thus we find Davy's "phlogistic" conjectures on the subject,[17] going back to Priestley but with a proper Romantic emphasis on polarity added; and the similar notions of Johann Ritter and Jacob Winterel, who believed water to be the basic substance, oxygen and hydrogen being differently-charged modifications of it.

The concern with forces rather than inanimate brute matter (which was excluded by Romantics who either denied that there was anything apart from forces, or else declared that matter was in some sense alive) fell in with a general distaste for mechanism. Thomas Carlyle in his review "Signs of the Times" in 1829[18] made the distinction between dynamics and mechanics, which can also be found in Coleridge's *Biographia Literaria* and was already a commonplace on the Continent. Dynamics, according to Carlyle, treats forces and energies, and particularly those of man: love, fear, wonder, enthusiasm, poetry, and religion, "all of which have a truly vital and infinite character". The dynamic is what grows and develops; or rather it is what has an Idea (rather than some mechanism) behind it. When applied to chemistry the term "dynamical" could imply some commitment to the idea of inorganic "development" or evolution, but it need not do so; in the authors we are considering, it does imply rejection of "mechanical" analogies, and interest in "forces" instead.

[17] J. Davy, *Memoirs of the Life of Sir Humphry Davy*, London, 1836, vol. I, pp. 403–6; see R. Siegfried, "The Phlogistic Conjectures of Humphry Davy", *Chymia*, IX (1963), 117–24.

[18] [T. Carlyle], "Signs of the Times", *The Edinburgh Review*, LIX (1829), 439–59.

In 1831 J. J. Waterson wrote an "exposition of a new dynamico-chemical principle", beginning with a quotation from Oersted.[18a] The introductory part refers to the "sublime grandeur and simplicity which so eminently characterises the works of nature", and to the conjecture that heat, light, electricity, magnetism, and perhaps gravitation "may all be particular modifications of one agent". We are still an enormous distance from knowing the primary cause of all these phenomena; but "Experiment, however ably conducted, has as yet shown nothing in heat, electricity and magnetism but simply and exclusively the existence of force". Heat is a repulsive energy existing between the constituent atoms of bodies; and electricity and magnetism "present still more curious instances of invisible forces which rival gravitation in the important parts they sustain in the economy of nature". The most remarkable feature in all these phenomena is the display of polar forces. Heat and electricity appear in chemical changes; electricity and magnetism are intimately connected; only gravitation seems to be left out. But this is probably because, like stallar parallax, the effects are too small to be detectable so far. It may well be that gravitation and heat are examples of the same elementary force.

Such notions are prevented from being of practical use, according to Waterston, because of our ignorance of the nature of *motion*. All attempts to account for motion so far fail "to convey any definite conception or satisfactory meaning to the mind". So philosophers have generally despaired. There has even been no general agreement over whether the quantity of motion in the world is constant or needs to be kept up. To prove either contention demands "an intimate acquaintance with nature. The constituent principle of the attractive and repulsive powers, and their mode of operation, may differ from all we can deduce in comparison". When we see things come to rest we suppose that their momentum is lost; but it may instead be reduced to an invisible state, according to fixed laws. On collisons, vibratory and undulatory motions are perceived: "These changes are influenced by the nature of the composing substance, which again is an immediate consequence of the peculiar molecular forces of the ultimate constituent particles. Since we have this reason to suppose that their molecular forces are, like gravitation, subject to fixed laws and are every way of like importance and universality, it becomes highly probable that they are alone the invisible agents which abstract the momentum of collision without any evidence of its existence being afterwards perceived".

If true, this would lead to a splendid simplification in which everything would really be reduced to the principles of matter and motion. Waterston

---

[18a] *The Collected Scientific Papers of John James Waterston*, ed. J. S. Haldane, London, 1928, pp. 531–536.

espoused an ætherial theory of gravitation, since he found action at a distance "inconceivable". He hoped for an account of molecular forces showing them to be analogous to gravitational ones, an objective which harked back to Laplace and eighteenth century workers, and looked forward to Lord Kelvin; but Waterston's "dynamical chemistry", as he later worked it out,[19] involved the reduction of chemistry to physics by means of corpuscular theories; which was not what early nineteenth century authors had meant by the term. The "dynamical" theory of gases, and A. W. Williamson's "dynamical" ideas of chemical equilibrium, fit into this tradition in so far as they explain apparently static states in terms of ceaseless movement; but hypothetical particles belong to another story.

William Whewell, in his *Philosophy of the Inductive Sciences*, took the idea of polarity as being perhaps the crucial one in chemistry, Faraday being singled out for commendation because he had expressed the polar character of chemical affinity without invoking particles with poles. He referred to the claim that Schelling had anticipated the suggestion in Davy's first Bakerian Lecture, of 1806, that chemical bonding is electrical; but declared that Schelling's remarks were too vague and ambiguous to count. Coleridge had in *The Friend* supported the idea of polarity[20]: "We anticipate the greatest improvements . . . from . . . those . . . who, since the year 1798, in the true spirit of experimental dynamics, rejecting the imagination of any material substrate, simple or compound, contemplate in the phenomena of electricity the operation of a law which reigns through all nature, the law of POLARITY, or the manifestation of one power by opposite forces". Whewell felt about chemical combination much as Coleridge and Faraday did[21]; that the two components "become one to most intents and purposes, and that the unit thus formed . . . is not a mere juxtaposition of the component parts". He rejected atomism, and the use of the term "attraction" for chemical affinity, which seemed to him to have no similarity to Newtonian "attraction", or gravitation.

The first edition of Whewell's book appeared in 1840, and Faraday's "Speculations" four years later. At about this time a number of translations

[19] *Collected Papers of Waterston*, pp. 565–87, 588–97. This other meaning of "dynamical", as found in Lord Kelvin, has been expressed by R. H. Silliman, *Isis*, LIV (1963), 465, as implying commitment to a programme of reducing "all natural phenomena to the laws of motion and impact". This was the traditional programme of atomists; and as such was anathema to those with whom this paper is concerned, who described it as "mechanical".

[20] K. Coburn, *Inquiring Spirit*, p. 259.

[21] W. Whewell, *The Philosophy of the Inductive Sciences*, 2nd ed., London, 1847, vol. I, p. 394; vol. II, p. 99.

of scientific works from the German began to be made. Richard Taylor's *Scientific Memoirs*, beginning in 1837, contained translations of papers mostly in the mainstream of scientific thought. Of less "respectable" works, Goethe's *Theory of Colours* was translated in 1840; Oken's *Elements of Physiophilosophy* in 1847; in 1849 Bohn began to publish the complete translation of Humboldt's *Cosmos*; and in 1852 he brought out Oersted's *Soul in Nature*. In 1848 Chapman, who two years before had published George Eliot's translation of Strauss' *Life of Jesus*, published J. B. Stallo's *General Principles of the Philosophy of Nature*, certainly the most readable account of *Naturphilosophie* to appear in English.

Oken's book was written "under a kind of inspiration" in 1810; and having been in its day influential, it was reacted against with some horror in the later nineteenth century. Coleridge read it in German, and while he had some rude remarks to make about Oken on colours, he followed him pretty closely in some notes on the differences between plants and insects.[22] Oken attached great importance to light; denied the inertness of matter; and stressed the polarities in nature. Unless the planets were in some sense alive, he believed, their orbits would be circular and not elliptical. There were three elements in the universe, which was composed of point atoms[23]: *"God being in himself is Gravity; acting,* self-emergent, *Light; both together,* or returning into himself, *Heat.* These are the *three Primals* in the world, and equal to the three which were prior to the world. They are the manifested triunity = *Fire"*. Hydrogen corresponded to caloric, oxygen to light, and carbon to matter.

No other elements were possible, for the world is composed on the principle of the zero or origin affirmed and negatived[24]; O, +, − : "All other bodies must be only different degrees of fixation of the above-mentioned bodies, or combinations of the same. Different degrees of carbon are without doubt the metals. Different degrees of oxygen are probably chlorine, iodine, bromine. Different degrees of hydrogen are probably sulphur. Nitrogen is probably peroxydized hydrogen, or an oxyd of hydrogen; this is indicated by its medium weight, and its perfectly azotic character". This all seems utterly lunatic,

[22] K. Coburn, *Inquiring Spirit*, pp. 250, 223.
[23] L. Oken, *Elements of Physiophilosophy*, trans. A. Tulk, London, 1847, p. 48. Davy's first paper, in T. Beddoes (ed.), *Contributions to Physical and Medical Knowledge*, Bristol, 1799, was concerned with the relationship between oxygen and light; and he returned to the question in his last work, *Consolations in Travel*, London, 1830, p. 196. Hydrogen, in the later phlogiston theory, was identified with phlogiston; to identify it with caloric must indicate slightly greater modernity. D. Low, *An Inquiry into the Nature of the Simple Bodies of Chemistry*, London and Edinburgh, 1844, considered hydrogen and carbon to be the only real elements.
[24] L. Oken, *Elements*, p. 61; see Coleridge, *Philosophical Lectures*, p. 403.

68

188

and yet it is not so very far from the speculations of Davy[25]; but Davy succeeded in using such ideas to generate experimentally testable consequences. We should notice that Oken's elements are essentially forms of energy rather than kinds of matter.

An intermediate position between Davy and Oken was taken up by Oersted, whose *Soul in Nature* contains a fair quantity of thoroughly *a priori* matter; and yet in reading it one cannot but be conscious that it was his Romantic conviction of the unity of all nature which led to his discovery of electromagnetism. His chemical theory was based on polarity[26]: "The antithesis between the process of combustion and the process of reduction, now becomes far more comprehensible, while we perceive that they depend on a preponderance of two opposite forces. That which we before named neutralisation, is no longer a secret to us, as we know that it depends on the equilibrium of exactly these opposite forces, only under a different form". He believed that: "Bodies . . . possess an inward power of acting, by means of which they occupy space. . . . What we most directly know of bodies, accordingly, is, that they are spaces filled with active powers". All bodies are probably composed of gaseous elements; material substances have been revealed by the progress of chemistry to be "mere phantoms of clouds and vapour". Rest, according to Oersted, is not an inactive state of existence, but a dynamic equilibrium; he had an Heraclitean partiality to waterfalls.

Elsewhere he declared that the atomic theory does not belong to science[27]: "Our investigation is indeed arrested before certain elements, which, for the present, we must consider simple elements; but science leaves no doubt that this is a mere passing idea. We may perhaps hit on some peculiar materials, which may be really acknowledged as the fundamental elements of matter, but even then our only power of distinguishing them is by the laws of their action. In short, matter is not an inanimate existence, but an expression of activity, by which the all pervading laws of Nature are determined and restrained".

This shows considerable parallels with Faraday's speculations, and also looks forward to Ostwald in its analysis of rest as dynamic equilibrium—Ostwald's example was an equally-loaded balance—and of matter as space occupied by forces. Oersted's book does not seem to have exerted any great influence in England; but Baden Powell, the Savilian Professor of Geometry

[25] H. Davy, "The Bakerian Lecture", *Philosophical Transactions*, XCIX (1809), 55–6, and elsewhere in this lecture.
[26] H. C. Oersted, *The Soul in Nature*, trans. L. and J. B. Horner, London, 1852, pp. 314, 4–5, 8.
[27] Oersted, *Soul in Nature*, p. 88.

at Oxford—who is best known as a contributor to *Essays and Reviews* and as the father of the defender of Mafeking—used arguments derived from Oersted in his *Essays* of 1855; arguing, from the principles of development and the unity of all sciences, in favour of the theses of *Vestiges*.[28] Baden Powell was associated with Sir John Herschel, and his account of the philosophy of science is generally more sober than Oersted's.

Humboldt's *Cosmos* represents an attempt to see nature whole, as a great panorama.[29] In general, Humboldt was rather severe towards speculators, taking in the introduction to the first volume of the Bohn translation what seems a very well-balanced position between empiricism and *a priori* science. In the Introduction to the fifth volume we do find some dynamical chemistry[30]: "Amid the boundless wealth of chemically varying substances, with their numberless manifestations of force—amid the plastic and creative energy of the whole of the organic world, and of many inorganic substances—amind the metamorphosis of matter which exhibits an ever-active appearance of creation and annihilation, the human mind, ever striving to grasp at order, often yearns for simple laws of motion in the investigation of the terrestrial sphere". The meaning of "motion" is clarified by a reference to Aristotle on qualitative motion, or metamorphosis, very different from mere mixture; such laws of motion would therefore explain the diversity of matter and its combinations in the manner we have already met in Coleridge and others.

Humboldt wrote a little further on[31]: "Metamorphosis, union, and separation afford evidence of the eternal circulation of the elements in inorganic nature no less than in the living cells of plants and animals"; and—a passage very like the conclusion of Davy's *Consolation*—"Our earthly sphere, within which is comprised all that portion of the organic physical world, which is accessible to our observation, is apparently a laboratory of death and decay; but that great natural process of slow combustion, which we call decay, does not terminate in annihilation. The liberated bodies combine to form other structures, and through the agency of the active forces which are incorporated in them a new life germinates from the bosom of the earth". In Humboldt therefore we find a view of the importance of *motion* which we shall meet again in Edmund Mills; and ideas of active forces in bodies, chemical change as

[28] Baden Powell, *Essays on the Spirit of the Inductive Philosophy, the Unity of Worlds, and the Philosophy of Creation*, London, 1855. My paper "Professor Baden Powell and the Inductive Philosophy" will appear in the March, 1968, number of the *Durham University Journal*.

[29] A. von Humboldt, *Cosmos*, trans. E. C. Otté, London, 1849–58. See M. P. Crosland, *The Society of Arcueil*, London, 1967, pp. 431–3.

[30] Humboldt, *Cosmos*, vol. V, p. 5.

[31] Humboldt, *Cosmos*, vol. V, p. 7.

metamorphosis, and the world as being in flux, characteristic of Romantic scientists.

From the attempt to illustrate a tradition from the display of parallels between a number of books in English, we may now pass on to two men who were indubitably connected together, J. B. Stallo, the German-born American philosopher of science, and his pupil, the chemist and geologist Thomas Sterry Hunt, a friend of Benjamin Brodie and Norman Lockyer. Because Stallo's *Concepts and Theories of Modern Physics* has recently been republished, edited by Percy Bridgeman, he is now quite well known as a positivist, who came independently to some of the same conclusions as Ernst Mach. His earlier book, *The Philosophy of Nature*, is undeservedly less familiar, and would seem equally to merit a new edition. Stallo wrote very well, and this exposition of *Naturphilosophie* is a *tour de force*. The first part of the book is Stallo's own; the latter part is devoted to an exposition of the systems of Kant, Fichte, Schelling, Oken, and Hegel—mostly the latter, who gets nearly two hundred pages. Hunt did not see any great clash between the doctrines of this book and the *Concepts and Theories*; positivism and *Naturphilosophie* were really not too remote from one another, as the opposition of Davy and Ostwald to atomism shows.

Stallo referred with approval to Goethe, who[32] "found the meaning of nature in nature's forms" and, abominating all mechanical views, treated the world as an organism. Stallo believed that science was animated by the search for unity, which had led to the conclusion that there was everywhere motion. Rest was only equilibrium, and could not even be conceived except as relative to motion; it could indeed be described as no more than an incident to motion. The substantiality of motion is the principle of all existence: the world presents an external, rigid, permanent *aspect*; but this is to be referred to an internal, vivifying, moving principle. All movement proceeds from vital forces, for action by impact is clearly absurd. Matter only arises from the essential self-exclusion of the vital principle; it exists only, for Stallo, in so far as the *Active* effects thereby its self-mediation. If matter may be said to exist in virtue of its inner vitality, it cannot consist of permanent atoms, the product of an irreflective and groundless theory, rejected by intelligent chemists "like Dr. Kane". Sir Robert Kane, an editor of the *Philosophical Magazine*, described Dalton's theory in his textbook, widely used in the United States, as[33]: "overlaid by a tissue of hypothesis so irregular, and so unnecessary, that for a long time the reeal dignity and excellence of the experimental laws was

[32] J. B. Stallo, *The Philosophy of Nature*, London and Boston, Mass., 1848, p. 18.
[33] R. Kane, *Elements of Chemistry*, 2nd ed., Dublin, 1849, p. 294. On the background to Stallo's speculations, see the review-article by George Wilson, *The Edinburgh New Philosophical Journal*, XXXVII (1844), 1.

underrated and misunderstood". Organisms, Stallo continued, frequently produce compounds of which the elements are not to be found in the medium in which they live. The value and being of the material sphere consists wholly in *relation*; it maintains the continuity of spirit, but is always limited qualitatively and quantitatively[34]: "Matter exists merely in virtue of its external limitation, and by reason of its internal relation to a process, it is finite, first in space and next in time".

Since dead matter is an absurdity, for everything that exists is alive; and since the atomic theory demands dead matter; it follows that the theory is nonsense, and that matter must be infinitely divisible. The Spiritual is the realm on which we should concentrate[35]: "The so-called different modifications of the Material are, in conformity with the above, so many different *phases* of the Spiritual in its outward existence". Ostwald's use of the term "phase" was not dissimilar, though Energy had replaced Spirit in his account.

The rest of Stallo's essay is taken up with a series of "evolutions", tracing the development of the Spiritual. Light plays a particularly important role[36]: "when one body becomes luminar for another, it states *itself*, its *external form*, AT the other". Light is not merely the symbol of mind, but also its representative in space. Nothing in Nature can be accidental or arbitrary; and that the metals all reflect light and are all—except potassium and sodium—heavy, proves that they are the bearers of "planetary" qualities. Gold is almost pure planetary matter, and hence unites only with chlorine. Everything else is more "individual", for every mass is "swayed" by two activities, one general and planetary, and the other peculiar and individual: "The terms of this new individual antithesis necessarily again intensate themselves to separate, unital existence, and as such they constitute the so-called CHEMICAL ATOMS OR EQUIVALENTS". The different elements are thus generated by a process of "special individualization".

[34] Stallo, *The Philosophy of Nature*, p. 34.
[35] Stallo, *The Philosophy of Nature*, p. 45.
[36] Stallo, *The Philosophy of Nature*, pp. 68, 82–3. For Davy's views on light, see J. Z. Fullmer, "The Poetry of Humphry Davy", *Chymia*, VI (1960), 102–26. It seems far from clear how Davy as a youth in Penzance formed ideas very like those current in Germany. J. Davy, *Memoirs of Humphry Davy*, vol. I, p. 38, writes that before Davy went to Bristol he had some acquaintance with the works of the Transcendentalists. From J. J. Tobin, *Journal of a Tour made in the Years* 1828–1829 *through Styria, Carniola, and Italy, whilst accompanying the late Sir Humphry Davy*, London, 1832, it appears that Davy's knowledge of German was negligible. A. F. M. Willich, *Elements of the Critical Philosophy*, London, 1798, would seem to be too late, and is far from easy to read. Elements of neo-Platonism Davy could have got from W. Enfield, *The History of Philosophy from the Earliest Times to the Beginning of the Present Century*, London, 1791, which we know he read.

Chemical combination is to be thought of, according to Stallo, in the way we have now come to expect, as a real synthesis[37]: "That all chemical combination is really an identification, and not merely an apposition of particles, is unwittingly taught by all the chemists, when they say that a chemical compound presents both components in the smallest possible molecule"—a term which Stallo allowed as sanctioned by long usage. An electric current is generated by the mere contact of dissimilar bodies, but continued by chemical action; and the "same chemical antithesis", or polarity, of which electricity was the ideal form, lay at the bottom of the differences of the chemical elements: "Had the Material formed itself in sole obedience to this antithetical power, we should have had numberless concrete individualities, but without qualitative distinctions". We have elements instead because the bondage of planetary life had interfered with the free working out of this scheme. Elective affinities represent an imperfect realization of this polarity.

We should therefore be very careful in our interpretations of chemical reactions, particularly in electrochemistry; for electricity is more likely to generate new substances than to analyse old ones[38]: "It is false, and certainly cannot be proved by experiment, that the chemical elements existed previous to electric action and to the individualizing process; this hypothesis again holds forth the radical error of assuming concrete, quiescent existences, and then *superinducing* movements, changes, and modifications. The logically preëxisting elements in special individualization were only the stages of planetary consistency; *from these the chemical elements were and are born in virtue of the individualizing and therefore antithetical electric action*". The Material, as we have already seen, cannot attain to permanence or enduring qualitative definition.

We should notice that Stallo used "evolution" in its then-usual sense to mean an individual process, the growth of an organism, and not to mean "development".[39] This next step was taken by his pupil Sterry Hunt, a Fellow of the Royal Society, in 1867, in a lecture on the Chemistry of the Primeval Earth delivered at the Royal Institution London. The idea that the various elements had developed from a common ancestor cannot have been entirely new, for in his famous review of *The Origin of Species* Asa Gray suggested that the widespread acceptance of theories of inorganic development might increase the readiness of readers to accept Darwin's conclusions.[40]

[37] Stallo, *The Philosophy of Nature*, pp. 87, 91.
[38] Stallo, *The Philosophy of Nature*, p. 91.
[39] T. H. Huxley, *Darwiniana*, London, 1893, chap. VI.
[40] A. Gray, *Darwiniana*, ed. A. H. Dupree, Cambridge, Mass., 1963, pp. 9, 44.

Much earlier, in 1853, Hunt had written of the "Theory of Chemical Changes and Equivalent Volumes".[41] He wrote that matter was subject to two forces, the effect of one of which was condensation, and of the other, expansion. Chemical change was to be viewed as "metagenesis"; a process in which two or more bodies unite and merge their specific characters in those of a new species. No notions of elements or preëxisting groups should have any place in chemistry: "It is to be remembered that our science has to do only with phenomena, and no hypothesis as to the noumenon or substance of a species under examination, based upon its phenomena, or those of a derived species, can ever be a subject for science, for its transcends all sensible knowledge". Chemical union is, as Kant taught, interpenetration rather than juxtaposition, and any atomic theory is quite mistaken. Hunt, in the same tradition, like Davy separated the sphere of biology completely from that of chemistry.[42] Crystals or organic tissues, he wrote, only become the subjects for chemical change when they have been destroyed. It was for this kind of vitalism that Liebig attacked *Naturphilosophie*; and indeed organic chemists seem to have been more ready to accept atomism than electrochemists and physical chemists.

In accordance with his view of chemical combination, Hunt believed that chemical changes should be studied in terms of volumes, which do merge, rather than of atoms, which do not; and he tried to apply Gay-Lussac's law to the combinations of solids and liquids. In 1854, in another paper[43] he suggested that solution is chemical union; an idea that lay behind Ostwald's attempt to discuss chemical combination in terms of the Phase Rule, which properly applies to solutions and alloys. Hunt remarked that there is often a definite limit beyond which the affinity of a liquid for water is satisfied; and that if any more is added a mechanical mixture results, which separates into two layers. Solution was a result of the tendency in nature towards unity, condensation, and identification. Hunt now found even the Kantian notion of interpenetration too mechanical as an account of chemical combination, for specific characters interpenetrate as well as volumes; and he therefore preferred the Hegelian view that the chemical process is an identification of the different

---

[41] T. S. Hunt, *Chemical and Geological Essays*, Boston, 1875, pp. 426–37. This essay had appeared in the *Philosophical Magazine* in 1853. *Cf.* Dumas' remarks on atomism in 1837; that the term *atom* should be abolished because it went beyond the facts, which chemistry should never do; J. B. Dumas, *Lecons sur la Philosophie Chimique*, Paris, 1837; G. Buchdahl, *British Journal for the Philosophy of Science*, X (1959), 120. Dr. W. H. Brock first called my attention to Sterry Hunt.

[42] Davy, *Consolations*, pp. 198–201; Schenk, *The Mind of the European Romantics*, p. 180.

[43] Hunt, *Chemical and Geological Essays*, pp. 448–58. Hunt's views on chemical change may also be found in his *New Basis for Chemistry*, London, 1887.

194

and a differentiation of the identical. In double decomposition, he believed, there was always first union, then division; and in this process affinities could sometimes be reversed by slight changes in conditions.[44]

In the lecture of 1867[45] he produced the idea that the elements might have developed from simple ancestors, drawing upon Deville's studies of dissociation. Brodie had at the same time hit upon a somewhat similar notion, and Lockyer also published views of this kind. In 1886 Crookes in his famous Presidential Address to the British Association made the idea widely known; but there seems no evidence that any of these workers, apart from Hunt, had any connexion with *Naturphilosophie*. In 1882 Hunt wrote another paper,[46] "Celestial Chemistry from the Time of Newton", suggesting that Newton's æther (from the Hypothesis touching Light and Colour) should be taken seriously. Newton "saw in the cosmic circulation, and the mutual convertibility of rare and dense matter, a universal law, and rising to a still bolder conception, which completes his hypothesis of the Universe, adds: 'Perhaps the whole frame of Nature may be nothing but various contextures of some certain etherial spirits or vapours condensed, as it were, by precipitation, much after the same manner that vapours are condensed into water, or exhalations into grosser substances, though not so easily condensible; and after condensation wrought into various forms, at first by the immediate hand of the Creator, and ever since by the power of Nature, which, by virtue of the command "Increase and multiply" became a complete imitator of the copy set her by the great Protoplast. Thus, perhaps, may all things be originated from ether'." Hunt added that Grove and Humboldt had employed a first matter not unlike this one, but that nobody had yet worked out these suggestions from a chemical point of view. But that some kind of æther was the food of the planets, the "material basis of life" constantly supplying the carbon which was continually being used up, seemed to Hunt more than likely.[47]

[44] There seems to be no explicit reference to Berthollet's views on chemical equilibria in Hunt; but in *Chemical and Geological Essays*, pp. 459–69, there is a paper in support fo Laurent.

[45] T. S. Hunt, "On the Chemistry of the Primeval Earth", *Chemical News*, XV (1867(, 315–7. On the speculations of Brodie and Lockyer, see W. H. Brock (ed.), *The Atomic Debates*, Leicester, 1967, pp. 11–4. W. Crookes, *Report of the British Association*, 56th meeting (1886), 558–76.

[46] T. S. Hunt, "Celestial Chemistry from the Time of Newton", *Chemical News*, XLV (1882), 74–6, 82–3; the passage from Newton may be found in I. B. Cohen (ed.), *Isaac Newton's Papers and Letters on Natural Philosophy*, Cambridge, 1958, p. 180.

[47] Hunt was hardly up to date in this suggestion; for earlier speculation on the subject, see R. Rigg, *Experimental Researches, shewing Carbon to be a Compound Body made by Plants*, London, 1844.

Sterry Hunt's suggestions do not seem to have fallen on very receptive ears, but at the Atomic Debate at the Chemical Society in 1869[48] two speakers, Carey Foster and Edmund Mills, made contributions which were very much in the tradition we have been looking at. Carey Foster proposed that chemical combination could be regarded as a kind of transmutation rather than as a juxtaposition of particles, though he seems to have felt (a trifle sadly) that such a change of heart on the part of chemists was unlikely.

Edmund Mills' protest against the atomic theory at the Debate seems to have taken up two points: first, that atoms were not a *vera causa*, as waves were in the undulatory theory of light; for we know about waves, whereas we have no analogy for an atom.[49] Second, as he put it in a paper of 1871[50]: "Surrounded on all sides with continuity, motion, and change, our most popular ideas relate to limits, repose, and stability". With these ideas the static atomic theory of Dalton is certainly in accord; but the sciences should be based on the most general idea existing at a given time, and that idea was then *motion*. Acidity and alkalinity, for example, should be explained in terms of polarity,[51] as Davy had correctly seen: "The acidity of any substance is a kind of resultant whose direction is hydrogen". In all combinations, one substance plays the role of an acid, and the other of an alkali; this antagonism constitutes affinity. Davy, Avogadro, Laurent, Graham, and Brodie—a somewhat strange group —had all contributed, in Mills' view, to the idea of polarity in chemistry. His discussion of the atomic theory[52] displays an almost Coleridgean erudition, with lengthy quotations from Sir Kenelm Digby, who had denied the existence of brute matter, and a range of more eminent natural philosophers to whom the atomic theory had been repugnant. Mills followed Berthollet in suggesting that definite proportions were in fact "tinged with continuity".

Ostwald's paper refined these ideas, and the "motion" and "force" of earlier writers became "energy". The polarity is harder to see, but the dynamical idea of equilibrium involves the idea of opposed forces, and one could interpret in terms of a balance of forces the stalactite model for the chemical elements. Certainly, for Ostwald chemical change was a blending, solutions with constant boiling point being the type of chemical compound, and rules derived from

[48] W. H. Brock (ed.), *The Atomic Debates*, pp. 19–26.

[49] *Chemical News*, XX (1869), 236; on *verae causae*, see J. Herschel, *A Preliminary Discourse on the Study of Natural Philosophy*, London, 1830, pp. 144–8.

[50] E. Mills, "On Statical and Dynamical Ideas in Chemistry"; Part III. "The Atomic Theory", *The Philosophical Magazine*, 4th Series, XLII (1871), 112–29. The quotation is from p. 123.

[51] E. Mills, "On Statical and Dynamical Ideas in Chemistry": Part I. "Acid, Alkali, Salt, and Base", *The Philosophical Magazine*, 4th Series, XXXVII (1869), 461. Part II of the series appeared in 1870.

[52] See above, note 50; the quotation is from p. 120.

alloys and solutions being applied to the explanation of definite proportions. There is also the objections to models from physics being used in chemistry; although the justification for this is much less high-flown than in earlier examples, it stil consists of setting the "dynamical" against the "mechanical". Any listeners who had read the literature of *Naturphilosophie* in English need not have been very surprised by this Faraday Lecture, and would not have regarded it as mere positivism.

If this be so, and if Romantic chemistry deriving from Kant (and from Renaissance chemists) can claim the founder of physical chemistry along with the founders of electrochemistry and electromagnetism, then clearly there is little need to provide a justification for taking this kind of material seriously. The opposition of Liebig can be set against the support of chemists equally important; although most of us might tend to feel the surprise of F. D. Maurice's unitarian[53] confronted with an orthodox saint, at the "plain and practical virtues" which "flowed out of the faith which he had been taught was so likely to beget immorality". It still remains to be proved in any detail that their philosophical views seriously influenced these chemists in their work; it is, of course possible that when they entered the laboratory they left all these ideas behind and proceeded by purely empirical methods. We have after all Einstein's dictum that one should look at what scientists do and not at what they say they do. But it appears unlikely that in these cases ideas derived from *Naturphilosophie* played no role outside the study and the lecture room; and in the case of Davy the recurrence of the ideas of polarity, of the unity of nature, and of the importance of light, would imply very strongly that they guided his experiments. It has already been demonstrated that the principle of conservation of energy was enunciated by, among others, those imbued with *Naturphilosophie*; and it need not surprise us, therefore, that those who introduced the concept of energy into chemistry drew on the same background.[54] From the time of Lavoisier down to our own century there was not only one paradigm available in chemistry, leading it inexorably towards objectivity; but at least two, one deriving from Lavoisier, and the other the one we have

[53] F. D. Maurice, *Theological Essays*, p. 154.

[54] See T. Kuhn's article in M. Clagett (ed.), *Critical Problems in the History of Science*, Madison, 1959, pp. 321–56. The influence of such ideas on Faraday is explored in L. P. Williams, *Michael Faraday*. For the situation in England, see W. F. Cannon, "History in Depth: The Early Victorian Period", *History of Science*, III (1964), 20–38, esp. pp. 23–4. The argument that in order fully to understand a work of art of another period we need to pursue philosophical arguments further than the artists had to, is presented in E. Wind, *Pagan Mysteries in the Renaissance*, 2nd ed., Harmondsworth, 1967, pp. 14–5; similar arguments can be applied to the history of science without committing the historian to the belief that science is really no more than an intellectual game.

examined. Lavoisier's was certainly the easiest to find in chemistry books; but the other was, as we have seen, not difficult to ferret out in print in English; and anyway chemists—and particularly those interested in theoretical chemistry —do not confine their reading to chemistry books.[55]

[55] J. J. Tobin, *Journal of a Tour*, records some of the books which Davy took on his last Continental journey, when he was writing the *Consolations in Travel*. Most of the other authors discussed show evidence of wide reading. On "Scientific Romanticism" generally, see W. Pagel's articles: "Religious Motives in Medical Biology". *Bulletin of the History of Medicine*, III (1935), 286–97; and "The Speculative Basis of Modern Pathology", *Bulletin of the History of Medicine*, XVIII (1945), 1–43. In a book of which I am the editor, *Classical Scientific Papers, Chemistry*, London, in press, there will appear in facsimile papers by Davy, Faraday, Whewell, Mills, and Ostwald relating to this topic.

VI

THE PHYSICAL SCIENCES AND THE ROMANTIC MOVEMENT[1]

"WHAT is man the wiser or the happier for knowing how the air-plants feed, or how many centuries the flint-stone was in forming, unless the knowledge of them can be linked on to humanity, and elucidate for us some of our hard moral mysteries."[2] "Socrates rose as a reformer and in the heat of reform he confined all philosophy to the knowledge of our own nature. . . . Plato, his great disciple, perceived that this were true if it were possible, but that the knowledge of man by himself was not practicable without the knowledge of other things, or rather that man was that being in whom it pleased God that the consciousness of others' existence should abide, and that therefore without natural philosophy and without the sciences which led to the knowledge of the objects without us, man would not be man."[3]

The historian of science cannot without unease read of Keats's toast of oblivion to the name of Newton, or his remark about the rainbow. A lack of respect for the achievements of the Romantic poets is a matter for pity rather than censure; and no doubt the same applies to Newton. At any rate the Newton we know, the last of the magi, might well have been more appealing to Romantics than the Augustan figure in whose shadow the natural philosophers of the eighteenth century seemed to poke about. But it is a serious matter for the historian to find those to whom as poets or philosophers he warms, and whose voices ring living through the passage of a century and a half, criticising in radical terms often those very aspects of the science of their contemporaries which have become the pillars of modern orthodoxy. While it is not the historian's job to judge those whose labours he describes—and the historian investigating science in the Romantic period cannot but assess theories by their coherence and consistency with then-known phenomena rather than whiggishly on their supposed correspondence with modern views—he would be a cold fish indeed if he were able to analyse trenchant and voluble writings on the value of science without taking them seriously.

In considering physical science in the Romantic period, there are two questions which must be kept distinct: the first is, What influences from the world-view of Romantics penetrated and transformed physics or chemistry? and the second, What impact did discoveries in these sciences have upon Romantics? In answering particularly the first, we must bear in mind the cautions against systems, notions, and hypotheses which all Romantics gave. It would be hard to frame a definition of a Romantic which would exclude, for example, Thomas Beddoes, Johann Ritter,

Hans Christian Oersted, or Humphry Davy; but it is another matter to establish to what extent, if at all, specifically-Romantic ideas played a part in their laboratory practice. This problem becomes acute when one tries to pursue philosophical doctrines across linguistic frontiers. Thus it would be impossible to falsify the contention that certain English natural philosophers were influenced by Kantian thinkers with whom their theories display some parallelism; but on finding that the chief candidates, Davy and Faraday, had no German, one wants strong proof of genetic connexion, of homology rather than analogy, if the result of the enquiry is to be more than a just-so story.[4] The historian, rather than searching for parallels which might indicate influences, should perhaps content himself with expositions, and might be well advised to explore the way in which certain terms, such as *force, power, energy, dynamical, etherial, hypothesis*, and *system* were employed in the thirty years or so either side of 1800.

The second question is perhaps one which has been less investigated, but which has interesting possibilities. In connexion with it, we may notice recent editions of Coleridge's *Friend*, and *Hints on the formation of a more comprehensive theory of life;* of *Omniana*, which he composed with Southey; of the English translation of parts of Goethe's *Farbenlehre;*[5] and of new translations of Hegel's *Science of logic*, and of his *Philosophy of nature*. In a period when the term 'scientist' had not been coined, when the natural philosopher was expected to develop and present a world-view,[6] and when most science seemed important as providing data for a cosmology rather than a basis for technology, the view of science held by Hegel or Coleridge is not unimportant. It is worth remembering here that while the pursuit of knowledge for its own sake now seems the foundation of academic freedom, and the intrinsic interest of pure science is apparent, to Hartley[7] the uninterrupted pursuit of truth was dangerous: "our Appetites must not be made the Measure of our Indulgences", in science or other amusements.

Let us turn, therefore, to our first question, on the impact of specifically Romantic ideas upon the physical sciences. We might begin by taking note of the recent study of L. R. Furst, *Romanticism in perspective*,[8] which emphasises the differences between the Romantic movements in Germany, England, and France. This and other works of literary history may provide a model for the historian of science venturing into this territory. Although the Romantic Movement spread across frontiers, we should remember, for example, that Coleridge[9] advises his reader to turn "with indignant scorn" from the "false Philosophy or mistaken Religion, which would persuade him that Cosmopolitanism is nobler than Nationality". It may be that physical science is a peculiarly cosmopolitan activity; but if we are to link it to Romanticism then we cannot forget national frontiers. In any nation, before invoking foreign influence without explicit documentation, the historian must exhaust the native possibilities; and he must remember that in its translation, dissemination, and assimilation any doctrine will undergo considerable modification. When ideas are transmitted through such individualistic and original authors as Coleridge and Carlyle, this effect will be even more strongly marked.

Examining Romanticism as a literary phenomenon, Furst remarks that in England it took a much less revolutionary form than in France or Germany. English Romantic poets are not separated by a great chasm from those of the eighteenth century; and Furst emphasises the importance of such pre-Romantics as Thomson, Akenside, and Young. In England there was less to rebel against. The classical tradition had not killed individualism; in Wordsworth, for example, Furst finds a compound of Augustan factuality and accuracy of observation with Romantic feeling and imaginative modification of phenomena. English Romantics were less thoroughgoing than the Germans, who had to outbid the *Sturm und Drang* of Goethe and Schiller, or the French, who had to destroy the Bastille of classicism; hence they form a much less coherent and formal group. It might be supposed that a detailed analysis along these lines might completely erode away the notion that there was an European Romantic Movement; but Furst recognises that there was this creative renewal, and finds a common factor among these Romanticisms in the acceptance of the creative imagination as the mainspring of the arts.

A version of this thesis might fit the history of science at the same period. Students of nineteenth-century chemistry should perhaps pay more attention to the eighteenth century. Wurtz's often-quoted remark that chemistry is a French science founded by Lavoisier has been taken too seriously even by those who have not believed it; and if Lavoisier is made to share his throne, his approximate contemporaries are put beside him. Before accepting too readily that the dynamical science of the nineteenth century is derived mainly from the dynamical philosophy of Kant, it is wise to investigage how far it resembles the dynamical science of the mid-eighteenth century. We are fortunate to have recent publications by Thackray and Schofield on Newtonian dynamical chemistry[10]; while a century ago William Odling, in a paper reprinted in the useful *Royal Institution library of science*,[11] drew attention to the theory of phlogiston as anticipating the application of the concept of energy in chemistry. Newtonians supposing that all the matter in the solar system might be contained within a nutshell would not have shared the slight surprise with which Frederick von Schlegel[12] declared that modern chemistry had "destroyed for ever, that appearance of rigidity and petrifaction which the corporeal mass of visible and external nature presents to our observation"; though they would probably not have shared his conviction that the so-called inorganic realm was alive. If it be Romantic to be interested in powers and forces, and to have high hopes that light and electricity were the key to the inner nature of things, then there were pre-Romantics among English Newtonians.

If the Romantic Movement in literature was a revolt against oppressive and moribund forms, then it might well be that in the sciences one would only find a comparable development in regions or in disciplines where paradigms had become inapplicable or oppressive. Goethe's *Farbenlehre* was a frontal assault upon Newton, while Young presented his optical papers as a development of Newton's; during the reign of Joseph Banks it could be that there was in England no need to reject overweening mathematicians. It may be doubted, though, whether any general Romantic Movement in science would emerge from such investigations; for while it is possible to find important physical scientists in Britain and in Germany who can be called Romantics, there seems to be none in France. And while it is plausible to urge that Augustan classicism had reached a dead end, the corpuscularian tradition still had plenty of life in it. Discoveries were made by unreflective or materialistic atomists, whereas important works of literature were not produced by latter-day Augustans; and only a small minority of physical scientists could be described as Romantics even in Germany, and certainly in England.

Before we abandon hope of finding any route across these quicksands —and it does seem implausible that a major convulsion in the arts would not have affected science—we should take note of another source of inspiration for Romantics which has been strongly canvassed by Kathleen Raine[13]; this is the so-called Platonic tradition, embracing particularly the Neoplatonists, the Hermetic authors, and such later figures as Boehme and Bruno. Miss Raine lays particular importance upon Thomas Taylor, to whom we owe the first English translations of writings of Plato, Proclus, Plotinus, Iamblicus, and Porphyry. Taylor was chiefly known, like the Emperor Julian, for his desire to restore the worship of the Ancient Gods; but rather shrilly in the introduction to a recent selection of his writings, and more soberly and persuasively in a paper in the *British journal of aesthetics*, Miss Raine makes out a strong case for the influence of his translations and writings on English Romantic authors, particularly in their choice of imagery and in their view of the intellect as active rather than passive. Although Plotinus was well known for his opposition to atomism, and might be held to be a precursor of the view that chemical combination is a matter of complete interpenetration of opposites leading to a new synthesis, nevertheless it would be a bold student who would set out to assess the influence of Plotinus or Proclus, with or without Thomas Taylor, on nineteenth-century chemistry or physics. Humphry Davy, converted (we are told) to immaterialism in his 'teens, composed at the end of his life *Consolations*[14] in which the doctrine of the transmigration of souls reconciled him to the loss of the

58

ardour of youth and to the prospect of death. This doctrine may have been derived from Neoplatonism, described with fascinated horror by Enfield in his *History of philosophy*,[15] which Davy read. It is curious that these *Consolations* went through many editions; the readership for this mixture of heterodox religion and vitalistic science was clearly larger than for the original contributions to chemistry in his *Elements*. Nuggets from the book appeared in German in 1856, and a French translation was brought out in 1869 which reached a ninth edition in 1883. We still have much to learn about the circulation and effects of popular science in its various species in the nineteenth century.

While there seems to be little doubt of the direct influence of 'Platonic' authors on Romantic poets, there is little to pin down in the scientific regions of the republic of letters. When we have the full-scale biography of Davy and the edition of his letters we may have more to go on; but the great desideratum is an edition of his notebooks. Any Neoplatonism in English science one might expect to have been derived from authors such as Cudworth, or through the Hermetic tradition. The Cambridge Platonists[16] seem currently to be in favour; Glanvill's *Saducismus triumphatus* has been reprinted, and other works are promised; Henry More's *Complete poems* have reappeared, and there is available a handsome facsimile of Cudworth's *True intellectual system*, as well as two competing volumes of selections from the writings of the group. Cudworth particularly seems to have enjoyed considerable prestige in the eighteenth century, though, no doubt, more for his ethics than his physics. His belief in atoms of brute matter and in immaterial substance seems to have been shared by Davy, and his argument that if matter cannot be destroyed, then *a fortiori* spirit cannot, appears in both Davy and Faraday. In any attempt to draw up a catalogue of pre-Romantics, one would need to know what was read during the later eighteenth century; a recently reprinted work of use to the student in this field is Christopher Wordsworth's *Scholae academicae*,[17] which gives disputation-subjects and lists of books in use at Cambridge. The lists are less confined to empiricists than their equivalents might be today; though perhaps by the end of the century Locke, Hartley, and Paley were dominant.

As to alchemy, it is curious to observe the interest taken in it by chemists opposed to atomism.[18] Of Peter Woulfe, the last Fellow of the Royal Society to be an alchemist, too little seems to be known; he delivered a rather factual Bakerian Lecture, invented a kind of wash-bottle, accompanied his chemical experiments with pious practices, and lived on until 1805. Goethe has been claimed as an adept; Faraday declared that the alchemists' ideas on transmutation were not false but a vision of truth, distorted; and Davy was interested in alchemy, though perhaps not to the extent implied by J. B. Dumas, who suggested that it formed a major part of Davy's view of the world. The Royal Institution

83

Library acquired during the second decade of the nineteenth century a considerable number of alchemical works. Later in the century another great experimentalist and opponent of chemical atomism, Berthelot, made his classical studies of alchemy, which have recently been reprinted. The widespread belief in the unity of matter, the view of chemical synthesis as a union of opposites, and the idea, implicit in Davy's *Consolations*, that the researches of the chemist can somehow cast light on the problems of the existence of God, freedom and immortality, all show an affinity with the alchemical scheme of things. It might be a worthwhile exercise to compare the attitudes of alchemy of those chemists who took an interest in it.

Plotinus apparently supplied the basis for the famous passage in Coleridge's Dejection Ode: "O Sara! we receive but what we give, / And in our life alone does Nature live / Our's is her Wedding Garment, our's her Shroud". The creative role of the artist's imagination was perhaps the central tenet of Romanticism; and any relationship between Romanticism and physical science might be illuminated by a study of the role which scientists believed the imagination might play in science. Nineteenth-century authors writing on this topic included Sir Benjamin Brodie, physiologist and President of the Royal Society, and John Tyndall. Both these are too late for our purposes, but at the very beginning of the century Davy had spoken out in favour of the scientific imagination; and it seems that his remarks had caused Wordsworth to modify the comments on science made in the preface to *Lyrical ballads*.[19] If science were to be more than fact-grubbing, an intellectual game or the basis of a technology, then it must involve the imagination as well as the intellect, and the reason—the capacity to perceive true analogies—rather than the fancy, which generates mere hypotheses. Science in which the imagination plays an important role might seem to be in danger of becoming subjective, but it will not be if the imagination is guided by the reason. This problem of reconciling the subjective and objective, of making science free from hypothesis and yet genuinely interesting, was an important one in the Romantic period. Polanyi's *Personal knowledge* and Torrance's *Theological science* may help to guide the student in this region[20]; but in the end the problems must be wrestled with by the historian in the writings of Romantics and their contemporaries. For the man of science, the question whether the pattern being imposed was true, or simply the outcome of convention, hypothesis or system was an acute one, especially for chemists about 1800. The search for certainty in the shadow of Hume and Kant would seem to have some bearing on Whewell's assault upon applied mathematicians as mere deductive reasoners in his Bridgewater Treatise, and on the development of the British school of so-called broad-and-shallow-minded physicists.

Studies along these lines are more likely to illuminate the context of

60

science in the Romantic period than to produce any dramatic examples of direct influences upon individuals, but there seems no reason to deplore this. In the same way it might be possible and fruitful to investigate the manner in which certain important words were used during these years. Thus it seems that "power" lost among scientists its evocative associations with principalities and entered the prosaic world of horses and machinery. In *Protean shape*,[21] Susie Tucker has shown how meanings of words changed during the eighteenth century. Most of her examples are not scientific terms, but she does cast some light on "hypothesis" and makes use of the *Essay* of the French chemical nomenclators. Dictionaries are invaluable for studies of this kind, and it is clearly of some importance to the historian to establish as far as possible what such phrases as "imponderable agencies" conveyed to the reader or listener. Thus Davy's Bakerian Lecture of 1806 is full of the terms "agencies" and "energies"; although it introduced polarity unavoidably into chemistry, it does not contain the words "polar" or "polarity". In an effort to understand exactly what Davy meant by the words he used one would have to delve among his notebooks and correspondence; but to establish at least what his audience took him to mean, one would consult dictionaries and standard works. In this particular case, we can conclude that the terms were not precisely understood and perhaps ambiguously intended, for Faraday collected a dozen incompatible theories of electrochemistry, all of which purported to be derived from Davy's lecture. One may also meet surprises; thus when Davy referred in a lecture to theoretical chemistry as "transcendental" it might be supposed that he had exposed himself to some Kantian influence, but in fact the same term is used in Jeremy Bentham's translation of Bergman's *Essay*.[22] It is therefore not diagnostic of any Romantic theory, but is used in the logical sense of not directly verifiable. Sometimes a word seems to have changed its meaning considerably; thus "dynamical" about 1800 implied a view of the world in which the phenomena were to be described in terms of forces, whereas in Maxwell's famous paper of 1859 it refers to a theory of the collisions of hypothetical elastic particles. "Energy", like "power", gradually lost its anthropomorphic flavour; it is probably impossible to pin down exactly what such words conveyed during such periods of flux.

Light is being cast upon the use of such terms as "powers" and "hypothesis" among Newtonians. "Hypothesis" for Romantics seems to have implied mechanical and atomistic explanations, mistaken attempts to penetrate behind the phenomena. Their non-Romantic contemporaries such as Wollaston and Berthollet, and later Liebig, made the same point about Daltonian atomism[23]; but to attack a theory because it defies the Reason is a different matter from remarking that it does not measure up to the criteria of a *vera causa*. It is possible to be opposed to hypotheses, and even to the same hypotheses, for very different reasons; allies after all seldom agree on everything. This can be illustrated from the writings of J. B. Stallo, whose *Philosophy of nature* was the clearest exposition of *Naturphilosophie* in English—not that that is any great compliment. In his later *Concepts and theories of modern physics* his position had changed to a view akin to that of Mach, but the targets of his wrath remained much the same. Among the few who did not use "hypothesis" in a derogatory sense was Thomas Exley, the compiler of an encyclopedia, who in his old age tried to develop the atomic theory of Mossotti; he claimed Newton as a devotee of the hypothetico-deductive method.

Recent work has been done upon the method in which works of Bergman and Scheele were translated into English; a Swede provided a literal translation, which was then polished up by Bentham or Beddoes.[24] How accurately theoretical terms stand up to this kind of treatment is uncertain, but the close study of translations should cast light on the use and meaning of words. By the time Ritter's discoveries got into British journals the original papers had been heavily abridged and passed through French on their way into English. If one were looking for transmissions of German Romantic physics, then it would be necessary to look very hard indeed at these English abstracts of Ritter to find clues indicating his theoretical framework; and even the standard dynamical textbook of Gren, which was translated fully, hardly gives the reader enough information about the Kantian scheme to encourage him to leave the path marked out by Newton, Lavoisier and Dalton. A selection from Ritter's voluminous writings now forms one of *Ostwalds Klassiker*, from which we can try to recover his view of electricity as a phenomenon occurring at an interface. But to the contemporary reader of English journals, his theory must have seemed very much akin to the phlogistic speculations of such authors as Priestley and Gibbes, who also denied that water was decomposed on electrolysis. There is much still to be learned about theories of galvanism in the opening years of the nineteenth century; there are strange survivals and curious twistings, and the serious study of galvanism before Davy and Berzelius brought some coherence and a measure of general agreement to the subject could hardly fail to be illuminating.[25]

While in English there seems to be little public reference by physical scientists to the conceptual scheme underlying the discoveries of their German Romantic confrères, in Cuvier's recently reprinted *Rapport historique*[26] there is a criticism occupying several pages of the chemistry of Winterl. Cuvier gives an account of his galvanic theory, and attacks him for invoking immaterial principles and publishing unrepeatable experiments. Thomas Thomson, in *Annals of philosophy*, did in the second decade of the century discuss Oersted's views, but actual translations of Romantic scientific writings, except for Oersted's famous paper, did not appear until the century was well advanced; thus Oersted's *Soul in nature*, which has recently been reprinted, appeared in 1852, and Oken's *Elements of physiophilosophy*, of which a reprint would be useful, in 1847. Oersted's third-person account of his great discovery appeared in Brewster's *Edinburgh encyclopedia*. An investigation of the various editions of the various encyclopedias might cast light on this question of transmission.[27]

With encyclopedias of this period we are approaching the border between science and popular science, and in this same uneasy territory are to be found the articles on science in the various reviews, at this period in their heyday. Sufficient attention has no doubt been given to Brougham's assault upon Young, and to the attacks on the English Universities, in the *Edinburgh review*, but a systematic study of the scientific reviews should prove worthwhile. In *The Romantic reviewers*,[28] J. O. Hayden gives a general discussion of the reviews and an analysis of their criticisms of the writings of Romantic authors, concluding that, despite glaring examples of prejudice and brutality, the critics performed their function on the whole commendably. Hayden, following his reviewers, divides the authors reviewed into the "Lake", "Satanic", and "Cockney" schools, though he is left with some over at the end and sometimes has to classify reviews as simply favourable or hostile. It should be easier for the historian of science, for the division into schools could probably be replaced by the more natural one into sciences,[29] the number of articles is less, and the criticisms do not so easily fall into a pattern of being for or against. With the Wellesley Index as guide, much might be discovered about the diffusion of science from reviews, which criticised not only books but also various papers in the *Philosophical transactions*, and clearly aimed at keeping their readers *au fait* with recent publications in science as in other branches of literature.

This has brought us to our second question: What was it about science at this period—say 1770 to 1830—that made it of interest to those philosophers, poets and men of letters who can loosely be described as Romantics? Such people were not on the whole interested in the technical details of new industries, nor in those heaps of miscellaneous information which constituted most of the multiple-volume sets of the standard Systems of Chemistry. Although Comte considered chemistry an erudition rather than a science, vast audiences flocked to hear Fourcroy, Garnett and Davy; and to Coleridge chemistry, "the striving after unity of principle, through all the diversity of forms", was "poetry, as it were, substantiated and realised".[30] Clearly Comte and Coleridge were looking at somewhat different aspects of the science; to investigate those features of a science which make it popular at a given time is rewarding because the science must then have engaged some deeper interest than intellectual curiosity or vulgar utility.

Light has been cast upon one small region of chemistry by Alethea

Hayter's *Opium and the Romantic imagination*.[31]   With the Romantic concentration upon phenomena went an interest in unusual states of consciousness and sources of imagery, and hence in drugs, particularly opium and hashish which were then readily available.   Further, Thomas Beddoes's adherence to John Brown's theory of excitability led to his recommending opium to those whose excitability had fallen too low; among his patients were Coleridge and De Quincey.   When Beddoes's assistant Davy discovered the anaesthetic properties of nitrous oxide, he had at hand friends already making subjective experiments with narcotics. His book on nitrous oxide contains some of the best descriptions of its effects ever written.   Davy, who himself wrote very vividly about his experiments, tried to persuade Southey to include a nitrous-oxide vision in *Thalaba*, and one did apparently find its way into *The curse of Kehama*. Southey's account of his respiring the gas which appears in Davy's book is prosy, but elsewhere he wrote of Davy having invented a new pleasure, for which language had no name.   And as we know from Gillray's famous cartoon, respiring nitrous oxide became a craze; it had the advantages of neither leaving a hangover like alcohol, nor producing the visions which terrify the opium addict.   Davy's first book must have recommended itself more on account of these reports than for its careful analyses of the oxides of nitrogen, which were important in the development of his own thought as revealing how different in properties different compounds of the same elements could be.[32]

    This begins to indicate how chemistry did not merely produce a new substance capable of yielding new pleasures, but was itself an exciting discipline promising to reveal the unity of matter—a sphere in which mechanics had merely scratched the surface.   Coleridge went to Davy's lectures to increase his stock of metaphors, and Burke had already used images from chemistry.   After 1800 evidence accumulated to show that metallic, acid or alkaline properties did not depend upon the presence in the substance of material principles.   Chemistry seemed to exemplify a drive towards unity and simplicity.   This can be followed in the "Proutian" speculations through the nineteenth century, to which Romantic and Darwinian evolutionary beliefs contributed.[33]   Among those sceptical about the doctrine of the unity of matter were Dalton, and also Beddoes, and Michael Donovan, who in discussing electricity tried to take the edge off Ockham's Razor.   Beddoes remarked that to postulate fewer real elements would entail the supposition of more complex relations between them, so that to prefer few or many basic entities was a matter of taste in world making.   Donovan argued that there was no good reason to suppose that the phenomena of electricity and galvanism were the effects of a single cause.

    But such remarks stand out as unusual.   Although there were opponents of premature reductions, and certainly of attempts to reduce chemistry

64

and biology to mechanics, nobody—least of all Romantics—could accept that in science there were unrelated facts or unrelated branches. What one might investigate is whether Romantic opposition to mechanical explanations constituted an outright hostility to true objective science, and what effect science had upon the literature and philosophy composed by Romantics. A work has recently appeared on *Romanticism and the social order*,[34] dealing particularly with the political doctrines of the British Romantics. Some of them had well-developed political views, while others seem to have taken an intermittent or desultory interest in politics and in social questions; and hence parts of the book are taken up with literary criticism only vaguely related to the social order. It should be possible to produce a similar work on *Romanticism and physical science*, drawing in the same way upon the writings of men of letters, and not diverging so far from one's main stream. What would be required in such a work is not that the authors concerned should have apprehended the way science was going, nor even been very exactly informed about current developments; there was a very wide variation in the political views and knowledge of Romantics too. What would be necessary is that they should not have been indifferent to science, and the effort would be worthwhile only if there was some agreement on principles to bring out. Sharrock has shown how Davy's inaugural lecture at the Royal Institution made Wordsworth modify his attitude to science, admitting it as a source of material for poetry and allowing parallels between poetry and science. And particularly in the writings of Wordsworth, Coleridge and Shelley much evidence has been uncovered of their direct indebtedness to scientists, most notably, perhaps, to Erasmus Darwin.

The most systematic study along these lines, investigating particularly the effects of pantheism on English Romantic poets, is H. W. Piper's *The active universe*.[35] Among foreign authors Piper is chiefly concerned with the French vitalistic materialists. His book has recently been supplemented by McFarland's *Coleridge and the pantheist tradition*, dealing at length with German pantheism. Piper shows how the pantheism and animism of Coleridge's Unitarian, and poetically-fruitful, period had roots in the materialism of d'Holbach and in the writings of Erasmus Darwin, Priestley and Hartley. Priestley is familiar to historians of atomism for the twist which he gave to Boscovich's theory in the process of popularising it; its value to him was that it enabled him to dispense with immaterial substance. For if we encounter not impenetrable bodies but simply forces, then matter is active and immaterial substance redundant; and hence such Platonic corruptions as the doctrines of the Immortality of the Soul and of the Trinity must be abandoned by all who accept Newton's Rules of Reasoning. French materialists, chiefly interested in the biological sciences, had come to similar views of the activity of matter by taking seriously Locke's remark that God might have given it the power of thinking. Piper seeks to demonstrate the importance of these ideas for both Wordsworth and Coleridge, and then to argue that, through Wordsworth's *Excursion*, they operated significantly upon the younger generation of English Romantic authors also. Among these, Shelley has been the subject of studies which have shown that the apparently random profusion of images in his poetry are not really unorganised but derived directly from his reading of science.

By the time Shelley was writing poetry the science of Erasmus Darwin was hardly new, and the attitudes of both Coleridge and Wordsworth to pantheism and materialism had become rather different. It seems clear that scientific thought did exert an influence upon Coleridge, in both his early and later writings. The study of Coleridge's attitudes as he shifted from the materialism of his early days towards what seems to be an acceptance of brute matter with powers not inherent in it which modify it, might provide an interesting topic for the historian of science. It might help to cast light on attitudes to the atomic theory in the opening decades of the nineteenth century, and it could hardly fail to illuminate, for example, Davy's *Consolations*, with its doctrine of the immortality of the soul and rejection of the activity of matter.

But before discussing Coleridge's philosophical works, it is worth noticing Southey as another Romantic interested in science, though Southey seems to have turned to it neither in search of metaphors for poetry nor as data for a cosmology, but out of pure curiosity. In his spirited *Letters from England*,[36] purporting to be a translation of an account by a visiting Spaniard, there are entertaining reports of animal magnetism and various quackeries connected with it; of Swedenborgianism, and Taylor's revived paganism; of agricultural changes; and of the Royal Institution. In *Omniana*, of which a new edition has recently appeared, the contributions by Southey greatly outweighed in bulk those by Coleridge; and as the editor remarks, the topics Southey discusses were very much like those which occupied the early Fellows of the Royal Society: "The English Romantics were heirs of their country's scientific revolution quite as much as the English inventors, who turned it into an industrial revolution". *Omniana* is a splendid bedside book; and the transcription and brief essays particularly of Southey often relate to curious matters of natural history or physical science. But they are not organised; Southey was prosy and donnish and did not perhaps, like Coleridge, succeed in transmuting and making his own much of the science which he thus recorded. Of other Romantics, De Quincey in his *Autobiography* records the troubles which the evolutionary doctrines of Lord Monboddo introduced into his geopolitical contests with his imperialistic elder brother, and refers to the writings of Boyle and Wilkins. Echoes of Keats's physiological studies may no doubt be found in his writings; but whether Byron would repay investigation by the historian of science is an open question.

66

In *Omniana*, Coleridge's contributions were on the whole matters of psychology and literature rather than of physical science. The primary materials in which the historian would search for the scientific views of Coleridge are first the *Notebooks*, and then the published works, of which *The friend* and the posthumous *Hints on the formation of a more comprehensive theory of life* are perhaps the most rewarding for these purposes.[37] The *Notebooks* are a splendid quarry of information on all sorts of things; they contain Coleridge's notes from lectures of Davy and other material of a scientific kind. In general, they display that breadth of curious and serious learning which ensure that nothing Coleridge wrote is without interest, and each volume is accompanied by a slightly-larger tome of notes by the editor. It is not necessary to praise this edition of the *Notebooks* but it is not certain that historians of science have yet probed into it as deeply as with profit they might. In *The friend* Coleridge's discourse on method appears; this was written for the *Encyclopedia metropolitana*—another valuable source for the history of science —which was organised in a logical rather than alphabetical manner. But the editors revised Coleridge's original discourse, and the version in *The friend* is clearly better. The new edition, the first volumes to appear of Coleridge's *Collected works*, is very handsome indeed. Though the historian of science may wish to quarrel with some footnotes, he cannot but find matters of interest frequently appearing; he will look forward to the appearance of equally-splendid volumes of the other works of Coleridge which have some bearing upon scientific questions.

It should be worthwhile, in tracing Coleridge's evolution from a pantheistic Unitarianism to orthodox Christianity, and from radical to conservative political views, to bear in mind, as well as the influence of German metaphysicians and Anglican divines, the transition from the materialistic chemistry of Priestley and Lavoisier to the immaterialism of Ritter, Oersted and Davy, and from the materialistic physiology of d'Holbach to the vitalism of Hunter. Passages from *The friend*, from the *Aids to reflection*, and from the *Hints* can illuminate Davy's *Consolations*; they indicate how chemistry and physiology might support a natural theology more exciting than the Paleyan variety. Whether science really supplied Coleridge with more than metaphors, on the other hand, is still open to doubt; that is, it is unclear whether analogies from chemistry became truths of the imagination or remained conceits of the fancy.[38] What seems certain is that while Coleridge was not incoherent in his views, he never incorporated his vast knowledge into an obvious and rigorous system; indeed, following the proper Romantic canons, he would have been wrong to try to pigeon-hole dead knowledge.

For a system which incorporates elements of what we would consider Romantic science, and in which this science is fitted into a complete framework including all other knowledge, we must turn to Hegel. Although his active life fell within the Romantic period, Hegel was not a Romantic. He exalted the understanding at the expense of the intuition; he favoured speculation, within limits; and he constructed a formidable system.   A new translation of his *Science of logic* has just appeared, followed by two translations of his *Philosophy of nature*, which formed the second part of his *Encyclopedia of philosophical sciences*, coming between the logic and the philosophy of spirit.   The *Logic* has been long known in the English-speaking world,[39] though probably not much read by historians of science; the *Philosophy of nature* must be generally unfamiliar.   Attempts such as that of J. B. Stallo to make the system known do not seem to have caught on, at least in Britain; and in Germany the reaction of scientists of Liebig's generation against *Naturphilosophie* must have produced a climate unfavourable to the systematic, speculative and dynamical science of Hegel.

The reader of Hegel's *Logic* cannot but be impressed by the range of knowledge there displayed, but while the *Logic* leads into the *Philosophy of nature* it does not contain detailed discussions of empirical science. The *Logic* presents us with a world in which extremes meet, in which opposites do not cancel each other out, but are to be grasped in their unity.   All being—except pure being, with which we begin, and which is indistinguishable, though clearly different, from pure nothing—involves limitation or determination, and negation; and hence the work is constructed in a dialectical pattern.   This pattern can readily incorporate a dynamical view of nature, and although Hegel is remarkably hard-headed, and critical of the "unbridled imagination and thoughtless reflection" to be found in *Naturphilosophie*, his science is, superficially at any rate, not very different.   The *Logic*, although Miller's translation is lucid, is not easy to read; the plunge into the unfamiliar dialectical world is a difficult one, the book is very long, and the argument abstract although relieved by wittily-presented arguments often from the sciences. The historian of science might nevertheless be well advised to tackle it before beginning the more obviously relevant *Philosophy of nature*, in order to see how the whole system fits together, though his attention may well be diverted by the penetrating asides from contemplation of the massive dialectical edifice.   It would be a pity were this so, because by comparison with the roughly contemporary *Logics* of Whately and Mill, Hegel's is powerful, bold and speculative, and such an attempt to incorporate physical science into a coherent and exciting system must command respect.

Looking in more detail at the text, we find Hegel anticipating Whewell in a caustic dismissal of the doctrine of a plurality of worlds, as part of a discussion of the spurious infinity of the astronomers; urging that there are nodes, discontinuities, or jumps in nature; praising Goethe's theory of colours, and criticising applied mathematicians; preferring

68

Newton's exposition of fluxions to Leibniz's of the differential calculus; and pointing to some tautological explanations—"witches' circles"—in astronomy and crystallography. Hegel will not allow causality in the organic and spiritual realms, and sees causes and effects as inextricably intertwined because the term "cause" implies "effect" and vice-versa. He separates the sphere of *mechanism*, in which isolated and self-contained bodies are involved in processes tending to rest, from *chemism* and *teleology*. In chemism—and his chemical remarks are supported by references to Berthollet and Berzelius—we find that objects can only be defined, as acidic for example, in relation to other objects. Mixtures belong to mechanism, and compounds to chemism; the former is the sphere of contingency, applied to things or persons, and the latter would include love and friendship. Affinity, in the inorganic or personal field, is an effect of tension leading to reciprocal adjustment and combination. In teleology, with which the *Logic* ends, we move from the sphere of finiteness into that of purposiveness, intelligence and freedom; here we find a discussion of the phenomenon of life, and a defence, both historical and philosophical, of the dialectical procedure.

Miller's translation of the *Logic* comes with only the briefest of introductions, and the reader must plunge in and sink or swim as luck befalls him. His translation of the *Philosophy of nature* has an illuminating but brief introductory essay by J. N. Finlay, but for an extended commentary on what Hegel was trying to do, and on his relationships with his predecessors, contemporaries and successors, one must turn to M. J. Petry's version of the *Philosophy of nature*.[40] Indeed, while Miller and the Oxford University Press are to be congratulated on making available a translation of Hegel's text with even a paperback edition, it could be argued that without considerable commentary and full notes the work is hardly intelligible, and Hegel's erudition is less than obvious. Petry's version also seems to read more smoothly. Its price will, unfortunately, put it beyond the reach of most individuals, but as a standard work which in text and notes illuminates both the science of the opening decades of the nineteenth century and the Hegelian method, it should find its way into all libraries. And certainly one of the translations should be on the shelves of the historian concerned with science, or philosophy of science, in the first half of the nineteenth century.

The first point to be borne in mind in looking at the *Philosophy of nature* is the obvious one that it was written not as a kind of scientific textbook but as a work of philosophy of science, with its emphasis on the organisation of scientific knowledge. That is, it should perhaps be compared with Herschel's *Preliminary discourse* or Whewell's *Philosophy of the inductive sciences*; and once one has plunged well in, it is not—at any rate in the physics and biological sections—bewilderingly abstract or obscure. Hegel was obviously well read, and in contact with some men of science; but he was not immersed in scientific circles as Whewell and Herschel were, and perhaps for this reason his *Philosophy of nature* seems to have had a negligible effect upon scientists—though more investigation might modify this judgment. But his blunders, which Petry lists, were few; and in general his judgment was acute, and his knowledge wide. Petry enumerates the sources, mostly very reputable, used by Hegel, and has succeeded in the notes in tracking down Hegel's remarks to their origin in a book or paper. Indeed, one of the valuable parts of this translation is the wide reference in text and notes to standard textbooks; for historians of science tend to concentrate too much perhaps upon the frontier rather than upon the consolidated science available at a given time in standard secondary works. But we should have missed the point, as Petry reminds us, if we simply looked for gobbets of information in Hegel, as one might in Fourcroy or Thomson; for the importance of the work lies in the way the matter is organised in hierarchies, levels and spheres. To dabble a toe into the *Philosophy of nature* is valueless; rewards only follow a total immersion.

Much philosophy of science has been concerned with discovery; either with describing how scientists make discoveries, or like Schelling's trying to anticipate discoveries by *a priori* reasoning.[41] Hegel, on the other hand, was concerned not with this but with the organisation of established knowledge. He could thus work in the main from secondary sources; and while there are speculative elements in his science—and he approved of what seems to us an alarmingly dynamical physics and teleological biology—his hierarchies are reasoned structures, and his work quite different from the outpourings of Oken, composed during a brief fit of inspiration. Indeed, Petry makes out a case for the continuing value of the Hegelian structure, in which the different sciences play complementary roles on different levels; and such a broad scheme is in many ways more satisfying and less speculative than the doctrine that all sciences must ultimately be reduced to mechanics. The work is thus not merely of interest to antiquarians, an Oken ossified and unimportant even in its day; though the levels, and the transitions between them, would have to be modified along with the empirical matter.

Hegel's *Encyclopedia* falls into three sections: the *Logic*, the *Philosophy of nature* and the *Philosophy of spirit*. This triadic and ascending pattern is also to be found in the *Philosophy of nature* itself, where the three great divisions are mechanics, physics (including chemistry), and organics (including geology). Error and confusion cannot but result if concepts appropriate to one level are applied at another. On Petry's view, Hegel misapplies his own principles in opposing Newton's theory of colours to Goethe's, to which he was a convert; he should have presented the Newtonian view in the *Philosophy of nature*, and Goethe's in the *Philosophy of spirit*, so that the truth in both would on different levels be preserved.

Hegel was interested in the connexion of the physical sciences, though not quite in the manner of Mary Somerville.[42] He is critical of those who were too ready to identify such phenomena as electricity and galvanism; the world displays differences as well as similarities, and the scientist must beware of glossing over them, of dispensing with determinateness. The contingency or caprice, and the blurring of classes in nature are not examples of freedom and rationality, but of nature falling short of rationality. The animal's inadequacy to its universal is the germ of death.

We begin mechanics with space and time; their union constitutes place and motion. Matter is the quiescent unity of time and space; its essence is to have weight, and gravity is matter's acknowledgement of its lack of independence. Gravity leads us to astronomy, and from astronomy we pass into physics—the sphere of matter qualified, rather than just having weight. Beginning with light, we pass to the four elements, and meteorology; then to cohesion and elasticity, and on through sound to heat. The physics is completed by the passage to magnetism, to crystallography, to colours, tastes, and smells; and then to electricity, galvanism, and chemistry. In the organics, the major transitions are from geology, the sphere of the terrestrial organism, through botany to zoology. Particularly in the physics, this arrangement seems at first view eccentric or perverse; but it is reasoned, and from such an unfamiliar viewpoint aspects of science in the early nineteenth century are revealed which are concealed from the main road. Hegel's is more than another of those classifications of sciences of which the nineteenth century was so fecund, though it might with profit be compared with some of them.[43]

Hegel's account of chemistry is of some interest, because there his views seem so remote from ours. That this should be so is an indication of the turmoil surrounding theoretical chemistry in the first two or three decades of the nineteenth century. For Hegel, chemical combination was emphatically not a matter of the juxtaposition of particles; but it was not simply a synthesis either, although it did involve the coming together of opposites in a medium. Really it was a transmutation, and this view, we may note, found adherents many years after Hegel's death.[44] Similarly, the growth of plants and animals involved the transmutation of water, as in the experiment of Cusa and Helmont; and the cycle of evaporation and rain is described in terms of the interconversion of air and water. A chemistry based ultimately upon the four ancient elements is perhaps startling at this period, though it is not far removed from the idea of one matter in four states as the basis of everything, which enjoyed contemporary favour in England; and anyway Hegel was not particularly interested in the starting materials or the *caput mortuum* but in the chemical process itself. Matter for Hegel is not dead; electric charge is an upsurge of anger in it, not an affair of some hypothetical electrical substance. The chemical process is the totality of the animation of inorganic individuality, the alteration of the entire material differentiation, the passing away of distinguishing characteristics. Electricity is a superficial interaction; for galvanism a medium is required; these are definite steps on the way to the chemical process, and none must be identified with one another. With chemistry we reach the point of transition to the sphere of self-sustaining and self-stimulating process—the realm of organism. In this sphere the laws of the chemical process no longer apply. They describe the flux of accidents when bodies are in process with one another, whereas organisms maintain their unity, with increasing effectiveness as we pass up the scale. Chemistry can cast light only on causes of death; similar chemical substances may be detected in bread and blood, but this tells us nothing about how bread is transformed into blood.[45]

These ideas of Hegel's are backed by references to authorities and deserve to be taken with some seriousness. There are occasions when one feels, as one does with Whewell, that Hegel is playing the gadfly, teasing somewhat unfairly; but in general he has a good eye for a weak point, and is resolute in his opposition to mere contingency or facile formalism in the structure of science. For vision and comprehensiveness, among roughly-contemporary English books only Whewell's *Philosophy of the inductive sciences*[46] can be compared to the *Philosophy of nature*; and Hegel's greater detachment from science, and obviously his familiarity with German sources, makes his perspective very different from Whewell's.

Petry's translation shows the value, given guides as well-read as Hegel and Petry, of taking as it were a horizontal cut across the history of science, of studying the whole spectrum of sciences at a given period rather than the development through time of one of them—following the *I ching* rather than the causal principle.[47] For there are relations between simultaneous events as well as between successive ones; certain ways of thinking and proceeding are characteristic of different periods in fields further removed than mechanics and botany. Before science became specialised the same people often worked in a wide range of sciences;[48] a horizontal study ensures that their contributions are not forgotten. And, returning to our main theme, in considering science and the Romantic movement, we must perforce follow this method. Hegel's *Philosophy of nature* shows that it can be done for a post-Romantic dynamical system; whether it would be worthwhile to do anything on this scale for any Romantic author may seem open to doubt. It ought to be possible to make a writer important in his own day come alive in ours, though to find the right tone may not be easy[49]; but non-specialists who make us stand back and look in the science of our own or another day for general questions deserve encouragement. What our general excursus through all this territory seems to show is that while there is a strong risk of desultoriness in attempts to show influences between

72

Romantic philosophers or poets and physical scientists, there may be signposts erected by historians of literature which can help guide the historian of science. While internal factors are no doubt greatly predominant in the development of the sciences, to investigate the interest non-scientists took in science would seem nevertheless an undertaking worthy of the attention of historians of science. To see chemistry as a source of metaphors, or as occupying a well-defined place in Hegel's structure, is refreshing, and a help in the discerning of important trends rather than details in the history of the subject.

REFERENCES

1. Review copies were received of the following works of Hegel: the *Science of logic*, transl. A. V. Miller (Allen & Unwin, London, 1969); the *Philosophy of nature*, transl. M. J. Petry (3 vols, Allen & Unwin, London, 1970); the *Philosophy of nature*, transl. A. V. Miller (Clarendon Press: Oxford University Press, Oxford, 1970).
2. J. A. Froude, *The nemesis of faith*, 2nd ed. (London, 1849), 86.
3. S. T. Coleridge, quoted in T. McFarland, *Coleridge and the pantheist tradition* (Oxford, 1969), 207.
4. This review will be chiefly concerned with sources in English. For an account of German Romantic science, see the paper by my colleague Dr B. S. Gower, forthcoming in *Studies in history and philosophy of science;* H. A. M. Snelders, "Romanticism and naturphilosophie and the inorganic natural sciences", *Studies in Romanticism*, ix (1970) 193–215. See also L. Pearce Williams, *Michael Faraday* (London, 1965) and *The origins of field theory* (New York, 1966); T. H. Levere, "Faraday, matter, and natural theology", *British journal for the history of science*, iv (1968) 95–107. For the view that Davy worked simply from soundly-based analogies see R. Siegfried, "Sir Humphry Davy . . . on the diamond", *Isis*, lvii (1966) 325–335.
5. *Goethe's theory of colours*, transl. C. Eastlake (London, 1840; facsimile, London, 1967). It is a pity that this reprint has no introduction. Other books mentioned will be described below. We might also notice H. C. Oersted, *The soul in nature* (London, 1852; facsimile, London, 1966). It is regrettable that Lorenz Oken, *Elements of physiophilosophy*, transl. A. Tulk (London, 1847), has not yet been reprinted.
6. H. M. Jones and I. B. Cohen (eds), *Science before Darwin* (London, 1963), 7.
7. D. Hartley, *Observations on man* (2 vols, London, 1749; facsimile, Gainsville, Fla., 1966), ii, 255.
8. L. R. Furst, *Romanticism in perspective* (London, 1969). For a more cosmopolitan view, see H. G. Schenk, *The mind of the European Romantics* (London, 1966); his account of *Naturphilosophie*, pp. 178–181, is perhaps a little harsh; see n. 4 above.
9. S. T. Coleridge, *The friend*, ed. B. E. Rooke (2 vols, London, 1969), i, 292–293. A recent valuable study of the problem of the collision of hard bodies, W. L. Scott, *The conflict between atomism and conservation theory 1644 to 1860* (London, 1970) emphasises the importance of national traditions in this branch of physical science. See also W. F. Cannon, "History in depth: the early Victorian period", *History of science*, iii (1964) 20–38; on a contact between Germany and Britain see A. Gillies, *A Hebridean in Goethe's Weimar* (Oxford, 1969).
10. A. Thackray, "Matter in a nut-shell: Newton's *Opticks* and eighteenth-century chemistry", *Ambix*, xv (1968) 29–53, and *Atoms and powers* (Cambridge, Mass., 1970); R. E. Schofield (ed.), *A scientific autobiography of Joseph Priestley* (London, 1966). See also J. E. McGuire, "Force, active principles, and Newton's invisible realm", *Ambix*, xv (1968) 154–208.
11. W. L. Bragg and G. Porter (ed.), *The Royal Institution library of science: Physical sciences* (11 vols, London, 1970), ii, 282–290. These handsome volumes are marred by a flashy binding; they do make available a wide range of papers on various levels, and are invaluable for the student of the dissemination of science after 1851.
12. F. Schlegel, *The philosophy of life*, transl. A. J. W. Morrison (London, 1847), 86.
13. K. Raine, *Defending ancient springs* (London, 1967), "Thomas Taylor, Plato and the English Romantic movement", *British journal of aesthetics*, viii (1968) 99–123; and G. M. Harper (ed.), *Thomas Taylor the Platonist: Selected writings* (London,

1969). See also F. Yates, *Giordano Bruno and the Hermetic tradition* (London, 1964);
A. Debus, "Alchemy and the historian of science", *History of science*, vi (1967)
128–138; and on Neoplatonists, A. H. Armstrong (ed.), *The Cambridge history of
later Greek and early medieval philosophy* (Cambridge, 1967). Those disposed with
Coleridge and Hegel to wrestle with the science of Jacob Boehme might note that
The aurora, transl. J. Sparrow, ed. D. S. Hehner and C. J. Barker (London, 1914)
has been reprinted (London, 1960); this is a seventeenth-century translation.
Cf. also Plotinus, *The Enneads*, transl. S. MacKenna (4th ed., London, 1970).

14. H. Davy, *Consolations in travel* (London, 1830); see J. Z. Fullmer, *Sir Humphry Davy's
published works* (Cambridge, Mass., 1969); cf. J. Davy, *Memoirs of the life of Sir
Humphry Davy, Bart.* (2 vols, London, 1836), i, 26–27. The German volume, not
in Fullmer, is W. Buchner, *Goldhörner aus dem literarischen Nachlasse eines christ-
lichen Naturforschers* (Erlangen, 1856). Fullmer promises us a biography, and an
edition of Davy's letters; see also Sir H. Hartley, *Humphry Davy* (London, 1966).

15. W. Enfield, *The history of philosophy* (2 vols, London, 1791). In a review of Willich's
account of the Kantian philosophy, in the *Monthly review*, xxviii (1799) 62–69,
appears the remark that the writings of the Platonists of Alexandria "do not differ
so widely in spirit, as is commonly apprehended, from those of the Königsberg
school"; this was not intended as a compliment.

16. J. Glanvill, *Saducismus triumphatus* (3rd ed., London, 1689; facsimile Gainesville,
Fla., 1966); *The vanity of dogmatizing* (London, 1661; facsimile, Hildesheim, 1970);
H. More, *The complete poems*, ed. A. B. Grosart (Blackburn, 1878; facsimile, Hilde-
sheim, 1969)—the "Democritus Platonissans" is perhaps the most interesting to
the historian of science; R. Cudworth, *The true intellectual system of the universe*
(London, 1678; facsimile, Stuttgart-Bad Cannstatt, 1964); G. R. Cragg (ed.),
The Cambridge Platonists (New York, 1968); C. A. Patrides (ed.), *The Cambridge
Platonists* (London, 1969). See also H. R. MacAdoo, *The spirit of Anglicanism*
(London, 1965). Approving references to Cudworth can be found, for example, in
J. Spence, *Anecdotes, observations and characters of books and men*, ed. S. W. Singer
and B. Dobrée (London, 1964), 41, and C. Maclaurin, *An account of Sir Isaac Newton's
philosophical discoveries* (London, 1748; facsimile, New York, 1968), 26. Coleridge
read him, but whether any scientists in the early nineteenth century did so is
uncertain. By 1809 the Royal Institution had a copy of *The true intellectual system*.
Given an edition of Davy's notebooks, perhaps a *Road to chlorine* could be written
in the manner of the *Road to Xanadu*; but it might be less entertaining.

17. C. Wordsworth, *Scholae academicae* (London, 1877; facsimile, London, 1968).

18. M. Faraday, *Lectures on the non-metallic elements*, ed. J. Scoffern (London, 1853). 2, 7.
The Royal Institution library catalogues of 1809 and 1821 survive. See G. C. Jung,
Psychology and alchemy (2nd ed., London, 1967); E. Ashmole, *Theatrum chemicum
britannicum* (London, 1652; facsimile, New York, 1968); H. M. E. de Jong, *Michael
Maier's "Atalanta fugiens"* (Leiden, 1969); M. Berthelot, *Les origines de l'alchimie*
(Paris, 1885), *Introduction à l'étude de la chimie des anciens et du moyen âge* (Paris,
1889; facsimiles, Brussels, 1966).

19. R. Sharrock, "The chemist and the poet: Sir Humphry Davy and the Preface to the
Lyrical Ballads", *Notes and records of the Royal Society*, xvii (1962) 57.

20. M. Polanyi, *Personal knowledge* (London, 1958); T. F. Torrance, *Theological science*
(London, 1969).

21. S. I. Tucker, *Protean shape* (London, 1967); M. P. Crosland, *Historical studies in the
language of chemistry* (London, 1962); D. S. L. Cardwell, "Early development of
the concepts of power, work and energy", *British journal for the history of science*,
iii (1967) 209–224.

22. B. Linder and W. A. Smeaton, "Schwediauer, Bentham and Beddoes: translators
of Bergman and Scheele", *Annals of science*, xxiv (1968) 259–273—see especially
the last page; T. Bergman, *A dissertation on elective attractions* (London, 1785;
facsimile, London, 1970); C. W. Scheele, *Chemical essays* (London, 1786; facsimile,
London, 1966).

23. D. S. L. Cardwell (ed.), *John Dalton and the progress of science* (Manchester, 1968);
D. M. Knight (ed.), *Classical scientific papers: Chemistry* (London, 1968); J. B.
Stallo, *The philosophy of nature* (London and Boston, Mass., 1848)—a reprint of
this would be useful—and *The concepts and theories of modern physics*, ed. P. W.
Bridgman (Cambridge, Mass., 1960). On Newtonian use of "hypothesis", see
A. Koyré, "Concept and experience in Newton's scientific thought" (1956), reprinted
in *Newtonian studies* (London, 1965).

24. See n. 22.

25. Such a study would require careful investigation of *Nicholson's journal* and the
Philosophical magazine among English journals—but it could not be made com-

74

pletely insular; cf. J. W. Ritter, *Die Begründung der Elektrochemie*, ed. A. Herman (Frankfurt am Main, 1968).

26. G. Cuvier, *Rapport historique sur les progrès des sciences naturelles* (Paris, 1810; facsimile, Brussels, 1968), 84 ff.

27. R. L. E. Collison, *Encyclopedias: their history throughout the ages, a bibliographical guide* (New York, 1964); S. P. Walsh, *Anglo-American general encyclopedias* (New York, 1968).

28. J. O. Hayden, *The Romantic reviewers* (London, 1969).

29. Though C. Babbage, *Reflections on the decline of science in England* (London, 1830; facsimile, Farnborough, 1969), discusses the Royal College of Physicians and the Royal Institution as schools or at least pressure groups. Other mildly-sinister later groups include the American Lazzaroni, described in N. Reingold, *Science in nineteenth-century America* (London, 1966); and the X Club: see J. V. Jensen, "The X Club: fraternity of Victorian scientists", *British journal for the history of science*, v (1970) 63–72; R. M. MacLeod, "The X Club. A social network in late Victorian England", *Notes and records of the Royal Society*, xxiv (1970) 305–322. On various scientific institutions, see B. H. Becker, *Scientific London* (London, 1874; facsimile, Farnborough, 1969).

30. S. T. Coleridge, *The friend*, ed. Rooke, i, 470–471. Whewell suggests that sciences are popular when they lend themselves to dramatic demonstration experiments, and seem easy to understand, but this does not seem to be the whole story.

31. A. Hayter, *Opium and the Romantic imagination* (London, 1968); there is in J. Cottle, *Reminiscences of Samuel Taylor Coleridge and Robert Southey* (London, 1847; facsimile, London, 1970), 464, a letter from Coleridge of 1803, sent with a parcel of Bang (Indian hemp) procured from Sir Joseph Banks, which ends: "We will have a fair trial of Bang. Do bring down some of the Hyoscyamine pills, and I will give a fair trial of Opium, Henbane, and Nepenthe."

32. R. Siegfried, op. cit. (n. 4).

33. W. V. Farrar, "Nineteenth century speculations on the complexity of the chemical elements", *British journal for the history of science*, ii (1965) 297–323, iv (1968) 65–67; D. M. Knight (ed.), *Classical scientific papers: Chemistry*, second series (London, 1970); T. Beddoes, *Contributions to medical and physical knowledge* (Bristol, 1799), 222; M. Donovan, *Philosophical magazine*, 4th series, iii (1852) 117–127. Donovan also wrote books on galvanism and on chemistry, the latter for the *Cabinet cyclopedia* of Dionysius Lardner. Hegel, as we shall see below, held that the unity of nature can only be appreciated when its diversity has been recognised.

34. R. W. Harris, *Romanticism and the social order 1780–1830* (London, 1969). Harris stresses the "scientific" interests of Constable and Turner (pp. 378–381). Studies of the relations between the sciences and the visual arts in the Romantic period might well be rewarding, though apart from studies of optics and colours, and perhaps pigments, this field would chiefly include illustrations of natural history and of technology. The rise of wood-engraving transformed the appearance of physics and chemistry books, putting illustrations on the page instead of at the back of the book, by about 1830. On the 'chemical' art of lithography, see M. Twyman, *Lithography 1800–1850* (London, 1970); on pigments, R. D. Harley, *Artists' pigments c. 1600–1835* (London, 1970).

35. H. W. Piper, *The active universe* (London, 1962); T. MacFarland, *Coleridge and the pantheist tradition* (Oxford, 1969). Piper seems to prefer Coleridge the unitarian and pantheistic poet, whereas MacFarland discusses the poetry and delights in the trinitarian philosopher, critic and theologian. MacFarland's point of departure is the enquiry into how Coleridge came to "borrow" such large chunks of Schelling when their basic beliefs were opposed. His discussions of plagiarism are rather tedious, but to follow Coleridge's ambivalent attitude towards pantheism has clearly been worthwhile. MacFarland has an appendix on "Coleridge and scientific thought", chiefly about *Naturphilosophie*, but there seems to be more to be said about this topic than he believes. On Erasmus Darwin and his influence, see R. E. Schofield, *The Lunar Society of Birmingham* (Oxford, 1963); D. King-Hele (ed.), *Essential writings of Erasmus Darwin* (London, 1968); P. C. Ritterbush, *Overtures to biology* (New Haven and London, 1964).

36. R. Southey, *Letters from England*, ed. J. Simmons (London, 1951); R. Southey and S. T. Coleridge, *Omniana*, ed. R. Gittings (Fontwell, Sussex, 1969), 18. De Quincey's problems increased when, his claims having been reduced to one minute, poor and isolated island which he had to maintain by diplomacy, his brother unfairly suggested that the inhabitants were so backward as still to have tails.

37. *The notebooks of Samuel Taylor Coleridge*, ed. K. Coburn (London, in progress); S. T. Coleridge, *Hints towards the formation of a more comprehensive theory of life*, ed.

S. B. Watson (London, 1848; facsimile, London, 1970); *The friend*—see n. 9 above; *Encyclopedia metropolitana* (29 vols, London, 1817–45)—see n. 27 above. E. L. Griggs, *Collected letters of Samuel Taylor Coleridge* (Oxford 1959) is, of course, also invaluable. For anyone interested in Coleridge, K. Coburn (ed.), *Inquiring spirit* (London, 1951) is a most useful introduction.

38.   Goethe's use of the term "elective affinities" to describe human relationships, and Hegel's similar grouping of love and friendship under "chemism", seem fanciful rather than imaginative; inverted, these comparisons might be constructive leaps of the imagination.

39.   Despite this, it is noteworthy that the headline over the review of Hegel's *Logic*, in Miller's translation, on the front page of *The Times literary supplement*, 19 June 1969, was: "Was Hegel a Great Philosopher?". For details of the Hegel translations, see n. 1 above.

40.   Petry's thoroughness may be gauged from his check from contemporary weather-reports that Hegel has reported an optical observation accurately.

41.   R. Harré, *The anticipation of nature* (London, 1965); F. W. J. Schelling, *On university studies*, transl. E. S. Morgan, ed. N. Guterman (Athens, Ohio, 1966) is probably the most accessible text of Schelling in English.

42.   E. C. Patterson, "Mary Somerville", *British journal for the history of science*, iv (1969) 311–339; H. I. Sharlin, *The convergent century* (New York, 1966). Gillispie's thesis in *The edge of objectivity* (Princeton, N.J., 1960) that science must be extricated from flux and process and from subjectivity would not have appealed to Hegel.

43.   A. M. Ampère, *Essai sur la philosophie des sciences* (Paris, 1834) has, for example, recently appeared in facsimile (Brussels, 1966).

44.   W. H. Brock (ed.), *The atomic debates* (Leicester, 1967), 23–24.

45.   Similar remarks are made by Davy in his *Consolations*, p. 198; see E. Mendelsohn, *Heat and life* (Cambridge, Mass., 1964).

46.   See R. Harré's remarks on Whewell, *British journal for the history of science*, iv (1969) 399. Two facsimile editions of the *Philosophy of the inductive sciences* have recently appeared.

47.   *I ching*, transl. R. Wilhelm and C. F. Baynes (3rd ed., London, 1968); C. G. Jung's foreword, p. xxiii.

48.   Of scientists of the Romantic period, W. H. Wollaston, for example, has probably suffered in posthumous reputation because of the range of his researches, from determinations of equivalent weights, and mechanics, to fairy-rings, sea-sickness, and why the eyes in portraits seem to follow one around. Many of his contemporaries similarly worked in a number of fields which to us, though perhaps not to them, seem almost completely separate.

49.   A recent example of an account of a minor but historically interesting literary figure, William Combe—who was *inter alia* ghost-writer for John Hunter—is H. W. Hamilton, *Doctor Syntax* (London, 1969). Combe's career casts light on many aspects of the world of letters; whether there are comparable figures in the scientific community who might provide a similar basis for a horizontal study in the Romantic period would be worth investigating.

VII

Chemistry, Physiology and Materialism in the Romantic Period

IN THE Conclusion to his *Aids to Reflection* Coleridge wrote:

I am persuaded, however, that the dogmatism of the Corpuscular School, though it still exerts an influence on men's notions and phrases, has received a mortal blow from the increasingly *dynamic* spirit of the physical Sciences now highest in public estimation. And it may safely be predicted, that the results will extend beyond the intention of those who are gradually effecting this revolution. It is not Chemistry alone that will be indebted to the Genius of Davy, Oersted, and their compeers: and not as the Founder of Physiology and philosophic Anatomy alone will Mankind love and revere the name of John Hunter.

This essay is intended as a gloss upon this passage, and the paragraphs which follow it; in which Coleridge sought to demonstrate that whereas the Frenchified, mechanistic science of his early years might have seemed to favour materialism, the dynamical chemistry and physiology of his maturity ruled it out of court.[2]

This enterprise has clearly some affinity to the natural theology of which so much had been produced in England;[3] but it is not the same. The argument from Design in the biological realm was revived by the great naturalist, and associate of the Cambridge Platonists, John Ray, in his *Wisdom of God Manifested*; and it received wide circulation in the *Physico-Theology* of his friend the parson-naturalist William Derham. Cotton Mather employed it in his *Christian Philosopher*; and it was polished and made more incisive by William Paley, who was well-informed and acute but could not be described as a naturalist. By the time he wrote his *Natural Theology*, the argument had long been shown by Hume to be less than compelling; but this does not seem to have affected the reception and influence of the work. Similar productions, by Bentley, Derham, and others were intended to make apparent the finger of God in the inorganic realm, particularly in the starry heavens. Indeed, much popular science in English in the eighteenth century, and the first half of the nineteenth, was written as natural theology; this approach at least made authors take a general view of their subject, and develop their cosmological notions.[4] But by the time the Bridgewater Treatises appeared, in the 1830s, it was becoming evident that too much weight was being put upon insecure foundations; and in some places the learned authors made themselves ridiculous. With our advantage of hindsight, such works cannot but be seen as the bold protestations of a bankrupt.

[1] S. T. Coleridge, *Aids to Reflection*, Conclusion 10; the edition I have used is ed. T. Fenby (Liverpool, 1874), p. 354. On the influence of the work, see D. Newsome, *Godliness and Good Learning* (London, 1961), pp. 14ff, 196–7: B. Willey, *Nineteenth Century Studies* (London, 1949), ch. 1.

[2] On pantheistic materialism, see H. W. Piper, *The Active Universe* (London, 1962). See also T. McFarland, *Coleridge and the Pantheist Tradition* (Oxford, 1969); he, unlike Piper, prefers the later, Trinitarian Coleridge; but his appendix on 'Coleridge and Scientific Thought' seems insufficient. See also R. G. Collingwood, *The Idea of Nature* (Oxford, 1945), pp. 142–52.

[3] J. Ray, *The Wisdom of God Manifested in the Works of the Creation* (London, 1691); C. Raven, *John Ray*, 2nd edn. (Cambridge, 1950); W. Derham, *Physico-Theology* (London, 1713); *Astro-Theology* (London, 1715); C. Mather, *The Christian Philosopher* (London, 1721); W. Paley, *Natural Theology* (London, 1802); D. Hume, *Dialogues concerning Natural Religion*, ed. N. Kemp Smith (Oxford, 1935); H. R. McAdoo, *The Spirit of Anglicanism* (London, 1965); see also the bibliography to chapter 3 of my *Natural Science Books in English, 1600–1900* (London, 1972).

[4] For the writings of Boyle, and the Boyle Lectures, see J. F. Fulton, *A Bibliography of the Honourable Robert Boyle*, 2nd edn. (Oxford, 1961); H. M. Jones and I. B. Cohen (ed.), *Science Before Darwin* (London, 1963); on the Bridgewater Treatises and their background, see C. C. Gillespie, *Genesis and Geology* (Cambridge, Mass., 1951); O. Chadwick, *The Victorian Church*, 2 vols. (London, 1966–70), I, 558–72, II, 1–35.

102

A different kind of natural theology had also been written by the Cambridge Platonists; this was concerned not with attempting to prove the existence of God but with using science to support a world-view in which religion and theology could flourish. In particular they sought to show that the atomic theory which was beginning to prove powerful in physics was inimical to fatalism and materialism. Henry More in his poem *Democritus Platonissans,* and Ralph Cudworth in the first chapter of his *True Intellectual System,* argued that if matter were composed of atoms possessing only the primary qualities, then immaterial substance must organize, arrange, and even move these inert and brutish particles. And if matter be composed of indestructible corpuscles, which are only rearranged and not created or destroyed in the various processes of geology, biology, and chemistry; how much more certain is it that immaterial substance cannot be destroyed. This might seem to prove more than was wanted—the pre-existence of souls as well as their immortality—but Cudworth argues that the creation of souls still goes on, even though that of brute matter does not. To improve his case, Cudworth produced an impeccable genealogy for theistic atomism, going back to Moses; and discussions among Newtonians of matter and the active powers which organized and moved it took place against this background.[5]

This brings us to our period; for in 1777 Joseph Priestley in his *Disquisitions relating to Matter and Spirit* presented an atomic theory which made matter inherently active, and spirit or immaterial substance unnecessary. His theory was derived from the Jesuit Roger Boscovich, who, believing in the law of continuity, argued that there could not be the sharp boundary between atoms and void required by ordinary atomism, and that atoms must be mathematical points—centres of force—rather than billiard balls. On this view, we only knew matter through its innate forces or powers; and matter occupies space only as soldierse occupy territory, the forces from each atom spreading out with increasing attenuation through all space. If immaterial substance is unnecessary, then by Newton's Rules of Reasoning it should not be admitted. Immortal disembodied souls are therefore impossible; and the doctrine of the immortality of the soul, along with that other Platonic corruption the doctrine of the Trinity, must be abandoned in favour of the primitive belief in the resurrection of the body. For Priestley, then, science favoured materialism, and materialism favoured true religion.[6]

Meanwhile in the biological sciences Cartesian mechanism had been carrying all before it. Although Descartes' physiology was as unsound as his astronomy, it did not receive the same amount of criticism. The *Physics* of the Cartesian Jaques Rohault was translated into English with notes by the Newtonian Samuel Clarke—famous for his correspondence with Leibniz. In the passages on astronomy and pneumatics, the footnotes from Newton and Boyle appended by Clarke often dwarf the text which they refute; but the biological part of the work is almost devoid of hostile annotation. Newtonian experimentalists such as Stephen Hales, Perpetual Curate of Teddington, performed vivisections which seem intolerable unless they supposed that animals were mere machines. And the famous *Homme Machine* of de la Mettrie placed man in the same category. Actual attempts to explain physiological processes in terms of

[5] R. Cudworth, *The True Intellectual System of the Universe* (London, 1678; facsimile, Stuttgart-Bad Cannstatt, 1964), ch. 1; H. More, *The Complete Poems,* ed. H. B. Grosart (Blackburn, 1878; facsimile, Hildesheim, 1969); R. H. Kargon, *Atomism in England from Hariot to Newton* (Oxford, 1966); A. Thackray, *Atoms and Powers* (Cambridge, Mass., 1970); J. E. McGuire, 'Force, Active Principles, and Newton's Invisible Realm', *Ambix,* XV (1968), 154–208; I. B. Cohen (ed.), *Isaac Newton's Papers and Letters on Natural Philosophy* (Cambridge, 1958), pt. III; A. R. and M. B. Hall, *Unpublished Scientific Papers of Isaac Newton* (Cambridge, 1962), pt. III.

[6] J. Priestley, *Disquisitions relating to Matter and Spirit* (London, 1777), ch. 1; R. E. Schofield (ed.). *A Scientific Autobiography of Joseph Priestley* (London, 1966), pp. 166–71; *Mechanism and Materialism* (Princeton, N.J., 1970); R. Boscovich, *A Theory of Natural Philosophy,* tr. J. M. Child (Cambridge, Mass., 1966); W. L. Scott, *The Conflict between Atomism and Conservation Theory 1644–1860* (London, 1970).

mechanics were less successful; in particular there was much research into the nature of respiration, which had been supposed a cooling process. Finally this was explained by Lavoisier and Crawford in chemical terms, although their account still left numerous questions for the physiologists. Breathing served not to cool but to maintain the body temperature; life was sustained not by the vital flame but by an oxidation process. Studies of the torpedo—an electric fish—and of frogs, and then of the 'galvanizing' of humans living and dead, seemed to show that the nervous processes were to be explained in terms of electrical science, which Franklin and Cavendish were erecting on the model of Newtonian natural philosophy.[7]

Chemistry had also progressed in a materialistic direction; although there was a dynamical tradition as well, particularly in England. According to the theory of phlogiston, a body was combustible if and only if it contained the substance phlogiston. Lavoisier's chemistry was as materialistic as that of his opponents; for a body to be acidic, it must contain oxygen—hence the name—and to be fluid or gaseous, it must be combined with the matter of heat, caloric. Lavoisier's list of simple substances included light and caloric, which while imponderable were not immaterial. Dalton's atomic theory also was more materialistic than that of Newton and Boyle in that it was an *homoiomeria*; which Cudworth had censured in Anaxagoras as spurious, bungling and counterfeit. In this atomism, chemical properties, instead of being explained in terms of the arrangements of atoms all of the same kind and possessing only the primary qualities, inhere in the atoms of hydrogen, oxygen, and so on. The properties of things depend upon their components; and the work of Lavoisier and Dalton led to the quantification of chemistry on the basis of weights.[8]

However by 1800 there was widespread dissatisfaction with this state of things. Mechanistic biology had been undermined by John Hunter; who had investigated for example the process of self-digestion. The living stomach is not attacked by the juices it contains, which rapidly break down any meat we eat; but in the corpse the stomach is digested by its own fluids. Such studies indicated that there was a gulf between the living and the dead; and it became a commonplace among chemists to remark that when inorganic matter entered into living form it obeyed new laws. The authority of Georges Cuvier supported a neo-Aristotelian biology, emphasizing teleology. From the mass of jumbled fossil bones discovered in the Paris Basin, Cuvier, employing his principle of correlation, was able to reconstruct the various animals to which they had belonged; ultimately his techniques were so refined that a palaeontologist could reconstruct a whole animal from one bone, even a broken one. No physiology based upon chance or mechanism, chemical or physical, could hope to account for the delicate adjustments of the parts found in the animal and plant kingdoms. The prestige of biology became such that particularly in Germany organic analogies were taken over into the physical sciences; and the very existence of a truly inorganic realm was

[7] J. Rohault, *System of Natural Philosophy*, tr. J. Clarke, ed. S. Clarke, 2 vols., 1723; H. G. Alexander (ed.), *The Leibniz-Clarke Correspondence* (Manchester, 1956); S. Hales, *Haemastaticks* (London, 1733; facsimile, New York, 1964); A. E. Clark-Kenedy, *Stephen Hales* (Cambridge, 1929); W. Shugg, 'Humanitarian Attitudes in the Early Animal Experiments of the Royal Society', *Annals of Science*, XXIV (1968), 227–38, seeks to defend these scientists from charges of wanton cruelty; J. O. de la Mettrie, *Man a Machine*, ed. G. C. Bussey (La Salle, Ill., 1961); E. Mendelsohn, *Heat and Life* (Cambridge, Mass., 1964); I. B. Cohen, *Franklin and Newton* (Philadelphia, 1956); L. Galvani, *Commentary on The Effects of Electricity on Muscular Motion*, tr. M. G. Foley, ed. I. B. Cohen (Norwalk, Conn., 1953); H. Cavendish, *The Electrical Researches*, ed. J. C. Maxwell (Cambridge, 1879), pp. 194–215.

[8] H. C. Oersted, *The Soul in Nature*, tr. L. and J. B. Horner (London, 1851; facsimile, London, 1966), pp. 300–324; A. Lavoisier, *Elements of Chemistry*, tr. R. Kerr (Edinburgh, 1790; facsimile, New York, 1965), p. 175; W. Whewell argued that all matter must be heavy; *Philosophy of the Inductive Sciences*, 2nd ed., 2 vols (London, 1847), II, 624–34; Papers on the introduction of chemical atomic theory are reprinted in facsimile in *Classical Scientific Papers—Chemistry*, first series, 1968; second series, 1970; of which I am editor; and are discussed in my *Atoms and Elements*, 2nd ed. (London, 1970).

denied. Chemists looked across to biology for schemes of classification, and later of evolution.[9]

In 1802 Schelling could remark that the key to the explanation of natural phenomena was supposed to be found in chemistry; but that 'the more all explanation of nature has been reduced to chemistry, the more difficult chemists have found it to account for chemical phenomena'. The juxtaposition of hypothetical particles could not explain chemical synthesis; in which opposites came together to form a product having new qualities. Hunter and others had shown how organisms maintain their unity despite the flux of their particles; and in the Heracleitian dynamical chemistry of the Romantic period everything endured only as columns of smoke or waterfalls do. Hegel advised concentrating upon the chemical process, not on the starting materials or residue, the *caput mortuum*. In place of mechanics, John Stuart Mill proposed chemistry as the paradigm for political science; for in chemistry the whole is more than the sum of the parts entering into the synthesis.[10]

Newtonians had sought for a dynamical chemistry on the model of astronomy; in which the science would be quantified by measuring not weights but forces. This proved a failure, but was revived with the rise of thermodynamics in the latter part of the nineteenth century. In the Romantic period, those who hoped for a dynamical, and hence philosophically and theologically respectable, chemistry derived comfort from the researches of Humphry Davy. Davy first achieved fame with his experiments which seemed to prove that heat was not a material substance but, as Bacon and Newton had believed, the effect of the rapid motion of particles. He then, in the course of demonstrating the intoxicating and anaestheic properties of nitrous oxide—laughing gas— showed how varied can be the products of two chemical elements. Oxygen and nitrogen on combination can generate a gas so delightful that inhalation of it became a craze and was depicted by Gillray; they can make ordinary air, then believed to be a compound because of the uniformity of its composition; and they can yield the acidic vapours, some brown and some colourless but all choking and poisonous, familiar to all schoolboys who have dissolved a penny in nitric acid. Clearly the properties of substances did not depend upon the presence or absence of material substances.[11]

At the Royal Institution in London, at which he became Professor of Chemistry, Davy carried on experiments on electricity; and in 1806 read to the Royal Society the Bakerian Lecture for which he received a prize from Napoleon. In this lecture polarity entered chemistry; for Davy demonstrated that the force which retains the particles in chemical combination is electrical in nature. He showed how an electric charge can make an inert metal reactive, or a reactive one inert; and in the following year he used a powerful electric battery, or 'voltaic cell', to decompose first potash and then

[9] J. Hunter, 'On the digestion of the stomach after death, *Philosophical Transactions,* LXII (1772), 447–54; W. Coleman, *Georges Cuvier* (Cambridge, Mass., 1964); G. Cuvier, *Essay on the Theory of the Earth,* tr. R. Kerr (Edinburgh, 1813); L. Eiseley, *Darwin's Century* (London, 1959); L. Oken, *Elements of Physiophilosophy,* tr. A. Tulk (London, 1847); see the paper by B. S. Gower on German Romantic physics forthcoming in *Studies in the History and Philosophy of Science;* T. H. Levere, 'Faraday, Matter, and Natural Theology', *British Journal for the History of Science,* IV (1968), 95–107.

[10] F. W. J. Schelling, *On University Studies,* tr. E. S. Morgan, ed. N. Guterman (Athens, Ohio, 1966), p. 131; G. W. F. Hegel, *The Philosophy of Nature,* tr. M. Petry, 3 vols. (London, 1971); J. S. Mill, *Autobiography* (London [*World's Classics*], 1963), p. 135: see my review. 'The physical sciences and the Romantic Movement', *History of Science,* IX (1970), 54–75.

[11] T. Bergman, *A Dissertation on Elective Attractions,* tr. [T. Beddoes], (London, 1785; facsimile, intr. A. M. Duncan, London, 1970); D. S. L. Cardwell (ed.), *John Dalton and the Progress of Science* (Manchester, 1968); H. Davy, *Collected Works,* ed. J. Davy, 9 vols. (London, 1839–40); H. Hartley, *Humphry Davy* (London, 1966): Davy (born 1778) had been a teenage materialist; J. Davy. *Memoirs of the Life of Sir Humphry Davy,* 2 vols. (London, 1836), I, 26; R. Siegfried, 'Sir Humphry Davy . . . on the diamond', *Isis,* LVII (1966), 325–35.

numerous other compounds into their elements in accordance with his theory—becoming one of the most prolific discoverers of new elements in the history of chemistry. Shortly afterwards he established that there was no oxygen in the strong acid then called muriatic acid—our hydrogen chloride—nor in the weak acids hydrogen sulphide and telluride. Here was fresh evidence that chemical properties did not depend upon a material constituent. And when Davy had married a wealthy widow, he burnt a diamond to put it beyond doubt that it was chemically identical with charcoal; here we have the same material displaying extremely diverse physical properties. Davy also demonstrated the parallels between ammonium salts and those of the metals sodium and potassium which he discovered. Ammonia was known to be a compound of two gases, hydrogen and nitrogen; experiments of Davy and Berzelius suggested that it might be possible to make an amalgam of 'ammonium' with mercury, proving it to be a true metal. All other metals might well be composed of the same or other gases; for Friedrich Schlegel chemistry had thus destroyed for ever the appearance of rigidity and petrifaction which the world presents, disclosing the living forces.[12]

A problem raised by these researches was that of the relation between electricity, magnetism, heat, and light on the one hand and matter on the other. For Schlegel these represented the inner life or soul of the Earth; and Schelling compared the relation of light and physical objects to that of soul and body. Indeed discussions of chemical synthesis show affinities with those on the union of soul and body to form a *tertium quid*, an integrated whole. Belief in the unity of all the forces or powers in nature, heat, light, electricity, magnetism, and chemical affinity, was strongly held by such Romantics as Ritter and Oersted, pioneers of electrical science, and was a factor in the postulation of the Principle of Conservation of Energy a generation later. Davy's view that electricity and chemical affinity were manifestations of one power is a step in this direction. Coleridge remarked on 'the striving after unity of principle through all the diversity of form' which characterized chemistry; giving it a 'strong hold on the imagination'. In chemistry, he wrote, 'we find poetry, as it were, substantiated and realized in nature: yea, nature itself disclosed to us . . . as at once the poet and the poem!' This refers to the old distinction between *natura naturans* and *natura naturata*, nature as artificer and nature as artifact; chemistry, if it is handled as a truly dynamical study, allows us to see both the process and the product in their unity.[13]

Davy was praised in his lifetime as England's answer to the gibe that chemistry was a French science; and for his opposition to materialism in particular. This is to be found in his lectures and chemistry books; but his main statement of it comes in his posthumously-published *Consolations in Travel*, of 1830. Davy wrote to console himself for the loss of creative imagination and energy; the dialogues of which the book consists were composed as, after resigning the Presidency of the Royal Society, he wandered around Italy and Austria in search of health. In the first dialogue we find Davy espousing the doctrine of Origen that the process of redemption will go on until all spiritual natures have passed through every mode of existence in their progress towards power and knowledge. But the fourth dialogue, 'The Proteus, or Immortality' is perhaps of most interest. The creature *Proteus Anguinus* was described in the Royal Society's *Transactions* for 1801, in the same number as Davy's first paper

[12] F. Schlegel, *The Philosophy of Life*, tr. A. J. W. Morrison (London, 1847), pp. 85–6; Oersted, *Soul in Nature*, p. 5.

[13] S. T. Coleridge, *The Friend*, ed. B. E. Rooke, 2 vols. (London, 1969), I, 470–1; see also p. 467, and p. 94, footnote; and p. 479, but ignore footnote 1 which should refer to the work of J. W. Ritter, reprinted, ed. A. Hermann, in *Ostwalds Klassiker*, new series, vol. 2 (Frankfurt am Main, 1968): on juxtaposition and mixing, see A. H. Armstrong (ed.), *The Cambridge History of Later Greek and Early Medieval Philosophy* (Cambridge, 1967), p. 358; Cudworth, *True Intellectual System*, p. 833: on the background to the postulation of the Principle of Conservation of Energy, see the article by T. S. Kuhn in M. Clagett (ed.), *Critical Problems in the History of Science* (Madison, Wis., 1959), pp. 321–56.

in that journal. It was described as an ambiguous, amphibious creature living in caves; it had eyes but seemed blind; its origin was unknown, but it seemed to have been flung from a subterranean sea; and nobody knew whether it was a perfectly-developed creature, or the larva of an unknown species. It was therefore an excellent symbol for man, the great amphibium.[14]

Davy insists upon the gulf between the organic and inorganic realms; chemistry cannot be reduced to mechanics and physiology cannot be reduced to chemistry. To argue that God might have bestowed the power of thought upon matter is to propose that He might have made a house its own tenant. Our ideas are undoubtedly derived from the senses; but they have no more in common with sense data than mathematical truths have with the symbols in which they are expressed. We gain thoughts through our present material instruments; but our existence is a succession of births and deaths, in which our powers increase as our bodies become each time more æthereal. Souls come not trailing clouds of glory, but wrapped in a refined æthereal matter to which is to be attributed instinct and conscience. Sharing Newton's preoccupations, Davy suggested that the relationship of God to the world might be the same as that of our soul to our body, except that the world does not run down; and as Newton had with gravity, Davy oscillated between æthereal explanations of the powers of matter—espoused in the *Consolations*—and ascribing them immediately to God. Such views might seem pantheistic, but need mean no more than that God can be apprehended through nature, as Henry More and Thomas Traherne believed. Davy added that the cold heavy doctrines of materialism necessarily lead to atheism; but that if we abandon the dissecting-room, a walk in the green fields shows 'in all the powers of matter the instruments of deity' and 'love as the creative principle in the material world.'[15]

With this Wordsworthian sentiment we can leave Davy and return to the passage from Coleridge with which we began. For Davy, Oersted and Hunter had compelled us to admit that the objects of our senses bear the same relation to the intelligible objects which the paper, ink, and lines in an edition of Homer bear to the *Iliad*. We could make a book about the eye out of the same chemical substances as the eye; when we refer to an eye, therefore, we cannot mean its components. Nor do we mean a particular combination of them; for a dead eye, a cold lump of jelly, is no more an eye than a heap of rubble in the desert is Babylon or Persepolis. In using the word 'eye' we necessarily include the Principle, or Operating Cause; and this the man of science can discern through the flux of material particles which compose the living organ. To suggest that the particles, through their own affinities, organize themselves is but to push the mystery back one stage; and is also to ignore the nature of organs, in which 'not only the characteristic Shape is evolved from the invisible central power, but the material Mass itself is acquired by assimlation.' In a machine, on the contrary, pre-existing parts are assembled.

Plants transmute carbon dioxide and water into grass or leaves; and 'the Organific Principle in the Ox or the Elephant exercises an alchemy still more stupendous. As the unseen Agency weaves its magic eddies, the foliage becomes indifferently the Bone and its Marrow, the pulpy Brain, or the solid Ivory'. What we see is not particles but flesh; it is the 'translucence of the invisible Energy', which soon abandons the matter to other powers. Reflection upon this cannot but make us reject the dualistic 'Mechanico-corpuscular philosophy' and both its offspring, materialism and subjective idealism:

[14] H. Davy, *Consolations in Travel, or the Last Days of a Philosopher* (London, 1830); see J. Z. Fullmer, *Sir Humphry Davy's Published Works* (Cambridge, Mass., 1969), for translations and later editions: C. Schreibers, 'Proteus Anguinus', *Philosophical Transactions*, XCI (1801), 241–64.

[15] Davy, *Consolations*, pp. 206–7, 210–5, 278, 219–20. Extracts from Davy's lectures are to be found in vol. VIII of his *Collected Works*;—on active powers, see pp. 336–7; G. Cuvier called the *Consolations* the work of a dying Plato; *Éloges Historiques* (Paris, n.d.), p. 354.

the one an abstraction, leading to a world of mere spectres and apparitions, and the other a mazy dream.[16]

The mechanico-corpuscular philosophy led to belief in a dead world, or to a pantheism in which the creator was degraded to an *anima mundi*; and then to an unwillingness to contemplate a Personal God, which must lead to the worship not of God but of Fate. Adherents to it search for 'evidences' for Christianity, instead of making men feel the want of it; and they become tinged with a literal-mindedness in which they are unable to appreciate symbols, and regard as mere metaphors spiritual truths which can only be spiritually discerned. To Coleridge, then, it seemed that a false philosophy of science had distorted theology, and had also led to a dead end in chemistry and physiology; an the effort to use for theological purposes the new dynamical doctrines of several important scientists was worth making, although in the event few were awoken from their Paleyan dogmatic slumbers in time.

[16] S. T. Coleridge, *Aids*, pp. 355–65; quotations are from pp. 357–8. The gritty factuality which Coleridge deplored is prominent in F. W. Newman, *Phases of Faith*, 6th ed. (London, 1860; facsimile, Leicester, 1970.)

VIII

"CONQUERING THE PREJUDICES ADOPTED FROM THE FRENCH SCHOOL
OF CHEMISTRY" : THE SCIENCE IN ENGLAND IN
GAY-LUSSAC'S TIME

Abstract :

   In the opening decades of the nineteenth century, the
French were believed, rightly or wrongly, to be setting the
pace in all the sciences . Nowhere was this belief more evident
than in Chemistry, where theory and nomenclature had both been
imposed by the French ; and it was against Frenchmen that men
of science in England measured themselves, as the title of this
paper, a quotation from DAVY, indicates . French and English
chemists not only came from countries at war for many years, and
were therefore spurred on by patriotism ; but their countries
also differed in their economies, in their political and social
relations, and in their attitudes to religion . Englishmen were
amateurs : none had the formal training that GAY-LUSSAC received
at the Ecole Polytechnique, and few confined their work to one
science . These diversities meant that Chemistry was seen a
little differently on the two sides of the Channel ; Englishmen
keen to make Chemistry at least in part an English science had
rather different general views of nature from those current in
France .

166

At the end of the eighteenth century and during the
opening decades of the nineteenth, the French were believed,
rightly or wrongly, to be setting the pace in all the sciences .
In pure and applied mathematics, in astronomy, in palaeontology,
in the applications of science, and even in shipbuilding, the
French were the leaders against whom Enghishmen measured themselves [1] .

The state of English mathematics was denounced in the
Edinburgh Review when it was all too apparent that there was
nobody in England competent to understand and to translate
LAPLACE's great work on astronomy ; salt was rubbed into this
particular wound when the work was very competently translated
by an American sea captain, Nathaniel BOWDITCH, between 1829 and
1839 . The Astronomer Royal, Pond, had translated only LAPLACE's
popular work, *The System of the World* . The French notation, and
with it a concern with up-to-date methods and problems in mathematics, was introduced into Cambridge about 1820 by a group of
enthusiasts - BABBAGE, HERSCHEL, PEACOCK, and WHEWELL - and as
late as 1830 BABBAGE could bemoan the "decline" of science in
England since the time of NEWTON, and point out that in France
things were better ordered and mathematicians properly recognised and honoured . Certainly, if one compares the names given to
capes and islands on the contemporary voyages to Australia of
BAUDIN and FLINDERS, one finds that while the French made much
use of the names of men of science, the English honoured members
of the Royal Family, and of the government and admiralty .

The state of palaeontology in the two countries can be
guaged from a glance at James PARKINSON's famous book, *Organic
Remains of a Former World*, which came out in three volumes between 1804 and 1811 [2] . The first of these is a leisurely discussion in the form of letters, as befitted a medical man chemist
writing in his spare time, establishing beyond doubt that fossils
really are the remains of animals or plants ; but in the two
later volumes, for which he had the benefit of the work of
LAMARCK and especially of CUVIER, these eighteenth-century preoccupations gave way to a more technical and modern approach to his
subject . It was during the First Empire that geology began as a
science in Britain ; but although Britain's prosperity depended
heavily upon coal-mining, the English lagged behind the French

in the application of geology . There was no school of mines
until the middle of the nineteenth century, and Henry de la BECHE
who was to become first Director of the Geological Survey, performed a service to his countrymen in translating a number of
papers from the *Annales des Mines* in 1824 . John BUDDLE the col-
liery "viewer", or manager, from WALLSEND owned a copy of this
publication ; and it was in his mine that the DAVY lamp, one of
the few examples of English applied science in this field, was
first tried . The lamp was contrived by DAVY in a burst of brilliant problem-solving that contrasts with the systematic acquisition and teaching of knowledge that one expects from a school
of mines .

If in other sciences the French were dominant, in chemistry their position seemed almost unassailable ; for the new
theory and its associated nomenclature had been imposed by Frenchmen, and the classic textbooks at the end of the eighteenth century were those of LAVOISIER, of FOURCROY, and of CHAPTAL [3] .
In these years, chemistry really was a French science . The antiphlogistic theory was normally referred to as "French" ; and as
BERTHOLLET, and then the younger generation including GAY-LUSSAC,
carried on the work it seemed as though the science might stay
French . Chemists in England pitted themselves against the French;
and as for twenty years the two nations were almost constantly at
war, national rivalry was an important factor in determining
Englishmen to break into the science . As it happened, by 1815
significant changes had been made in the structure of chemistry
by Englishmen, notably by DALTON and WOLLASTON in atomic theory,
WOLLASTON in crystallography, and DAVY in electrochemistry ;
while Thomas THOMSON in Scotland had written a textbook that even
received the accolade of a French translation . Despite the war,
communications between England and France seem to have been
quite good ; chemistry in the two countries was not developing
independently, but the different situations in which the rivals
found themselves helped determine the course of their investigations .

During the 1790 s, the struggle between Britain and
France had an ideological character . The English have always
been prone to distrust charm, polish, and wit, seeing in them
evidence of insincerity and levity, and identifying them, along

112

with an idolatrous devotion to reason, as Gallic qualities ;
their ideal is instead the plain blunt man, whose science should
be empirical, inductive, and Baconian, in contrast to that of the
French who were all too likely to be carried away by fashionable "systems" [4] . To the generation (including DALTON, WOLLAS-
TON and DAVY) who came to maturity in the years of the French
Revolution, it seemed that a mixture of levity with rationalism
had led to the collapse of religion and to the end of hopes of
ameliorating rather than abolishing the *ancien régime* . A proper
revolution would have followed the pragmatic and bloodless course
of the Glorious Revolution of 1688 in England .

It could not be denied that the French Revolution had had
intellectual origins, and men of science in England in the early
nineteenth century were as anxious to dissociate themselves from
the "Enlightenment" as their predecessors had been to prove that
they were no allies of HOBBES . Materialism particularly came in
for assault because it seemed to be here that *philosophes* like
HOLBACH had used science to attack society ; though the excessive reliance of some Frenchmen on mathematics, a purely deductive
system, was also blamed as a cause of infidelity . Curiously
enough, PRIESTLEY, the most notable English chemist of the late
eighteenth century, had been a materialist, although he was also
a fervent if unorthodox Christian ; but his support of the French
Revolution, which led to his laboratory being sacked in the last
"Church and King" riot so far in English history, ensured that
Christian materialism would have few converts in the next generation [5] . Indeed, as late as 1819 when William LAWRENCE published the lectures on physiology which he had delivered at the
Royal College of Surgeons, the book was denounced by another prominent doctor, John ABERNETHY, as materialistic ; and this charge was so grave that LAWRENCE decided to suppress the work .
When an unauthorised reprint appeared in 1822, he tried to suppress that ; but the legal position was that an author had no
rights in "blasphemy" - which in this case meant too sympathetic
an awareness of the work of French physiologists, and an avowal
that John HUNTER's "vital principle" was of little use .

In chemistry, the poet and philosopher COLERIDGE praised
DAVY and some of his contemporaries (including WOLLASTON) for
leading the science away from materialism into dynamics [6] .

Whereas for LAVOISIER, as indeed for most believers in phlogiston, the properties of bodies depended upon their constituents (most notably acidity upon oxygen) ; for DAVY, converted to immaterialism in his teens, properties were the outcome of forces or powers and not of matter . His earliest work was on the oxides of nitrogen, of which he believed air to be one because of the constancy of its composition ; and he demonstrated that their very different properties, chemical and physiological, were the result simply of different proportions of the same components, and were to be attributed to different balances of power or force . The nature of this force was made apparent in his work on electricity, which he took to be a different manifestation of the same power as chemical affinity ; and he found that the chemical reactivity of elements depended upon their electrical state . Thus positively-charged silver was reactive, and negatively-charged iron inert ; this principle DAVY later used in his scheme to save the copper bottoms of ships from corrosion by attaching to them lumps of more active metals, zinc or iron . Heat and light in reactions, and acidity, were for him not the result of the presence of oxygen, but were simply manifestations of force ; and when he established the chemical identity of diamond and graphite, this again proved that very different properties were the outcome of arrangements of matter under powers .

The real task then of the chemist was not to weigh things, but to study the powers that modify matter . This view chimed in with what was being taught as *Naturphilosophie* in Germany, which may help to understand COLERIDGE's enthusiasm for it [7] ; but in England generally, German speculative chemistry aroused no more enthusiasm than it did in CUVIER who made acid remarks about it in his *Rapport Historique* . Really, DAVY's seems to be a world-view going back to the Cambridge Platonists of the seventeenth century, and to LOCKE and NEWTON ; and its survival and revival in England owed no doubt something to patriotism, but much to the strong persistence of religious beliefs among men of science in England . Just as HUNTER's vital principle seemed to support the doctrine that man had a soul as well as a body, so DAVY's researches seemed to demonstrate that brute matter alone could not account for the phenomena of the inorganic realm .

170

The first half of the nineteenth century saw the publication of a surprising number of works of natural theology, demonstrating the existence and benevolence of GOD from nature as BOYLE
had done [8] . The most famous of these was PALEY's *Natural Theology* of 1802 ; PALEY's writings were required reading for undergraduates in a number of universities in Britain down to the
middle of the century, and no doubt helped to prevent the emergence of "two cultures", where those trained in science have
little in common with those who have read for an arts degree .
Not only did natural theology popularise science for those entering upon public life or the professions, but it also helped to
delay the divorce of science from morality . Opposition to vivi-
section as practised by MAGENDIE, and opposition to evolutionary
schemes which seemed to make some races of men closer to apes
than others and thus permit slavery, were features of this moral
commitment . A generation later, with the publication of DARWIN's
theory, men of science like T. H. HUXLEY did their best to expel
non-experts from the discussions, because by then the professional scientist had begun to emerge ; but in the early nineteenth
century he is not to be found .

Natural theology played a smaller role in chemistry than
in astronomy or biology, although in 1834 William PROUT published
a volume on this topic . Public lectures, on the other hand [9] ,
such as those delivered by DAVY, FARADAY, and J. M. GOOD, were
replete with references to GOD, and indicate an acceptance in the
doctrine of providence - indicated not only in the victories of
NELSON, but more seriously in the disposition of coal and minerals for the use of man, and even in the provision of chemical
properties to the elements . Such teleological beliefs fitted
very happily with an emphasis upon applied science ; WOLLASTON's
discovery of malleable platinum, and DAVY's work on fertilisers
and on the miners' safety lamp, showed how the useful properties
given by GOD to things could be found out . It was perhaps no
accident that the standard work of the late eighteenth century
on pure and applied chemistry, Richard WATSON's *Chemical Essays*,
1781-7, was written by a Bishop .

DAVY and FARADAY held forth in the Royal Institution,
which was during the First Empire a kind of temple of applied
science supported by the landed gentry [10] . In a sense, DAVY

was the first professional chemist in England ; but the amateur
tradition was strong enough for his ambition to be the achievement of the status of a gentleman . He was happy to accept large
fees for delivering lectures, but refused to patent his inventions and looked down on WOLLASTON for profiting thus from his
science . DAVY's research was thus the outcome of collective
patronage, by the members of the Royal Institution ; and he turned to them for particular needs, such as a larger battery in his
electrical experiments . England was still after all governed by
an *ancien régime* of a kind ; and DAVY's career was the result of
enlightened patronage by a number of people, including Sir Joseph
BANKS, President of the Royal Society and landowner, and the most
powerful patron in the world of science .

While DAVY was supported by landowners anxious to increase their own wealth, and the wealth and stability of their country, through the application of science ; DALTON in Manchester
was associated with the Literary and Philosophical Society in
that booming centre of the cotton trade . Manchester was not in
DALTON's day an intellectual centre, and indeed it may be that
the stubborn DALTON could develop his atomic theory there whereas in more sophisticated places it would have been stifled at
birth because it seemed so crude [11] . The 'Lit. and Phil.' seems
to have been founded by doctors and clergy to disseminate culture, and to have been supported by manufacturers conscious of
their lack of social grace . Acquaintance with science was easier
to acquire than taste in literature or the fine arts, and hence
had great appeal ; and DALTON was thus supported, though meagre-
ly, by the Society and eked out his income with some private
teaching .

DAVY had begun as an apprentice to an apothecary and
surgeon, and DALTON had hoped at one time for a medical career ;
for medicine was a way of getting a training in chemistry and in
biology, and of supporting oneself while one later did research
on the side . It was however a profession with first and second
class members ; the physicians, who were University men and were
consulted chiefly by the well-to-do were the patricians, and the
apothecaries and surgeons, who learned by apprenticeship, the
plebians . Unless one could find a patron, as HUXLEY later found
Sir John RICHARDSON, one had little hope of emerging into the

world of science from the lower strata [12] . DAVY's English
contemporary Foreign Members of the Institut were WOLLASTON and
Thomas YOUNG, both of them trained as physicians . WOLLASTON
hated medicine, and used his income from his platinum process
as a way of escape from it, while YOUNG continued to be supported by his medical work .

    Both of these men surprise us by the range of their work,
which is another indication of the amateurism of English science.
WOLLASTON was an important crystallographer, and the inventor of
the reflecting goniometer for measuring crystal angles ; in his
theory of crystals, he revived the idea of HOOKE that the forms
were built up of spherical or spheroidal particles, rather than
the "integrant molecules" of HAUY . Thomas THOMSON found CHENE-
VIX's support of the French theory evidence of a lack of proper
patriotism, in a blistering attack in his new journal, Annals of
Philosophy, in 1813 [13] . WOLLASTON also wrote about fairy rings,
and (with Sir Thomas LAWRENCE, the painter) on why the eyes in
portraits seem to follow one around ; both these essays were
published by the Royal Society .

    YOUNG tried to quantify elective affinities in the eigh-
teenth-century manner ; he delivered and published some exceedingly important lectures on natural philosophy ; he introduced
the wave theory of light into the mainstream of physics ; he
began to compile a medical bibliography ; and he worked on the
Rosetta Stone . Both in optics and in Egyptology, his work was
fragmentary, and both these endeavours were brought to fruition
by Frenchmen, FRESNEL and CHAMPOLLION, who has perhaps a more-
specialised approach to science . BERZELIUS described DAVY's
work as "brilliant fragments" [14] , and the same epithet could
be applied to that of YOUNG . A lack of the opportunity for formal training, the absence of a formal and professional scientific community, and the need for the man of science to make his
way rather than follow a career-pattern, were no doubt important
in keeping able scientists unspecialised ; so that chemists easily, and often fruitfully, strayed into what would now be physics,
and were often sporadically rather than continuously occupied
in research . When they were elected to the Royal Society, instead of an Academy they would find a group of amateurs, few of
whom had published in science, but who would all agree about

maintaining religion and the British constitution . The world of
GAY-LUSSAC was in many ways more modern .

N O T E S

1   The chapters on science in different countries in J. T.
MERZ, *European Thought in the Nineteenth Century*, 4 vols., Edin-
burgh, 1904-12, I, chapters 1 and 3, are still of interest . See
also M. P. CROSLAND (ed), *The Emergence of Science in Western
Europe*, London, 1975 . *Edinburgh Review*, 11 (1808) 249 ff ;
N. REINGOLD (ed), *Science in Nineteenth-century America*, London,
1966, pp. 11 ff ; J. M. DUBBEY, *The Mathematical Work of Charles
BABBAGE*, Cambridge, 1978 ; C. BABBAGE, *Reflections on the Decli-
ne of Science in England*, London, 1830 ; S. F. CANNON, *Science in
Culture*, London, 1978, on HERSCHEL and the "Cambridge Network" .
On names, see E. H. J. & G. E. E. FEEKEN, & O. H. K. SPATE, *The
Discovery and Exploration of Australia*, Melbourne, 1970 ;
M. FLINDERS, *A Voyage to Terra Australis*, 2 vols., London, 1814 ;
N. BAUDIN, *Journal*, tr. C. CORNELL, Adelaide, 1974 .

2   On PARKINSON, see my paper in *Ambix*, 21 (1974) 78-85, and
J. C. THACKRAY's bibliographical study in *Journal of the Society
for the Bibliography of Natural History*, 7 (1976) 451-66 .
J. DAVY, *Memoirs of the Life of Sir H. DAVY*, 2 vols., London,
1836, II, p. 11 ; my copy of the *Annales* volume was formerly
BUDDLE's .

3   M. P. CROSLAND, *The Language of Chemistry*, London, 1962 ;
W. A. SMEATON, *Fourcroy*, 1962 .

4   On materialism, and on the philosophical background to
the chemistry of this period in England generally, see my *Trans-
cendental Part of Chemistry*, Folkestone, 1978, esp. chapter 3 .
On science in relation to the social and economic structure, see
M. BERMAN, *Social Change and Scientific Organization : the Royal
Institution, 1799-1844*, London, 1978 .

118

5   On PRIESTLEY, see J. McEVOY, "Joseph PRIESTLEY", *Ambix*, 25 (1978) 1-5, 93-116 (continuing) ; R. E. SCHOFIELD, *Mechanism and Materialism*, Princeton, 1970 ; A. THACKRAY, *Atoms and Powers*, Cambridge, Mass, 1970 . On LAWRENCE and ABERNETHY, see *Dictionary of National Biography* . On HUNTER's principle, see my paper "The Vital Flame", *Ambix*, 23 (1976) 5-15 ; a version of this in French is in *La Tourbe des Philosophes*, 1 (1977) 15-18, 2 (1978) 7-18, tr. E. DELMAS .

6   T. H. LEVERE, "S. T. COLERIDGE : a poet's view of science" *Annals of Science*, 35 (1978) 33-44, and *Affinity and Matter*, Oxford, 1971 .

7   B. S. GOWER "Speculation in Physics", *Studies in the History and Philosophy of Science*, 3 (1973) 301-56 is the standard work on *Naturphilosophie* . G. CUVIER, *Rapport Historique*, Paris, 1810, pp. 84 ff .

8   On natural theology, see my *Natural Science Books in English*, London, 1972, chapter 3, and the Ph. D. thesis by Richard YEO, Sydney University, 1978 ; and see the chapter on Science as a Career in my *Nature of Science*, London, 1977 .

9   For DAVY's lectures, see J. Z. FULLMER, *Sir Humphry DAVY's Published Works*, Cambridge, Mass., 1969 . For FARADAY's, A. E. JEFFREYS, *Michael FARADAY : a list of his lectures and published writings*, London, 1960 . J. M. GOOD, *The Book of Nature*, New York 1833 : this has a sketch of his life at the front .

10   M. BERMAN, note 4 ; C. A. RUSSELL et al., *Chemists by Profession*, London, 1977, pp. 24 ff ; J. DAVY, *Fragmentary Remains of Sir H. DAVY*, London, 1858, p. 310 . On BANKS, see W. R. DAWSON, *The Banks Letters*, London, 1958 ; A. M. LYSAGHT, *Sir Joseph BANKS in Newfoundland and Labrador*, London, 1971 ; H. B. CARTER, "Sir Joseph BANKS and the plant Collection from Kew ..." *Bull. Br. Mus. Nat. Hist.* (hist. ser.), 4 (1974) 283-385 .

11   A. THACKRAY, *John DALTON*, Cambridge, Mass, 1972 ;
D. S. L. CARDWELL (ed), *John DALTON and the progress of science*,
Manchester, 1968 ; and my *Atoms and Elements*, 2nd ed., London,
1970, chapter 2 .

12   On DAVY, DALTON, WOLLASTON and YOUNG, see C. C. GILLSIPIE
(ed), *Dictionary of Scientific Biography*, New York, 1970 (continuing) . R. E. JOHNSON, *Sir John RICHARDSON*, London, 1977, describes another contemporary of GAY-LUSSAC who from a surgeon
became an eminent ichthyologist and achieved fame as a Polar
explorer ; and was then patron of HUXLEY and of Joseph HOOKER .

13   T. THOMSON, *Annals of Philosophy*, 1 (1813) 251-58 ; see
also M. C. USSELMAN, "The WOLLASTON/CHENEVIX controversy over
the Elemental Nature of Palladium", *Annals of Science*, 35 (1978)
551-79 . S. H. MAUSKOPF, "Crystals and Compounds", *Transactions
of the American Philosophical Society*, new series, 66 (1976)
5-82, is concerned with French work .

14   H. HARTLEY, *Humphry DAVY*, London, 1966, p. 148 .
On science in Britain, see R. PORTER "Gentlemen and Geology : the
emergence of a scientific career, 1660-1920", *The Historical
Journal*, 21 (1978) 809-36 . Part of YEO's thesis (note 8) is to
appear in *Annals of Science* . For what alarmed the English in
Science, see R. W. BURCKHARDT, *The Spirit of System : LAMARCK
and evolutionary biology*, Cambridge, Mass., 1977 .

IX

SCIENCE AND PROFESSIONALISM IN ENGLAND, 1770-1830

When in 1792 an Embassy under Lord Macartney was sent to China, the letter which he took from King George III to the Emperor described the King as a patron of science,[1] who "had taken various opportunities of fitting out ships, and sending, in them, some of the most wise and learned of his own people, for the discovery of distant and unknown regions ... for the sake of increasing the knowledge of the habitable globe, of finding out the various productions of the earth; and for communicating the arts and comforts of life to those parts, where they had hitherto been little known." It added that "he had since sent vessels, with animals and vegetables most useful to man, to islands and places where, it appeared, they had been wanting, /and/ had been still more anxious to inquire into the arts and manners of countries, where civilisation had been improved by the wise actions and virtuous examples of their sovereigns". On the staff of this embassy was John Barrow, who later became Secretary of the Admiralty and a great patron of exploration; and who in 1818 wrote that "The reign of George III will be referred to by future historians as a period not less distinguished by the brilliant exploits of our countrymen in arms, than by the steady and progressive march of the sciences and the arts."

On the whole, this is not the view which historians of science have taken, though they would admit the progress in 'arts', meaning techniques. But perhaps because historians of science have been too ready to accept Rutherford's famous dictum that science is either physics or stamp-collecting, they have come to the conclusion that there was little science worthy of the name in George II's England. Charles Babbage's famous polemical work of 1830, On the Decline of Science in England, with its plea for more money and honour for applied mathematicians, has fitted very well into this scheme.[2] But we have recently had re-evaluations of American science,[2] attempting to replace denigration or encomium by real historical understanding; and it seems worthwhile to attempt the same task for science in Britain. We must try to see what constituted 'science' between about 1770, when Captain Cook arrived at Botany Bay in Australia, and 1830 when Babbage's book was written; to see how and why this activity, or group of activities, was supported by Government or by individuals; and to see what

54

opportunities there were of pursuing a career in science. We must try to see 'science' as contemporaries saw it; remembering that science has intellectual, practical, and social aspects.

In the official account of Macartney's Embassy, praise was given[3] to one who "stored with the knowledge of whatever former naturalists had observed, and making, for the sake of further researches, a voluntary sacrifice of the enjoyments of fortune, at an early age, to encounter the extremities of opposite climates, and the perils of unknown routes, succeeded in enriching the history of nature in its several departments." Readers would have recognised this portrait of Joseph Banks; who after his return from his voyage with Cook became in 1778 President of the Royal Society. He held this office until his death in 1820, that is through most of our period; so long was his reign, that his immediate successor, Humphry Davy, had been born after Banks became President. The impression is sometimes given, that in this high office Banks was amateurish and high-handed; but recent studies do not confirm this view, although they do indicate that what was expected of a President of the Royal Society in 1800 was different from what would be expected or desired today. In 1958 there appeared a guide to Banks' correspondence. The editor estimates that Banks probably wrote 100,00 letters, of which (mercifully) only something over 7,000 survive. These show Banks' finger in every pie; indicate the extent of his foreign acquaintance; and show him a zealous defender of the Royal Society, and a keen natural historian.

Banks' competence as a natural historian has been in doubt, chiefly because he published so little. Indeed it is only in the last few years that we have had an adequate edition of his journal kept on his voyage with Cook[4]; the fishes he and his associates collected have at last been published, and the flowers are to appear; and we have had a splendid edition of his Newfoundland journal and other papers. In his own day, his knowledgeability and the value of his collections were generally recognised; he kept open house, and while his materials were unpublished they were generally available. At his death, they passed to the British Museum, which he and Davy had hoped might become something more like Cuvier's Museum in Paris; and indeed at the British Museum a team of professional classifiers did develop even in the years before Natural History was separated from the rest of the collection in the 1850s.

One of the greatest botanists of his day, Robert Brown, was Librarian to Banks at the end of his life, and remained in charge of his collections when they were transferred to the British Museum. Banks had opened Brown's career in science when he had arranged for him to sail with Matthew Flinders in his survey and circumnavigation of Australia in 1801-3; Brown collected many new species, but also wrote an essay on the

Australian flora which was full of interesting comparisons and
generalisations.[5] Indeed, many natural historians were supported either by the Government or by some society such as the
African Association, usually in either case on Banks' nomination; or supported directly by him. In the letter to the
Emperor of China, the transference of useful animals and plants
from one country to another was given prominence; this was an
old practice of navigators, but particular stress had been laid
upon it on Cook's voyages, and Banks had been interested in
bringing merino sheep to England, and in taking breadfruit
trees from Tahiti to the West Indies. This latter project was
to have been carried out by Captain Bligh in HMS Bounty; but
after their stay at Tahiti the crew mutinied and set the Captain and those loyal to him adrift in an open boat. He navigated the boat across several thousand miles of ocean, and
survived to perform the task satisfactorily in another vessel.
The Bounty had been specially fitted up to take the tubs with
breadfruit trees in them; and a gardener, David Nelson with an
assistant William Brown, had been appointed on Banks' recommendation. In the mutiny, Nelson went in the boat with Bligh,
while Brown joined the mutineers. What is more to the point
is that in such men as these we can recognise that important
figure in modern science, the technician.

Indeed, by the end of the eighteenth century, natural
history required a number of such people: artists to draw the
specimens, or work up the drawings made on the spot by the
botanist or zoologist[6]; engravers, to transfer the artist's
picture onto copper so that it could be printed; collectors,
to search out new specimens, particularly those of economic or
decorative importance to introduce into public gardens, like
those at Kew, or to enrich private gardens, menageries, or
collections; and gardeners and curators to look after the
specimens collected. The lines between these various classes
of people, and the 'philosophical' gentlemen who supported
their activities, cannot be drawn with any exactness; on the
one hand, social boundaries were fluid, and on the other the
man of science unless he were as rich as Banks might well find
himself having to draw and even perhaps engrave his own plates;
in the latter part of our period, the invention of lithography
cut out this last stage because drawings could be made on the
stone and printed directly. The career of a plant collector,
George Caley, has recently been described, mostly in the words
of his own journals and correspondence with Banks. He was
sent out to New South Wales during the first decade of the
nineteenth century, and when government support was not forthcoming for him, Banks employed him to collect objects of natural history. Fairly soon after his return to England, he was
appointed Superintendent of the Botanical Garden at St.Vincent,

in the West Indies. Beginning in humble circumstances, he had
raised himself through his natural history to what could have
been a position of security and usefulness; but he was a tact-
less man, and soon threw up the post after a series of dis-
agreements. A man living by his science, as Thackray has shown
in his study of John Dalton, had at this period to shape a
career for himself; but in natural history there were the beginnings of a career structure.

Caley's career was made in New South Wales and in the
West Indies; and the connexion between the growth of the
British Empire in these years and the prosecution of natural
history is close. In nations with an empire to describe, descriptive science will be prominent. Banks had been one of the
discoverers of New South Wales, and was one of the chief advo-
cates of settlement there. He was also a supporter of the
African Association, which was sending explorers into North
and West Africa during our period.[7] On his return from a
voyage to New Guinea, undertaken on behalf of the East India
Company, Thomas Forrest met Banks in London in 1780, and introduced to him William Marsden, later the historian of Sumatra,
who described the introduction as the most important of his
life. Stamford Raffles, the founder of Singapore and of the
Zoological Society of London, was like Marsden a keen natural
historian; and he used his power to support Thomas Horsburgh,
the zoologist who first made known in the West the Malay Tapir,
and Joseph Arnold, a botanist. In India itself, the East India
Company supported the Botanical Garden at Calcutta, where
William Roxburgh supervised the drawing of splendid pictures
of the Indian flora; and attempts were made to introduce tea
plants into India. Servants of the 'Honourable Company' engaged on diplomatic missions, exploration, or administration
were expected to describe natural history, particularly if the
animals, vegetables or minerals they saw were of economic importance; and competence at natural history no doubt favoured
one's career in the Company, though it did not usually make
one anything like a professional scientist.

In the armed services too, a capacity for science could
lead to promotion. This was especially true in those branches
devoted to survey; the Ordnance Survey in Britain, and its corresponding service in India, and the Admiralty Hydrographic
Service.[8] These did offer careers to soldiers, sailors, and
civilians; and represent substantial government investment in
science. The sailors, soldiers, and surgeons who conducted
the first fleets to New South Wales and to Van Diemen's Land
found an avid public for their descriptions of Australasia; and
the writings of Governor Philip, Governor Hunter, Governor King,
David Collins the first Judge Advocate, and John White the first
Surgeon General, abound in natural history and are illustrated

with good pictures of the creatures described. While natural history had economic value, and sometimes survival value, in the new colony, it was to be some time before a scientific career was possible in the Antipodes; though the botanist Allan Cunningham, who had been trained at Kew and had begun as a collector eventually became Colonial Botanist in 1835 - but soon resigned because the most important part of his duties seemed to be growing vegetables for the Governor.

In the Royal Navy, his unrivalled capacity as a surveyor raised James Cook to the rank of Post Captain; and throughout our period, but especially in the ten years either side of the Revolutionary and Napoleonic Wars, naval officers who were good at astronomy and ship's surgeons who were capable natural historians could hope for rapid promotion. On his voyages, Cook had taken civilian astronomers and natural historians; though he had himself made observations of the Transit of Venus on the first voyage, and was clearly an excellent observational astronomer; as a navigator needed to be in the late eighteenth century, because even when he had a chronometer it needed to be 'rated' - that is, to have its timekeeping checked and its rate of gain or loss per day determined - and this could only be done by making accurate lunar observations. Setting up the observatory was always one of the first things which had to be done when the ship came to land.

On Cook's voyages civilian scientists were taken, the implication being that the ship's officers were not themselves men of science; in the late seventeenth century, the same belief had led to the civilian astronomer Edmond Halley being given command of a survey ship. Among the junior officers whom Cook trained was George Vancouver, who in 1790-5 himself commanded a survey on the West coast of North America.[9] He was experienced enough he hoped to make the observations "at least, in an useful manner" and he addedthat "it was with infinite satisfaction that I saw, amongst the officers and young gentlemen of the quarter-deck, some who, with little instruction, would soon be enabled to construct charts, take plans of bays and harbours, draw landscapes, and make faithful portraits of the several headlands, coasts, and countries, which we might discover; thus, by the united efforts of our little community, the whole of our proceedings, and the information we might obtain in the course of the voyage, would be rendered profitable to those who might succeed us in traversing the remote parts of the globe that we were destined to explore, without the assistance of professional persons, as astronomers or draftsmen."

Not only were civilian astronomers unnecessary, but a surgeon was chosen for the voyage who was qualified to add to the stock of botanical information; this was Archibald Menzies, who did indeed collect various new plants, and who provides an indication that the status of ship's surgeons was improving.

We have already noted that Flinders, who was also trained by Cook, did take a trained botanist, Robert Brown, on his voyage around Australia in the opening years of the nineteenth century; and P. P. King, on his surveys between 1818 and 1822 was accompanied by Allan Cunningham, whom we have already met. The contemporary expedition to the Congo, under Captain Tuckey, had various civilian scientists on board; at the end of our period Charles Darwin sailed as a civilian on the Beagle; and on the famous cruise of the Challenger from 1872-6 there were a number of civilian scientists. But on the whole, during our period the necessity for carrying civilians decreased. This seems to be most interestingly shown on the various Arctic voyages, sent to determine the possibility of a North-West Passage from Europe to the Far East.

The great proponent of these voyages was Sir John Barrow, a protégé of Banks', who was Second Secretary of the Admiralty from 1803-45 and was a Fellow of the Royal Society although he was nothing like a professional scientist. In voyages with a scientific purpose, we see both professionalisation beginning with the posts made available to astronomers and natural historians, and its opposite in the training of naval officers and surgeons to do some science in addition to their other duties. What is striking about the voyages is that in Vancouver's words "the united efforts of our little community" were required. Much of the science of about 1800 was not a team activity; but voyages were 'big science' both in the sense that they required a large investment and in the sense that a voyage naturally required the co-operation of numerous more-or-less specialised people.

The first of this series of Arctic expeditions to make dramatic discoveries were those of W. E. Parry in 1819-20 and John Franklin in 1819-22.[10] Parry took with him Edward Sabine as astronomer; he was not a civilian, but was a Captain in the Royal Artillery, who went on to rise to the top of the tree both in his profession and in science, becoming a General and President of the Royal Society. It was his task to make astronomical observations, and also pendulum experiments to determine the figure of the Earth; that is, to measure the flattening at the Poles. Parry also took with him young Naval Officers who were to be trained in the making of observations, both astronomical and magnetic; for terrestrial magnetism was something else that could be profitably studied near the Poles. Parry's expedition successfully wintered in the Arctic Circle at Melville Island; he had two well-found ships, and his object was to get round the northern part of North America by sea.

Franklin's expedition went overland, to reach the Arctic Ocean down the Coppermine River and to explore the coast. He

took with him John Richardson, a Naval Surgeon who had recently
obtained an Edinburgh MD; they reached the Polar Sea, and to the
amazement of contemporaries navigated upon it in birchbark canoes.
They then had a desperate journey back, and very nearly starved
to death. They returned to a hero's welcome; Richardson became
a leading ichthyologist and authority on Arctic fauna, and
Franklin was subsequently elected to the Council of the Royal
Society. Richardson was later the patron of such Naval Surgeons
as T. H. Huxley and J. D. Hooker, who like him went on to become
very distinguished scientists; indeed, one could say that to go
on a voyage, as a civilian like Banks, Brown, or Darwin, or as
a surgeon like Menzies, Richardson, or Huxley, was the best
route to professional success in the biological sciences in
Britain in the early or middle years of the nineteenth century.

On Parry's second voyage to the Arctic, he again took an
astronomer who was not a naval man; in this case he was the
Rev. George Fisher, who acted also as Chaplain. But on his
third voyage, Lieutenant Henry Foster[11] went "nominally as
assistant surveyor, but in fact to perform the duties of Astronomer to the Expedition, for which he was fully qualified."
Foster was indeed awarded the Copley Medal of the Royal Society-
its highest award – for his astronomical observations made on
the voyage; and was immediately promoted and given command
of a vessel, HMS Chanticleer, to take on a cruise for scientific purposes. A committee of the Royal Society drew up
some recommendations; but the voyage terminated sadly when
Foster was drowned in a river in Panama. This rapid promotion
was the result of the Duke of Clarence, later William IV,
occupying the office of Lord High Admiral and determined to be
a patron of science. Such patronage produced a slightly sour
reaction from that excellent surveyor Captain W. F. W. Owen,
who was conducting a much more dangerous and less glamorous
survey than the Arctic voyagers, namely that of the rivers and
ports of Africa. He wrote that "superficial observers suppose...
that the art of critical navigation is only to be acquired by
the learned few. The affectation of extreme minutiae and of
reasoning on new hypotheses to account for all possible effects,
and to make the Royal Society a stepping-stone to the honours
and benefits of our service, has certainly produced more injury
by discouraging the unpresuming man of real professional merit,
than it has done good by raising talent from obscurity."
Captain Edward Belcher also commented after visiting Foster's
mean grave that it showed "how little is thought of the pet of
science, when his services are no longer available."

When Arctic voyagers, or indeed those who had surveyed
other regions, returned, they were often elected into the Royal
Society;[12] and the leader of the expedition usually wrote up

60

the voyage for publication. Although the authors usually apologise for their lack of a polished style, their plain unvarnished histories are a model. Sometimes they contain systematic accounts of the magnetic, optical, and astronomical work done, with tables of the results and descriptions of the specimens of natural history collected.[13] Thus in John Ross' Voyage to Baffin's Bay we find scientific appendices, one of which - by Ross himself - was concerned with magnetic deviation, following up work by Flinders. This had been read to the Royal Society and would, we are told, have appeared in its Philosophical Transactions if Ross had not published it with the Voyage; the Royal Society always refused to publish second-hand material. But more usually, the scientific appendices appeared in separate volumes from the narrative; and for the publication of these, which could not be expected to sell widely and were often very handsome, government grants were forthcoming as a rule. These were not sufficient to pay the various authorities consulted, but only to cover the costs of printing and publication; national prestige was thus involved both in the dispatch of the expeditions, and in the publication of their results.

A wide range of authorities were consulted, or entrusted with writing up accounts of the specimens. Thus Michael Faraday analysed air from the Arctic brought back on Parry's second voyage, which seemed (curiously enough) to contain less oxygen than the air of London. Geologists who described collections included William Jameson of Edinburgh, William Buckland of Oxford, and John MacCulloch, of the Royal Military Academy at Woolwich who became President of the Geological Society; botanists included Robert Brown and William Hooker; and zoologists included J. G. Children, of the British Museum, Joseph Sabine, and John Curtis. John Richardson drew together the various observations made on the voyages into the Fauna Boreali-Americana, a sumptuous government-supported publication which superseded Pennant's Arctic Zoology. The best practice of such expeditions was later enshrined in the Admiralty Manual of Scientific Enquiry, which was edited by Sir John Herschel.[14]

Government supported these voyages, on the urging of Banks, Barrow, and others, partly for foreign policy reasons, and partly because of their general utility to navigation. Thus Vancouver's voyage was sent partly to survey, and partly to ensure that the Spaniards evacuated their settlement on what is now called Vancouver Island. The North-west passage voyages were partly planned to counter the Russian presence in what is now the North and West of Canada; and the African surveys were made necessary by the laws against the slave-trade

and the need to enforce them.[15] Surveys of South America were
necessary for British trade there as countries gained their
independence from Spain; and surveys around Australia and
Indonesia shortened passage times and diminished the chance
of shipwreck. In Britain, it was generally expected that
science would be useful; the distinction between pure and
applied science does not seem to have been made, and, as in
America, the distinction made was between pure empiricism or
rule of thumb on the one hand, and soundly based scientific
knowledge on the other. This state of affairs goes well with
having scientific, or 'philosophical', navigators, agriculturalists, or industrialists, but does not lead to the emergence
of a class of men calling themselves scientists and thinking
of themselves as a professional body.

We are taught to expect that large enterprises involving
the government and the armed services may lead to technological spin-off; and it is worth asking if these expeditions
did so. Tinned food was prepared for these voyagers, and was
a help in the conquest of scurvy;[16] but because the principles
of bacteriology were not understood there was a high chance
that the food would be tainted as indeed McClure's was as late
as 1850. The importance of interchangeable parts was seen as
early as 1821 by Parry, who took on his second voyage two
vessels as far as possible identical, so as to economise on
spare parts – and also to keep company better. Instruments
were developed for Arctic use; and prefabricated, and finally
inflateable, boats were invented. Much was learnt in these
extreme conditions about central heating and about nutrition.
In the realm of science, much coastline was charted, and much
learned about the distribution of plants and animals, and the
formation of icebergs and of ice at sea. Arctic expeditions
provided access to a low-temperature laboratory which could
not be matched in temperate latitudes; and much information
was gained about terrestrial magnetism, the shape of the Earth,
and the phenomena of atmospheric refraction and of the Aurora
Borealis.

Such voyages represent a kind of 'normal science'; that
is, a discipline in which the questions are difficult to answer
rather than difficult to ask. In such spheres, government
or other support is crucial, because to transport a magnetic
or astronomical observatory around the world was beyond the
means of individuals. And the magnetic or pendulum experiments
needed to be done very accurately or not at all;[17] that they
were seen as important contributions to science is shown by
the award of the Copley Medal of the Royal Society to both
Sabine and Foster within ten years of one another for their
work on Parry's voyages. In America, Joseph Henry classed

those who had thus "braved the terrors of a polar winter"
among the "speculative" scientists "who possess the genuine
spirit of investigation and who have tasted the satisfaction
arising from an advancement in intellectual acquirements".

We have seen that for sailors, science might offer a
hope of professional advancement; but it did become almost
possible to make a career in 'naval science'.[18] The skills
required on voyages of survey and exploration were not
formally taught, but were learnt on voyages - that is, by a
kind of apprenticeship system. Cook had begun the tradition
of training his junior officers in this way; and the prime
example of the tradition which grew up might be Graham Gore,
who died on Franklin's last disastrous expedition to the
Arctic and whose father and grandfather (who had sailed with
Cook) were distinguished hydrographers. Franklin had sailed
as a midshipman with his uncle Matthew Flinders (who had
sailed with Cook) and George Back learnt his trade with
Franklin. Beechey was trained by Parry, and in turn he trained
Belcher. By the 1820s, commands in survey voyages and promotions into survey ships were as a rule being given to those
who already had learned the tasks involved as junior officers;
and a corps of scientific sailors was growing up. But they
never became completely separated from the rest of the Royal
Navy, and many distinguished hydrographers only obtained
promotion when they distinguished themselves on active service.

To sailors then, in general, science was a way of
furthering one's career rather than a career itself. Science
played in the life of Banks a not very different role; his
wealth and social position were assured, but his pursuit of
science gave him a sphere of considerable power and usefulness.
We have seen that within natural history, the patronage of a
man such as Banks could open a career to a young man of ability
who wanted to live by the pursuit of science; there was no
formal training, until in the 1820s the Horticultural Society
of London began a scheme for training gardeners.[19] But much
natural history could be learned by a formal or informal
apprenticeship system, particularly on a voyage; and then by
use of the herbaria and libraries of the Linnean Society and
of Banks. Voyages and travels also gave one the chance to
meet foreign men of science; who were also to be met in London
at Banks' house. The international character of scientific
voyages was indeed referred to in the account of Macartney's
Embassy:[20] "Enterprizes such as these were so much above the
usual course of things, and the motives of ordinary actions,
that in the midst of war, they were held sacred by an admiring
enemy; and without solicitation, were excepted from the danger
of the hostile attacks, to which every other English property

ınd person were exposed." But perhaps the most famous example
of science being above war in this period was the award to
Humphry Davy of Napoleon's prize for electrical discovery, and
his visit to Paris to collect it.

Davy's career was very different from that of Banks, whom
he succeeded as President of the Royal Society. Davy was a
professional scientist in that he lived by his science; at any
rate down to the time of his marriage, at the height of his
fame, to a wealthy widow. But Davy's training in science was
informal, and he hardly provided an example of a career in
science in the modern sense; his life only shows that an excep-
tionally able young man who pursued science could rise from
humble and provincial origins to the highest place in the
scientific world, and gain thereby an entrée into society. In
his own last work, Consolations in Travel, which appeared in
1830[21] he wrote: "Accident opened to me in early youth a philosophical career, which I pursued with success. In manhood,
fortune smiled upon me and made me independent."
On being asked, in the dialogue in that book, why he did not
embark upon a professional career in law or politics, he replied
that he was not ambitious or reptilian enough, adding that
"to me there never has been a higher source of honour or distinction than that connected with advances in science." He
added a complaint that so few of the aristocracy now pursued
science; urged the utility and dignity of science, and particularly of his own science of chemistry; and suggested that the
best preparation for a life of science was "a liberal education,
such as that of our universities." To men like Barrow, Davy
seemed too much the scientist (and also too much dazzled at the
eminence he had attained) to be a good President of the Royal
Society; but they need not have worried that Davy did not share
their view of science as a useful activity for men in professions,
or of independent means, and not in itself a profession.

In the last years of Banks' Presidency, in 1818, there
happened what Barrow[22] called "the first attempt, that I remember,
to open salaried office to men of science; and small as the
boon was, it was greatfully received; and the system worked with
great advantage both to the Admiralty and the public. But in
a subsequent fit of economy, this poor pittance was withdrawn
from science." This was the appointment of Commissioners to
the Board of Longitude; and the episode was seen in a very
different light by Charles Babbage, in his Decline of Science
in England of 1830. To him it seemed a piece of jobbery; but
his accents are those of a disappointed man. At all events,
although as Babbage wrote fairly large sums of money were
available for those lucky enough to get them, the experiment
did not last long and the Board of Longitude never became

64

anything like the Scientific Civil Service offering full-time
posts and a career structure. What was offered was some extra
emolument to men like Thomas Young, W. H. Wollaston, and
Michael Faraday, who were chiefly dependent for their income
upon their science.

There were few men like these in England in our period;
and not only in the Navy but also in Medicine, in the Church,
and even in the Law, men might have leisure to pursue some
science,[23] perhaps even at an advanced stage; and with the
hope of being useful to their fellow men, and of advancing
their own career – for a reputation for scientific knowledge
could do one nothing but good in any profession. This situation no doubt led to a state of relative weakness in
theoretical and mathematical science, where the problems
were more abstruse than those required for navigation. On
the other hand, it led to a wide diffusion of science within
the educated community; there was no question of "two cultures"
in England at this time. We should remember also that The
Philosophical Magazine, Nicholson's Journal, and the Royal
Institution's Journal, and the Annals of Philosophy all
flourished at this date; at any time between 1800 and 1830
there were at least two of these semi-popular journals
appearing, and publishing articles of some importance and
information about science at home and abroad. That there was
in general no scientific profession did not mean that little
science was done, or that few people were interested in it;
in some ways the absence of professionalism, like the reviews
of scientific books in literary periodicals and the large
audiences at the Royal Institution, indicates a feeling that
science was too important, useful, and interesting to be left
to the experts.

65

1   G.Staunton, _An Embassy to the Emperor of China_, 2 vols.
    Dublin, 1798,I,37-8. J.K. Tuckey, _An Expedition to_
    Explore the River Zaire, ed. J. Barrow, London, 1818,i;
    C. Lloyd, _Mr. Barrow of the Admiralty_, 1764-1848, London,
    1970.
2   A. P. Molella & N. Reingold, "Theorists and Ingenious
    Mechanics", _Science Studies,3_ (1973)323-51. See also my
    Sources for the History of Science,1660-1914, London,1974.
3   G. Staunton, _Embassy_,I,20; A. M. Lysaght, _Joseph Banks in_
    Newfoundland and Labrador, 1766, London, 1971; W.R.Dawson
    (ed.) _The Banks Letters_, London, 1958 and supplement in
    the _Bulletin of the British Museum (Natural History)_,
    Historical Series,3 (1962-9)41-93.
4   J. C. Beaglehole(ed.), _The Endeavour Journal of Joseph_
    Banks, 2 vols, Sydney, 1962; P.J.P. Whitehead(ed.),
    Drawings of Fishes from Captain Cook's Voyages, London,
    1968; B. Faujas St.Fond, _Travels in England_, 2 vols.
    London, 1799, I, contains an encomium on Banks; J. Davy,
    The Life of H. Davy, 2 vols. London, 1836,II,342ff.
5   Brown's essay is in M. Flinders, _A Voyage to Terra Australis_,
    2 vols + atlas, London, 1814, and is reprinted in Brown's
    Miscellaneous Botanical Works, 3 vols., London, 1866-8.
    The botanical artist F. Bauer accompanied Brown; and his
    Australian Flower Paintings, ed. W. Blunt and W.T. Stearn,
    has been announced in a sumptuous edition, London, 1974.
    J.C. Beaglehole (ed.), _The Journals of Captain James Cook_,
    4 vols in 5, Cambridge, 1955-74; W. Bligh, _A Voyage to the_
    South Sea, London, 1792.
6   W. Blunt, _The Art of Botanical Illustration_, London, 1950;
    F.Klingender, _Art and the Industrial Revolution_, ed. A.Elton,
    London, 1968; A. Coats, _The Book of Flowers_, London, 1973;
    P. Mitchell, _European Flower Painters_, London, 1973;
    M. Twyman, _Lithography 1800-1850_, London, 1970. Works on
    natural history are listed in my _Natural Science Books in_
    English, 1600-1900, London, 1972, chapters 5 & 9.
    J.E.B. Currey (ed.), _George Caley's Reflections on the_
    Colony of N.S.W., London, 1967; On James Douglas, another
    collector, see W. Moorwood, _Traveller in a Vanished Land-_
    scape, London. 1973
7   R. V. Tooley, C. Bricker, & G.R. Crone, _A History of Carto-_
    graphy, London, 1969; T. Forrest, _A Voyage to New Guinea_,
    intr. D.K. Bassett, Kuala Lumpur, 1969, introduction pp.18-9;
    W. Marsden, _The History of Sumatra_, intr. J. Bastin, Kuala
    Lumpur, 1966; J. Bastin, "Dr. Joseph Arnold", _Journal of the_
    Society for the Bibliography of Natural History,6(1973 305-72.
    M. Archer, _Natural History Drawings in the India Office_
    Library, London, 1962; W. Roxburgh, _Icones Roxburghianae_,
    Calcutta, 1964-.

66

8 C. Close, The Early Years of the Ordnance Survey, ed.
   J.B. Harley, Newton Abbot, 1969; A. Day, The Admiralty
   Hydrographic Service, 1795-1919, London, 1967;
   G.S. Ritchie, The Admiralty Chart, London, 1967; M.Deacon,
   Scientists and the Sea, 1650-1900, London, 1971.
   E.H.J. & G.E.E. Feeken & O.H.K. Spate, The Discovery and
   Exploration of Australia, Melbourne, 1970; facsimiles of
   early works are available in the Australiana series pub-
   lished by the State Library of South Australia.

9 A journal of Halley's voyage is being prepared for publication by the Hakluyt Society. G. Vancouver, A Voyage
   of Discovery, 3 vols + atlas, London, 1798,I,xiv; on
   ship's surgeons, J.J. Keevil, C. Lloyd & J.L.S. Coulter,
   Medicine and the Navy, 4 vols, Edinburgh, 1957-63, III
   & IV; C. Lloyd, The British Seaman, London, 1968.

10 On these voyages, see L.H. Neatby, Search for Franklin,
   London, 1970; reprints of the accounts of their voyages
   by George Back, Franklin, Parry, Richardson, J. Ross,
   J.C. Ross, and others are available.

11 W.E. Parry, Journal of a Third Voyage, London, 1826,xi;
   W.H.B. Webster, A Voyage... in H.M. Sloop Chanticleer,
   2 vols, London, 1834; W.F.W. Owen, Narrative of Voyages,
   2 vols, London, 1833,I,231-2; E. Belcher, A Voyage round
   the World, 2 vols, London, 1843,I,14.

12 In 1827 there were 35 Fellows of the Royal Society who were
   Naval men or had been on Arctic Voyages; see the printed
   lists of Fellows. Naval men were also well-represented
   on the Council of the Society.

13 J. Ross, A Voyage of Discovery, London, 1819; on the Royal
   Society and publication, M. Faraday, Selected Correspondence, ed. L. P. Williams, 2 vols, Cambridge, 1971, letter
   329; on government support for publication, F.W. Beechey,
   A voyage to the Pacific, 2 vols, London, 1831,I,xvi-ii.

14 J.F.W. Herschel (ed.), Admiralty Manual of Scientific
   Enquiry, London, 1849 and later eds.; a reprint of the
   2nd ed., introduced by me, is in press, London. Brown did
   the botany for J. Ross, Voyage, 1819; Hooker for Parry,
   2nd Voyage, Appendix, 1825, and for Beechey, Botany, 10
   parts, 1830-41.   Jameson did the geology in Parry,
   3rd Voyage, 1826; MacCulloch in J. Ross, Voyage, 1819;
   Buckland in Beechey, Voyage, 1831. Sabine did the zoology
   for Franklin's Narrative, 1823; Children for Back's Arctic
   Land Expedition, 1836; and Curtis the insects for J. Ross,
   2nd Voyage, 1835. J. Richardson, Fauna Boreali-Americana,
   4 vols, London, 1829-37. Systematic study of who described
   what would cast light on the organisation of science at this
   period.

15 G. S. Graham, <u>Great Britain in the Indian Ocean</u>, Oxford, 1967. On agriculture, see M. Berman, "The Early Years of the Royal Institution", <u>Science Studies,2</u>(1972)205-40.

16 On tinned meats, see W. E. Parry, <u>2nd Voyage</u>, 1824, pp. vii-viii; P.P. King, <u>Australia</u>, 2 vols, London,1827, II,91; J. Ross, <u>2nd Voyage</u>, 1835,I,108,619-10; R.Le M. McClure, <u>North-West Passage</u>, London, 1856, pp.121,131. Parry, <u>2nd Voyage</u>, iii-iv on identical ships; and see the descriptions of instruments and tables of observations in the various <u>Voyages</u>. Franklin took portable boats on his <u>2nd Journey</u>, London, 1828, pp.xiv-xvi; Richardson, <u>Arctic Searching Expedition</u>, 2 vols, London, 1851, had an inflateable boat. On heating, see Parry, <u>Voyage</u>, 1821, pp. 103ff, clxxiii; J. Ross, <u>2nd Voyage</u>, I,210ff, and his remarks on nutrition, pp.199ff.

17 P.E. de Strzelecki, <u>New South Wales</u>, London, 1845, pp.47ff. N. Reingold (ed.), <u>The Papers of Joseph Henry</u> Washington, D.C., 1972-,I,395.

18 See Day, Deacon, & Ritchie, note 8 above. But on promotions, see W.F.W. Owen, <u>Voyages</u>,II,224; and note that Darwin's Captain Fitzroy was an 'outside' promotion; H.E.L. Mellersh, <u>Fitzroy of the Beagle</u>, London,1968,p.29.

19 H.R. Fletcher, <u>The Story of the Royal Horticultural Society, 1804-1968</u>, London, 1969,pp.83ff.

20 G. Staunton, <u>Embassy</u>,I,21

21 H. Davy, <u>Consolations in Travel</u>, London, 1830, pp.224-5, 249; a short biography is H. Hartley, <u>Humphry Davy</u>, London, 1966.

22 J. Barrow, <u>The Royal Society</u>, London, 1849, p.62; C. Babbage, <u>The Decline of Science</u>, London, 1830,pp.66ff.

23 On Geology and Churchmen, see M. Rudwick, <u>The Meaning of Fossils</u>, London, 1973; Wollaston, Young, and James Parkinson were medical men who made a reputation in the sciences; Henry Brougham and W.R. Grove were lawyers interested in science. On general problems raised by science, see my "Chemistry, physiology and Materialism in the Romantic Period", <u>Durham University Journal,64</u> (1972)139-45.

X

Agriculture and Chemistry in Britain around 1800

Summary

This paper is concerned with the application of science to a practical activity. The story begins in the late eighteenth century, a period of agricultural innovation, with various authors urging that definite chemical knowledge should replace rule of thumb in the application of fertilisers. In the work of Archibald Cochrane, ninth Earl of Dundonald, we find this exhortation beginning to give way to descriptions of actual chemical experiments, and interpretations of equilibria in the soil. But it is only with Davy's *Agricultural chemistry* of 1813 that we get clear descriptions of soil analyses that could be undertaken by a farmer, accompanied with a certain amount of biochemical information on the growth of plants. Davy's recommendations were essentially conservative; he provided support for the best practices already being recommended by innovators. His book is interesting too, for the light it casts upon his more theoretical writings.

1. Introduction

The latter part of the eighteenth century in Britain was a time of agricultural innovation, as landowners and farmers introduced new methods of tillage such as seed-drills and horse-hoes, new crops such as turnips and ' artificial grasses ' like clover, and new breeds of stock. The value of manure as a fertiliser was increasingly appreciated, and various inorganic fertilisers such as nitre, marl, and common salt were employed in various places. Different soils were distinguished largely on physical grounds, as clayey, sandy, or chalky, and as light or heavy; and different treatments and crops were recommended for different soils. Towards the end of the century, we find authors who hoped that chemistry might provide better distinctions than these, and thus put agriculture upon a firm basis of science.

At first, we find little but exhortation. Bishop Watson, in his famous and successful *Chemical essays*, described some inorganic fertilisers, refuted the notion that snow was better for vegetation than rain because it was more ' nitrous ', and discussed rather inconsequentially the idea that water might be transmuted into earth by boiling it, or in the growth of plants as in the famous experiment of Helmont.[1] This would have helped the farmer very little; and Watson's ' Preliminary observations ' in a volume on the Agriculture of Westmorland again contain little useful chemistry, his main recommendations being that landowners should plant more larches, and should not allow the tar from wood to go to waste in making charcoal. However, in the preface

---

[1] R. Watson, *Chemical essays* (5 vols., 1793–96, London; the first three vols. of this set are 6th ed., the fourth is 5th ed., and the fifth is 2nd ed.), vol. 4, vii, xi.

138

to the fourth volume of *Chemical essays*, we find some exhortation in what was to become a familiar pattern. Watson wrote: ' My own notion, indeed, of National Improvement Security, and Happiness, tends not so much to the extending of our commerce, or the increasing the number of our manufacturers; as to the increasing of an hardy, and, comparatively speaking, innocent Race of Peasants, by making Corn to grow on Millions of Acres of Land, where none has ever grown before '. Good farming, Watson believed, would follow if all gentlemen were in youth ' instructed in the *Principles of Vegetation*, in the *Chemical Qualities of soils*, and in the *Natures and uses of different manures* ', and he hoped for an institution where this could be done. The emphasis upon agriculture as a source of social stability is something we meet again and again, particularly in works written after the outbreak of the French Revolution.

The agricultural propagandist Arthur Young was in France in 1789, and he recorded in his diary a dinner on 1 August:

> Dined with M. de Morveau ... Even in this eventful moment of revolution, the conversation turned almost entirely on chymical subjects. I urged him, as I have done Dr. Priestley more than once, and Mons. La Voisier also, to turn his enquiries a little to the application of his science to agriculture; that there was a fine field for experiments in that line, which could scarcely fail of making discoveries; to which he assented; but added, that he had not time for such enquiries.[2]

In 1793 the Board of Agriculture was set up, with Sir John Sinclair as its President and Young as its Secretary, with the aim of improving agricultural practice and expanding output. At once the Board began commissioning authors for a series of volumes on the agriculture of counties. Some of these were well-known men, familiar with the practices of their localities and qualified to comment upon them, while others were not; and the series is uneven. What is interesting is that even the best of the surveys did not use any chemistry. Thus Young's *Oxfordshire* contains a discussion of whether dunghills should be left to ferment, or be spread on the land when fresh or ' long '; but the evidence is purely based upon empirical case studies rather than systematic experiment, and no chemistry is brought in at all.[3]

In other surveys the story is much the same; thus Bailey and Culley, describing Northumberland, were simply practical, and so was Vancouver on Devonshire. The importance of manure was recognised—as it had been by Gilbert White of Selborne who ' borrowed ' loads of it—but there were no chemical discussions of how it worked, or of how crops enriched or impoverished the soil. Sometimes there were not even genuflections towards chemistry.

---

[2] A. Young (ed. M. Betham-Edwards), *Travels in France during the years 1787, 1788, 1789* (2nd ed., 1889, London), 222.

[3] A. Young, *General view of the agriculture of Oxfordshire* (1813, London; reprinted 1969, Newton Abbot); C. Vancouver, *General view of the agriculture of the County of Devon* (1808, London: reprinted 1969, Newton Abbot); J. Bailey and G. Culley, *General view of the agriculture of Northumberland, Cumberland and Westmorland* (1805, London: reprinted 1972, Newcastle-on-Tyne) with a useful general introduction and bibliography by D. J. Rowe. Bishop Watson's observations are at p. 280 ff. There is a full list of County Reports in my *Natural science books in English 1600–1900* (1972, London), 123–124. A useful introduction to farming at a slightly later period is H. Stephens, *The book of the farm* (1844, London and Edinburgh), abridged and edited with the new title *Victorian farming* by C. A. Jewell (1975, Reading), with a useful introduction by E. J. T. Collins. The standard guide to our subject is E. J. Russell, *A history of agricultural science in Great Britain* (1966, London).

Thus William Marshall, who was disappointed not to get the Secretaryship of the Board of Agriculture, and was perhaps a better agriculturalist than Young, described in his *Review and abstract of the County Reports* of 1808 the qualities of the ideal agricultural reporter.[4] He should be familiar with both agriculture and estate management, and also with

> the different SCIENCES which are intimately connected with rural subjects: particularly Natural History, as it relates to fossils and vegetables ... and with mechanics, to enable him to appreciate, with greater ease and certainty, machines and implements in use. Some knowledge of the higher branches of the mathematics, to form his mind to method, and to teach him to think with precision, and decide with clearness, on the subject before him.

There is no mention of chemistry at all, although—or perhaps because—Davy had been giving lectures on agricultural chemistry at the request of the Board of Agriculture for some years. Davy was not even the first to try to link chemistry and agriculture, rather than simply to urge that there ought to be some kind of link; for in 1795 Archibald Cochrane, Ninth Earl of Dundonald, had published *A treatise, shewing the intimate connection that subsists between agriculture and chemistry*.[5]

2. Cochrane's work

Cochrane was born in 1749, and died in 1831. His more famous son was one of the most dashing sea-captains of the Napoleonic Wars, and later of the wars of independence in South America, but he himself has a place in the history of chemical industry. He was involved with the setting up of the Walker Alkali Works on Tyneside to carry out the le Blanc process, and he took out ten patents between 1781 and 1812, most of them for chemical processes. Having coal mines on his estates, he tried to get coal tar accepted for caulking ships; but in this as in his other enterprises, he was not successful (although companies such as Walker flourished under different management) and he died, in Paris, impoverished. It is odd to find in the writing of this chemical manufacturer the remark that an ' unwise preference ' had been given by government to manufactures and to commerce. Perhaps because of his unfortunate experiences in industry, he now believed that husbandry was a ' sober and healthful employment ', not associated with infidelity and revolutionary opinions. Works on agriculture were written to be read by members of the powerful landed interest, and are not radical social documents; nor do there seem to be in them any explicit reference to the doctrines of Malthus, but rather a conviction that up-to-date agriculture could feed an expanding population.

[4] W. Marshall, *The review and abstract of the county reports to the Board of Agriculture* (5 vols., 1818, York; reprinted 1968, Newton Abbot), vol. 1, xxxi. G. White (ed. J. Clegg), *The Garden Kalendar, 1751–71* (1975, London), 219v.

[5] On Cochrane, see *Dictionary of national biography* (63 vols., 1885–1900, London) vol. 11, 160; and for his processes, B. Woodcroft, *Alphabetical index of patentees of inventions* (1854, reprinted 1969, London) under ' Dundonald '—the pagination is confused. I have used A. Cochrane, *A Treatise, shewing the intimate connection that subsists between agriculture and chemistry, addressed to the cultivators of the soil, to the proprietors of fens and mosses, in Great Britain and Ireland; and to the proprietors of West India estates* (2nd ed., 1803, London), 2. On Tyneside chemical industry, see W. A. Campbell, *A century of chemistry on Tyneside* (1968, Newcastle-on-Tyne), and *The chemical industry* (1971, London).

190

From a man familiar with chemical processes in industry, Cochrane's book is surprisingly short on detailed chemical discussion. Indeed, it marks a mid-point between the merely exhortatory remarks of Watson and Young and the experimental chemistry of Davy. The book is initially organised as Plot's *Oxfordshire* had been over a century before,[6] in terms of Earth, Air, Water, and Heat; but this framework then gives way to a discussion of the chemical nature of vegetables, soils and rocks, and fertilisers. The second half of the book is practical, with specific advice for improving soils after chemical tests have determined the proportions of components. The recommendations are cautious and empirical; fertility was dependent upon a due admixture of components, and in general ' the success of most operations, but most especially those of a chemical nature, greatly depends upon a regular and due observance of circumstances apparently trivial '.[7]

The introductory parts of the book make some telling general points in favour of the study of chemistry: ' there is no operation or process, not purely mechanical, that does not depend on Chemistry, which is defined to be a knowledge of the properties of bodies, and of the effects resulting from their different combinations '. Suitable substances for fertilising have been neglected, and attempts have not been made ' to explain on chemical principles the operation of manures and substances now in use. Had such researches been prosecuted to effect, they would have led to the discovery and application of other substances capable of being employed, with equal, or perhaps superior advantages '. Here Cochrane is surely right in general; that in applied science the way to begin is to see how devices or processes in current use work, and then try to improve or replace them.

The problem with agriculture was that the processes are complicated, and that the number of variables in any situation is large. Cochrane was aware of the importance of substances such as iron, which ' although it is only found in small quantities in vegetable and animal substances, still its effects in promoting vegetation may, in a chemical point of view, be much greater than can possibly be accounted for by the very small proportion of iron found in organic bodies '—perhaps because it can readily be reduced and again oxygenated. He recognised a succession of dynamical processes going on in the soil, and urged that ' by a thorough knowledge and application of chemistry to agriculture, the several substances in soils may be made to undergo a varied change of new combinations, tending to promote the greatest of all objects, a more plentiful supply of food '. Cochrane's book is partly posing questions for an able chemist to answer by research, and partly exhorting farmers to rely on chemical knowledge rather than mere empiricism; but the chemical questions raised are too complicated to be answered by a farmer, to whom the book would have been of little practical value. The textbook-writer Samuel Parkes expected in 1815 that the farmer would be able to carry out soil analyses, and interpret the results—but this seems utopian.

[6] R.Plot, *The natural history of Oxfordshire* (2nd ed., 1705, reprinted 1972, Chicheley, Bucks.); and W. Borlase, *The natural history of Cornwall* (1758, reprinted 1970, London) with a useful introduction by F. A. Turk and P. A. S. Pool. See also my *Sources for the history of science* (1975, London), 161 ff.

[7] Cochrane (footnote 5), 112; the quotations which follow are from pp. 1, 3, 53, 58. S. Parkes, *Chemical essays* (5 vols., 1815, London), vol. 1, no. 1, is also mentioned.

141

By 1793, when Cochrane wrote his book, chemists had at last reached a point when they might hope to improve agriculture because they understood something of how plants grew. Out of the ' pneumatic chemistry ' of the later eighteenth century came the discovery of photosynthesis, the process whereby plants in sunlight absorb carbon dioxide from the air. Plants were therefore seen to gain their nourishment primarily from the air and from water; and it became a problem to understand why some soils were so fertile, and others so sterile. Inorganic analysis had reached a level where mineral constituents of plants could be determined, and minerals were indeed found in plants, though often in variable quantities. Organic analysis was not sufficiently developed to do more than determine the elements and their proportions, and could cast little light upon structure. To perform meaningful and valuable chemical experiments in agriculture was therefore not a straightforward matter, for the interpretation of a soil analysis was difficult, and there was no readily testable theory such as Liebig was to provide in the 1840s. Davy, the first major chemist to involve himself deeply in agricultural chemistry, came to it from work on gases, and approached the subject in a judiciously empirical manner, rejoicing when his analyses confirmed what farmers had found about the fertility of soils.

3. Davy's *Agricultural chemistry*

It was Humphry Davy who made agricultural chemistry a subject to be taken seriously; as we can see from the sales of his book, which first appeared in 1813 and was still in 1840 in sufficient demand for it to be split between two volumes (7 and 8) of the *Collected works*, so that this edition should not interfere with sales of the one-volume version, then in its sixth edition.[8] Davy was born in 1778, and had made his name with his researches on nitrous oxide done at Thomas Beddoes' Pneumatic Institution in 1800. In 1801 he had come to the Royal Institution in London, first as Lecturer but from May 1802 as Professor. Dr. Berman has shown the prominent part played by landowners interested in the improvement of their estates in the early Royal Institution; he indicates that it was thus no accident that Davy, who was to be the Apostle of Applied Science, should be set to work on tanning and on agriculture rather than on industrial chemistry.[9] In 1802 Davy duly began a course of lectures on agricultural chemistry, sponsored by the Board of Agriculture; and it was these lectures, given each succeeding year, which were eventually written up into his book. In 1803 he was elected a Fellow of the Royal Society, and in 1805 awarded its Copley Medal primarily for his researches on tanning, in which he had proposed various substitutes for oak bark.

As one would expect in a work by Davy, the *Agricultural chemistry* abounds in broad generalisations often only loosely connected with agriculture; but also in manipulative details, so that from the book it would be possible for an

[8] J. Z. Fullmer, *Sir Humphry Davy's published works* (1969, Cambridge, Mass.), 70 ff. The best life of Davy is H. Hartley, *Humphry Davy* (1966, London); and see also the short life by me in C. C. Gillispie (ed.), *Dictionary of scientific biography*, vol. 3 (1971, New York), 598–604.
[9] M. Berman, ' The early years of the Royal Institution ', *Science studies*, **2** (1972), 205–240.

192

otherwise untrained person to do soil analysis. Everywhere Davy saw cyclic processes, harmony and equilibrium:

> And by the influence of heat, light, and chemical powers, there is a constant series of changes; matter assumes new forms, the destruction of one order of beings tends to the conservation of another, solution and consolidation, decay and renovation, are connected, and whilst the parts of the system continue in a state of fluctuation and change, the order and harmony of the whole remain unalterable.[10]

Davy drew upon a number of good authorities for his botany and chemistry, referring to experiments on plant growth by de Candolle, de Saussure, Gay-Lussac and Thenard. His heaviest dependence for his material on plant physiology was upon Thomas Andrew Knight, but what is interesting is his historical perspective, for Grew, Tull, Hales and Linnaeus are all drawn upon as current authorities—indeed, his plant sections are redrawn from Grew, whose *Anatomy of plants* had appeared in 1682, 130 years earlier.[11]

Davy had done some work on the chemistry of vegetation himself, some of his earliest researches being on photosynthesis, while later he had discovered silica in the outer layers of various grasses.[12] The *Agricultural chemistry* contains an appendix describing experiments on various grasses undertaken at Woburn by George Sinclair, gardener to the Duke of Bedford; they had been grown on various small similar plots, and were analysed by Sinclair with comments by Davy. Altogether 108 species were investigated, including the ' artificial grasses ' lucerne and clover, on various types of soils; we are sometimes apt to forget that the improvement of pastures was as important at this period as the more spectacular improvement of stock. We get an idea of the accuracy of Davy's analytical procedures from one of his soil experiments, where from 400 grains of sandy soil he found 379 grains of products, losing 21 grains or rather over 5% in the process and commenting that ' the loss in this analysis is not more than usually occurs, and it depends upon the impossibility of collecting the whole quantities of the different precipitates; and upon the presence of more moisture than is accounted for in the water of absorption, and which is lost in the different processes '. Davy collected carbon dioxide in a submerged bladder in estimating calcium carbonate, and measured the water displaced; this method was unsound, because it took no note of the increased pressure in the bladder, but it would have been relatively easy for farmers to use.

Davy was able to use his chemical knowledge to give definite answers to some questions of agricultural practice; notably whether compost or dung

---

[10] H. Davy, *Elements of agricultural chemistry* (2nd ed., 1814, London), 9; compare pp. 308–309.

[11] N. Grew, *The anatomy of plants* (1682, reprinted 1965, New York); Davy's plates 8, 9 and 10 are from Grew's 28, 33 and 29. On Knight, see H. R. Fletcher, *The story of the Royal Horticultural Society* (1969, London) 63 ff. Davy was not alone in still using Grew's plates; see R. Brown (ed. J. J. Bennett), *Miscellaneous botanical works* (2 vols., 1866, London) vol. 1, 384.

[12] The references for this paragraph are as follows. H. Davy (ed. J. Davy), *Collected works* (9 vols., 1839–40, London), vol. 2, 45–49 on photosynthesis, in his first (highly speculative) published paper; his papers on silex in vegetation were in (*Nicholson's*) *J. nat. phil. chem. arts*, 3 (1799), 55–59, 138. George Sinclair wrote an independent work on grasses entitled *Hortus gramineus woburnensis* (1816, London), with plates done by the newly-invented process of lithography. Davy (footnote 10), 373–479 discusses the grasses, and 172 the analytical experiment.

should be applied fresh or fermented.[13]   His conclusion was that dung should be applied as fresh as possible, because as it decomposes, it ' throws off its volatile parts, which are the most valuable and most efficient. Dung which has fermented, so as to become a mere soft cohesive mass, has generally lost from one third to one half of its most useful constituents. It evidently should be applied as soon as fermentation begins, that it may exert its full action upon the plant, and lose none of its nutritive powers '.   He believed that ' water and the decomposing animal and vegetable matter existing in the soil, constitute the true nourishment of plants'; and although he recognised that sometimes mineral matters, such as silica in grasses, did enter into the composition of vegetation, he believed that only salts of the alkalies and alkaline earths could be really described as ' fossil manures '.   Other inorganic substances could only improve the texture of the soil—which was important because good soils were those which held plenty of water—or neutralise some objectionable component already present.   Thus Sir Joseph Banks had passed on a sample of sterile soil from Lincolnshire, which Davy found to contain sulphate of iron; he recommended lime, ' which converts the sulphate into a manure '— that is, a slightly soluble calcium salt.   Dundonald had similarly recommended magnesian limestone as an antidote to acidity caused by the decomposition of iron pyrites.

Davy was himself tempted, as we know by what he wrote elsewhere, to believe that all matter was ultimately the same, and that the chemical elements were different arrangements of one prime matter.[14]   Bishop Watson had discussed at some length experiments in which plants had apparently generated new elements from water in which they grew, though admitting that the evidence was inconclusive either way.   If transmutation were simply rearrangement of corpuscles, as Hooke, Boyle, and Newton seemed to suggest, then there was nothing immediately implausible in the suggestion that plants might transmute elements.   For various reasons, belief in the unity of matter survived through the nineteenth century, and was in a sense justified in the work of J. J. Thomson, Rutherford, Moseley and Bohr; but by 1813, when the *Agricultural chemistry* came out, the evidence for unity of matter from plant physiology had been seriously weakened by experiments done by Davy and others.   He had shown, for example, that distilled water usually contained salts; this had been demonstrated in the course of his work on the electrolysis of water, and the conclusion provides a link between what might be called his work in ' pure ' and in ' applied ' chemistry.   Plants grown in distilled water which were found on combustion to have contained salts had therefore probably not made them from water, but simply absorbed them.

[13] H. Davy (footnote 10), 6, 184, 315, 203.   We might notice that Davy's agricultural chemistry differed from Liebig's in that the latter recommended insoluble inorganic fertilisers, fearing (unnecessarily) that soluble matter would rapidly be washed out of the soil; see E. J. Russell (footnote 3), 97 ff., and the paper by Dr. W. V. Farrar, of the University of Manchester Institute of Science and Technology, read at the meeting mentioned in the acknowledgements below.

[14] W. V. Farrar, ' Nineteenth century speculations on the complexity of the chemical elements ', *British j. hist. science*, **2** (1965), 297–323, **4** (1968), 65–67; my *Atoms and elements* (2nd ed., 1970, London) and *Classical scientific papers, chemistry*, 2nd series (1970, London) are concerned with this problem.   See also R. Watson (footnote 1), vol. 4, 257 ff; H. Davy (footnote 10), 311, and ' The Bakerian Lecture, on some chemical agencies of electricity', *Phil. trans. Roy. Soc. London*, **97** (1807), 1–56 (pp. 6–8); and C. A. Russell, ' The electrochemical theory of Sir Humphry Davy ', *Annals of science*, **15** (1959), 1–25, **19** (1963), 255–271.

194

In his seventh lecture, devoted to the question of mineral manures, Davy referred to

> enquirers adopting the sublime generalization of the ancient philosophers, that matter is the same in essence, and that the different substances considered as elements by chemists, are merely different arrangements of the same indestructible particles, have endeavoured to prove, that all the varieties of the principles found in plants, may be formed from the substances in the atmosphere; and that vegetable life is a process in which bodies that the analytical philosopher is unable to change or to form, are constantly composed and decomposed.[15]

But the evidence was that such rearrangements happened only among the organic compounds, and involved no transmutation; plants were found on analysis to contain no salts that had not been present in the soil or medium in which they had grown. Davy thus concluded:

> that the different earths and saline substances found in the organs of plants, are supplied by the soils in which they grow; and in no cases composed by new arrangements of the elements in air or water. What may be our ultimate view of the laws of chemistry, or how far our ideas of elementary principles may be simplified, it is impossible to say. We can only reason from facts. We cannot imitate the powers of composition belonging to vegetable structures; but at least we can understand them: and as far as our researches have gone, it appears, that in vegetation compound forms are uniformly produced from simpler ones; and the elements in the soil, the atmosphere, and the earth absorbed and made parts of beautiful and diversified structures.

It is first of all curious that in his *Elements of chemical philosophy*, published in the previous year (1812), Davy had devoted a chapter to the arguments from analogy for the unity of matter, cautiously favouring the doctrine.[16] His alternations between bold analogical speculations and careful adherence to facts make him a fascinating figure, and he would no doubt have considered himself in this to be following the example of Newton, who combined a refusal to feign hypotheses with a readiness to advance queries, again upon different occasions but at much the same time. It is also curious that believing this about mineral substances, and recognising minerals in many vegetable structures, Davy should have attached relatively little importance to mineral fertilisers. It is this which particularly distinguishes the agricultural chemistry of Davy from that of Liebig. Davy no doubt supposed that there would always be enough siliceous matter in soil to support grasses; but the other experiments to which he refers seem to indicate that plants absorbed certain salts, but not in general that these were essential to their growth—except for some which grew near the sea, and did not flourish without salt. In Saussure's experiments, described in tabular form by Davy, pines, rhododendrons and whortleberries were grown on limestone in the Jura, and on a granite mountain, Breven; when the quantities and proportions of various minerals in their ashes were very different. Organic fertilisers had been shown, to the satisfaction of any intelligent farmer, to benefit the land; and it was easy to understand that animal or vegetable matter could help in the production of

---

[15] H. Davy (footnote 10), 310, 314 (for the next quotation). Some agriculturalists continued to believe in the synthesis of elements; see W. V. Farrar (footnote 13).

[16] H. Davy, *Elements of chemical philosophy* (1812, London), 478–489; reprinted in my *Classical scientific papers, chemistry* (1968, London), 166–177. Davy (footnote 10), 116–117 tabulated Saussure's results; other analyses are on p. 150. He recognised that leguminous plants fix atmospheric nitrogen (p. 357), which some of his successors doubted.

new crops, whereas there was little experimental or chemical evidence for the efficacy of minerals. The great value of bones, as a source of phosphate, was appreciated later, and led to Liebig's overriding emphases on inorganic fertilisers despite his own background of work in organic chemistry, especially analysis.

Davy analysed various manures, including guano which had been sent from South America in 1805, and concluded that ' from its composition it might be supposed to be a very powerful manure '.[17] On plant nutrition generally, he wrote that soluble vegetable matter in the soil was the food of plants, along with water and carbon dioxide; that fluid matter in the soil was taken up by the capillary action of the roots; and that the sap, modified by exposure to heat, light and air in the leaves, was the source of new organized matter—it ' is thus, in its vernal and autumnal flow, the cause of the formation of new parts, and of the more perfect evolution of parts already formed '. Following Hales, he believed the flow of sap to be a purely physical process; writing that there was nothing *immaterial* in the vegetable economy, and that ' we must not suffer ourselves to be deluded by the very extensive application of the word *life*, to conceive, in the life of plants, any power similar to that producing the life of animals. In calling forth the vegetable functions, common physical agents alone seem to operate; but in the animal system these agents are made subservient to a superior principle '. Irritability and animation ought to be excluded from botany along with Dryads and Sylphs. In arguing thus, Davy was setting himself against strong traditions in eighteenth-century biology where animal-plant analogies were emphasised; and also against traditions in speculative German thought of the time, to be emphasised in his friend Coleridge's posthumous *Hints towards the formation of a more comprehensive theory of life* (1848, London).

Davy was a believer in the presence of an immaterial part in man, supporting this conviction from the researches of John Hunter and his disciples;[18] and his own posthumous work, *Consolations in travel* (1830, London) is full of this belief. To oppose the dynamic materialism and pantheism of the late eighteenth century, it was not unreasonable to break the Chain of Being and deny the analogies between plants and animals; and indeed in the *Consolations* Davy went further in asserting the distinctness of man—which in the days of slave emancipation was an important assertion, making all men equal rather than some akin to apes—and was naturally in no sense original to Davy. Even in a practical and ' applied ' work, we can thus see traces of his speculative interests.

[17] H. Davy (footnote 10), 296, 9, 250. On the animal/plant analogy, see P. C. Ritterbush, *Overtures in biology* (1964, New Haven). On *Naturphilosophie*, see T. H. Levere, *Affinity and matter* (1971, Oxford); B. S. Gower, ' Speculation in physics; the history and practice of *Naturphilosophie* ', *Studies hist. phil. science*, **3** (1973), 301–356; and T. H. Levere, ' Samuel Taylor Coleridge and Humphry Davy: natural philosophy and natural science ', forthcoming in the same journal.

[18] Recent work on vitalism takes as its point of departure J. H. Brooke, ' Wohler's urea and its vital force ', *Ambix*, **16** (1969), 84–114; see also E. Benton, ' Vitalism in nineteenth-century scientific thought: a typology and reassessment ', *Studies hist. phil. science*, **5** (1974), 17–48, and my paper ' The vital flame ', forthcoming in *Ambix*, which examines Davy's vitalism. On man's distinctness, see J. C. Prichard (ed. G. W. Stocking), *Researches into the physical history of man*, 1813 (1973, Chicago), xlix ff. On the difficulty or impossibility of strictly separating pure and applied science in the early nineteenth century, see N. Reingold (ed.), *The papers of Joseph Henry* (1972–, Washington, D.C.) vol. 1, 368 ff, 380 ff.

146

Davy's *Agricultural chemistry* was thus a conservative work in the main, recommending upon chemical grounds practices which the best farmers were already using.[19]   When he did make really new recommendations, they did not always turn out well, and there are few that survived from the lectures to the book.   He did provide workable systems of soil analysis, and a reasonable account of plant nutrition as generally understood, and he presented it in an attractively readable form.   His book met a need, that of feeding an expanding population in a time of world war; and no doubt, if it propounded few novelties, it helped to circulate knowledge of the best practice.   It was not simplicistic, like some subsequent writings on agricultural chemistry have been; for Davy recognised that agriculture was very complicated, while chemistry dealt with simplified laboratory situations.   Therefore agricultural chemistry was still not in a regular or systematic form, and Davy was therefore less likely to hit upon over-simple answers to complex problems than were his successors faced with the difficulty of feeding a restive population in the Hungry Forties.

Acknowledgments

This paper was read at a meeting of the Society for the History of Chemistry and Alchemy in London on 25 October 1975.   I would like to thank John Brooke, of the University of Lancaster, and Alec Campbell, of the University of Newcastle upon Tyne, for their comments on how a draft of this paper should be improved.

[19] Note Davy's demonstration ((footnote 10), 184) that chemical analysis confirmed that soils from farms where the rent was high were in fact better than those where it was low.   For a recommendation for a fertiliser that did not work, see Berman (footnote 9), 227; Davy (footnote 10), 218 in fact had the candour to admit his failure in this respect.   On Liebig's work generally see W. V. Farrar, ' Science and the German university system ', in M. P. Crosland (ed.), *The emergence of science in Western Europe* (1975, London), 179–192 (pp. 184 ff.); in the same volume, J. R. R. Christie, ' The rise and fall of Scottish science ', 111–126 (pp. 119 f.), describes Cullen's interest in agricultural chemistry, thus providing some background to Cochrane's work.   On agricultural chemistry in France, see F. W. J. McCosh, ' Boussingault versus Ville: the social, political and scientific aspects of their disputes ', *Annals of science*, **32** (1975), 475–490.   See also M. W. Rossiter, *The emergence of agricultural science: Justus Liebig and the Americans, 1840–1880* (1975, New Haven).

XI

TYRANNIES OF DISTANCE IN BRITISH SCIENCE

One of the major themes in this collection is distance in science, and the problems and opportunities presented by dependence, independence and interdependence; another is the relationship of the central or metropolitan and the peripheral. These things are easy to see in the context of the USA, politically but not yet culturally independent in the nineteenth century, and Australia, still formally a colony into the twentieth century; but they can also be seen within Britain, in the relationships between those in or near London (especially in Oxford and Cambridge) and those in the provinces. The position of somebody in the north of England, like John Dalton, was not so very different from that of a colonial; when he bade farewell to his audience after a course of lectures in London to return to 'comparative retirement' in Manchester he sounded like someone going home to a colony.[1] The Australian or American experience of being snubbed or patronized happened to Englishmen all the time; and the sort of people who wrote disparagingly about the domestic manners of Americans or Australians reacted in the same way to those of manufacturing districts in their own country. This indeed is classic metropolitan behaviour.

Australia did not loom very large in British consciousness throughout the century, unlike Ireland which rocked Church establishment, underwent famine, produced terrorism and split a great political party. Britons were happy when it suited them to applaud the progress made by their Anglo-Saxon cousins in subduing the wilderness and replacing savage races, regarding both Americans and Australians as honorary Englishmen of some kind. There was a steady flow of emigrants to both countries, more going to the USA; but the cultural centre for these people, however much they may have disliked the country they had left, remained Britain. For intellectuals like Henry James and Ernest Rutherford, south-east England was the place to go. The British system, if one can dignify it by such a term, of supporting science through patronage, and breaking into tight groups, meant that few American scientists (after the efflux of

R.W. Home and S.G. Kohlstedt (eds.), International Science and National Scientific Identity, 39–53.
© 1991 *Kluwer Academic Publishers. Printed in the Netherlands. Reproduced by kind permission of Kluwer Academic Publishers.*

40

Loyalists at Independence brought W.C. Wells and Rumford to London) came to Britain permanently. Australians and Mancunians found it easier, but heterodoxy made it difficult as we shall see with William Swainson, and Dalton like some colonials preferred to be a big fish in a small pond. We shall look at Humphry Davy, leaving Cornwall to make a career in London, receiving and then dispensing patronage; and at Swainson, whose bid for metropolitan recognition did not in the end come off and who emigrated in a huff to the colonies. Neither of them was particularly happy in the end, but that is perhaps less our concern.

Britain is a small country, but this was less obvious in the days before steamships and railways. A geologist in Devonshire[2] or a chemist in Manchester could seem very distant from those in London, Cambridge or Edinburgh; and geographical isolation could be a real problem. This was one of the ills that the British Association for the Advancement of Science, founded in 1831 just on the verges of the railway age, was meant to relieve.[3] By meeting each year in a different provincial city, the Association did give everybody from time to time a chance to see and hear the eminent, perhaps to be seen and heard, and at least to meet the likeminded. Even now, the north-east of England seems a long way from the centre of things; for the nineteenth century it is not easy to distinguish robust provincial pride from protesting too much about impotence and distance. One reason is that social and geographical distances go together.

Londoners then and since take it for granted that everything important happens there,[4] though in intellectual life Oxford and Cambridge might count as extensions of London. Provincials are not taken seriously unless they settle down and become Londoners, while for anybody brought up there it even remains difficult to work on scientists of the past who never left the provinces. Provincials and their institutions are perceived as unsophisticated and slightly comic; they might be expected to feel inferior, and perhaps bluster about it. This is a tradition that goes back to Shakespeare; and the achievements of Manchester in the last century cut little ice in London. To the ambitious in the nineteenth century, London was a magnet. There it first became possible to make a career for oneself in science, as Davy, Faraday, Tyndall, Huxley and Crookes did. Here were the Royal Society and the Royal Institution, the great museums, Greenwich Observatory, and the centre of government not merely of an island but of an empire. London was the Rome or Constantinople of the nineteenth century, anyway to Londoners; to be tired of it was to be tired of life. Cobbett saw it as the great wen, an excrescence or ulcer draining the life blood of Britain. Others hated its fog and filth. But it was the place to be for anybody who wanted to keep up and get on, especially in the intellectual world: and in this world the nineteenth century was the age of science.

An intellectual distance, and a difference in outlook, separates Londoners from those living further away. Equally important, however, was difference in social rank. Gentlemen who had estates in the country came up to London for the season, spending the summer and early autumn at home, and the meetings of the Royal and Linnean Societies (and the other specialised societies that came to join them as the century wore on) were arranged to fit in with this programme.[5] Science in Britain in the early nineteenth century was an occupation for gentlemen, after all, or for those in the learned professions – clergy, physicians or lawyers – who with a degree from an ancient university had the right to describe themselves as Esquire and thus to acquire the status of gentlemen. Even at Oxford and Cambridge, there were great social distances between the gentlemen commoners, there for a finishing school, and those from the less affluent classes, mostly training to be clergymen.

Nationalism does not seem to have been especially important for science in nineteenth-century Britain, especially because France at first and then Germany was clearly the centre of things and Britain always rather an outsider; it is after all an offshore island. Only from time to time, as with geology in the 1820s and '30s and electromagnetism later, were Britons in the lead. British inferiority was often manifest, especially perhaps in chemistry and physiology, and a rhetoric of decline played an important part not only in the first years of the BAAS but throughout the century. Patriotism is a different thing. It was not science that gave the English an image of themselves, but the splendid Constitution issuing out of the Glorious Revolution of 1688, imitated fairly well by Brother Jonathan in 1776 (though American democracy was often seen as a dreadful warning) and very badly by the French in their overrated revolution of 1789. Although aristocrats in Britain feared a revolution on the French model, especially in the thirty years after Waterloo, gradual political changes, organised religion and economic prosperity combined to ensure political stability. Regimes without the long standing of British governments, and without natural frontiers, might promote a nationalism based upon scientific achievements to bring them legitimacy. There was no great need in Britain to promote those episodes where Britons had done exciting things in science – though nobody was allowed to forget that Newton was an Englishman. Germans, coming like Scots from a poorer country with a better educational system than that of England, played important roles in pure and applied science[6] in the late nineteenth century in their adopted country, just as they were very important in Australasia. The English exported to Australia a pattern of scientific activity in which social distinctions were extremely important and social distances preserved, but not one in which foreign links were to be despised or rejected.

Those living in provincial towns in England found, and still find, what must also be familiar to Australians; that the most eminent intellectuals who grew up there went to the London area to make their career. Only around London was there a big enough intellectual community. This was true also for those from the other nations within the British Isles: in London and Cambridge, Irishmen like Tyndall and Stokes and Scots like Brown, Maxwell and Ramsay all settled down, just like those from the English provinces.[7] Snobbery is more important than nationalism in structuring British science in the nineteenth century and except for the few of independent means, most scientists depended upon patronage, especially in the first half of the century. Recognition abroad might help, early on (as for Davy) from France, and later, especially in chemistry, from Germany; and those of unsure status might therefore be more cosmopolitan. No Englishman in the interests of nationalism ever seems to have thought of refusing membership of an overseas academy, as Rod Home suggests Australians might more recently have done; indeed, they generally listed such things on their title-pages. Whether coming from a colony was worse than coming from the industrial North is an interesting question, to which it would be very hard to give a definite answer in general terms.

Banks had made his reputation with his voyage to Botany Bay, but his status depended upon his land and family; the two together led to his Presidency of the Royal Society from 1778 to 1820.[8] During this time he exercised patronage, notably in voyages of exploration such as that of Flinders to Australia, and in the wool trade, but also at home in the Board of Agriculture, where he promoted Davy's career. Davy had grown up in Cornwall, then a backward corner of the country, part of the Celtic fringe, where his father had been a wood-carver when in work and his mother when widowed a milliner. In the manner characteristic of the day, the Davys liked to think of themselves as a good family fallen on bad times; but Henry Mayhew was one of those struck by the social mobility Davy achieved through science.[9]

With the patronage first of James Watt, whose son Gregory had stayed with the Davys in the hopes of curing his tuberculosis in the mild climate of Penzance, and then of other members of the Lunar Society when he worked as assistant to Beddoes[10] at Clifton, Davy was propelled with the success of his work on the oxides of nitrogen to a post at the Royal Institution in London. His lectures here, including a series on agricultural chemistry, and then his electrochemical research, made him in the course of a few years one of the best-known scientists of the day; and his Bakerian Lecture to the Royal Society of 1806 was crowned by the award of a prize by the Institut in Paris. He became a great diner-out, and was clearly a spirited talker; and he put behind him his apothecary's apprenticeship in rural Cornwall, which he hardly ever revisited. He was happy to be in the society of aristocrats but relaxed only in the company of professional men, particularly doctors, with whom he went fishing.[11] Social mobility has a high price in unease.

His first Continental trip, to Napoleon's Paris to receive his prize, was followed by others. He was abroad when he heard of Banks' fatal illness, and hurried back as a candidate for the Presidency of the Royal Society: Banks had been elected a few days before Davy's birth. Banks and others had hoped that Wollaston would succeed him, but Wollaston withdrew rather than face a contest with Davy, who emerged as the only serious candidate. It was astonishing that one from his background should have risen to a position of such eminence, and some of his elders and betters were appalled. That the part of Banks should now be played by Davy, who had a reputation as a philistine bored by the pictures in the Louvre and whose cocked hat was usually askew, was mildly shocking.

In the event, Davy's Presidency was not a great success. His own research on protecting the copper bottoms of ships from corrosion was unsuccessful, and the expectations of the hostile camps[12] between which he found himself could not be satisfied. He tried to keep up the style of Banks, having acquired the means by an unhappy marriage to a wealthy widow, but found social climbing a lonely activity. A president neither born to the purple nor a university graduate was not in an easy position socially, and Davy's fast-working mind, his rhetorical skills and his conviction of his scientific merits did not necessarily endear him to everybody. Even with Faraday, to whom he became a father in science, he proved an oppressive parent: Faraday learned from Davy's example to avoid exposed positions of great eminence, though he followed Davy's

44

research programme with even greater success than Davy himself.[13]

One might have expected that a President with very different interests from Banks's might have propelled British science in a different direction, towards the laboratory rather than out into the Antipodes. But during Davy's time, voyages of exploration in the tradition of Cook and Banks continued, especially to the Canadian Arctic and to South America. And just as Banks had spent much time and effort in acclimatising the Merino sheep in Britain, so Davy was with Raffles a founder of the London Zoo. He hoped that some exotic creatures might be made economically important in Britain; but English grasslands were never graced with alpacas or kangaroos and the Zoo became a place where, as at the Royal Institution, entertainment and scientific study went on in uneasy union.

Bank's position had made it easy for him to exercise patronage, though he missed out on some promising men – for instance Faraday. Patronage is bound to be hit and miss, perhaps like any kind of selection. Davy used his patronage to obtain for his friend J.G. Children a post in zoology at the British Museum.[14] Children did not know any zoology but was a mineralogist and chemist of independent means, who had worked with Davy on gunpowder and also on the vile-smelling hydrogen telluride. When he fell upon hard times, Davy secured him a job, which Children did not at first particularly want, at the Museum. In the event he proved a competent zoologist, but the British Museum was not transformed as Davy had hoped it might be into something resembling the Musée d'Histoire Naturelle in Paris; that only happened when the natural history collections were moved to South Kensington in the 1880s,[15] under Owen who had strong Australasian connections. Children's social status was such that he could deal with the Museum's Trustees as an equal rather than as an employee, which was an advantage: but his appointment was controversial, and the prestige of the Museum was dented by the attacks of an unsuccessful candidate for the post, William Swainson, and his friend Thomas Traill. Collections made on national expeditions were sometimes sent elsewhere, like Darwin's from HMS *Beagle*. Mismanaged patronage seriously damaged an institution that Banks had successfully promoted.

Davy's successor in 1827 was the Cornish MP, Davies Gilbert (formerly Giddy); but after a brief reign he stood down and there was a bitterly contested election between John Herschel and the Duke of Sussex, the most intellectual of Queen Victoria's uncles. This raised in acute form questions of rank, professionalism and the status of science in general and the Royal Society in particular.[16] Even an affable Royal Duke was unlikely to become the close friend of working scientists but his brother, soon to be King William IV, had as Lord High Admiral sponsored scientific voyages, and the Duke himself took a genial interest in the advance of knowledge. He won the election; and we get a clear picture of social distance in letters written to Swainson by Peter Mark Roget, a Secretary of the Society (the other being Children) and author of the famous *Thesaurus*.[17] Roget was a doctor, and was responsible for the new journal that became *Proceedings of the Royal Society*, less formal than its *Philosophical Transactions*; he also wrote a Bridgewater Treatise on the goodness and wisdom of God.

The Duke had been elected on 30 November 1830, and on 5 December Roget described his first 'interview (I should perhaps say an audience) with our new President'. Clearly, the Duke succeeded in charming Roget, who earnestly wished 'he may eventually succeed in obtaining the cooperation of those who now assume so hostile an attitude, & threaten to secede from the Society, & perhaps establish another of their own'. His misgivings were less about the Duke than about the 'very imbecile Council' elected at the same time, because those who 'arranged' it 'evidently took pains to retain those of the former Council who were the most inefficient members, & particularly those who have never once attended: while on the other hand they excluded the really valuable members, & put into their places persons who, with one or two exceptions, will be totally inactive'.

By 2 March 1831 Roget was finding the Duke less assiduous than he might have been: 'he has not yet commenced either his evening or his morning conversazione's (sic)' and 'has been very little with us of late: & tomorrow is the first day, since the 20th of January, that he will take the chair at the meetings: having been during almost the whole of the interval at Brighton', where his brother George IV had built a palace of more than oriental splendour. To Roget's disappointment, the friends of Herschel did not respond to ducal overtures and the prospect of a return to the days of Banks when science had basked in the approval of the Establishment: 'The Duke takes every means in his power that is at the same time consistent with the dignity of his office, to conciliate them: but as yet without the smallest success', Roget wrote on 12 March 1831. The President had proposed revision of the Society's Charter, and 'We have taken care to nominate on that committee the principal reformers, both moderate and radical: in order that their plans may undergo full & fair

46

discussion. Of this number are Babbage, Baily, Beaufort, Brown, Fitton, Herschel and Warburton. These have, by letter, declined attending, without assigning any reason: and their places must be filled up by others. This is, indeed, to use your words, "carrying matters much too far"'.

Roget still hoped 'that the anticipated good from having a Royal President may eventually arrive'. In December 1830 he had indeed felt that if party spirit were to go on running so high, he would resign as Secretary in November 1831; but in the event he continued until 1849, by which time the Royal Society was transforming itself from a club into an Academy of Sciences. What the episode shows is that considerations of rank and status were closely involved with the moves towards specialisation and professionalisation, and that social distances were formidable, especially just at the time of the great Reform Bill of 1832.

Roget's letters were to William Swainson, a man who failed to cross these distances and who as a consequence later sailed as an emigrant to New Zealand, and for a time worked in Australia. Swainson did not have so far to travel as Davy or Faraday, who came from the artisan class. He came from Liverpool where his father had been a Customs Officer, and during the Napoleonic Wars he served in the Mediterranean with the Commissariat. On his return, after a voyage to Brazil, he took advantage of the new process of lithography as a cheap way of printing natural-history plates, especially of birds and shells. He was a very talented artist and supported himself and an increasing family by publishing illustrations, and by buying and selling specimens. We can learn the details from his correspondence.[18] But Swainson rated an illustrator some distance below a naturalist, and longed to shine as a writer.

His chance came with government support for publication of the natural history of the northern voyages, and then with the publishing revolution of about 1830, associated with the 'March of Mind', that brought cheap case-bound books to the British public. Swainson undertook to write the natural history for Dionysius Lardner's series, *The Cabinet Cyclopedia*, a series of informative volumes of which the first, Herschel's *Preliminary Discourse*, is the best-known.[19] Swainson wrote a *Preliminary Discourse* himself, on natural history, which is interesting for its descriptions of the generally rather turgid meeting practices of scientific societies, and of natural history publishing. In this and his other works he adopted the Quinary System of classification, an attempt to find an order that was both natural and tidy. He much admired the French and saw 'decline' in British science, but he did not follow Cuvier's lead where theory was concerned. His attempt to close the distance between the artist and the theorist did not work, because his persistence in unorthodoxy (the quinary system went with his strongly high-church views) kept him apart from most zoologists; McLeay, the other notable quinarian, being in Australia. But Swainson was also not quite a gentleman born and he was of a very quarrelsome disposition, suitable perhaps to a generation that had grown up in a war so long that peace after 1815 seemed abnormal and military metaphors natural.

The Quinary System never lost its heretical taint; but small groups that have had to fight for status and for truth as they see it remain defensive and distanced even after victory. Thus the Darwinians who formed the X Club, some like Huxley and Hooker with Australian connections, seem to have seen themselves as an embattled minority within the 'Church Scientific' in their struggles to cut British science free from religious ties and amateurism.[20] It is difficult here to separate social and intellectual distances; although the men of the X Club held the highest offices in science, like Davy they never felt easy in the Establishment. He tried to be a Gentleman of Science, as Faraday and the X Club members did not; but nobody in nineteenth-century Britain (and to some extent still) could avoid questions of social rank and the distances associated with it.

One way in which this showed itself was in attitudes to science as a profession. In writing about Faraday, Tyndall referred to his giving up professional work (meaning routine analyses for a fee) in favour of research into electricity.[21] Davy had refused in a gentlemanly way to patent his safety lamp, comparing himself favourably to Wollaston, born a gentleman, who had made a great deal of money from work with platinum. In Davy's time, pure and applied science were not distinguished: science was opposed to practice, or mere rule of thumb.[22] By the second half of the century, however, a gulf had opened, especially in chemistry: professional chemists earned their living generally by analyses, while academic or learned chemists increasingly began to find posts in the expanding educational system especially after 1870. The distance between the two groups, both social and intellectual, led to a schism in the Chemical Society of London, which continued as a learned body while the Royal Institute of Chemistry represented the professional interests of those in industry, whose status in the nineteenth century remained low. It was a century before this disruption was healed.[23]

By this time there were also distances within the scientific community: the sciences not only had a pecking-order changing over time, but also became increasingly specialised. Davy, Wollaston and Young worked across a spectrum of sciences. Herschel was one of the last who could refuse to specialise,[24] and by the end of the century most practitioners saw themselves not as men of science but as, for example, organic chemists. To Rutherford, 'chemist' meant damn fool; the remark had a particular context, but it reminds us of the distances between the sciences that had become a feature of the specialised life of the nineteenth century.

In the 1880s the *Proceedings of the Royal Society* were divided into parts A and B, A including the physical sciences and B what my physics teacher used to call 'the less exacting discipline of the more descriptive sciences'. Journals like *Nature* were required to publish each week the latest work and also to keep those in one specialism aware of what was going on in others, something that in more leisurely days Mary Somerville, Herschel and others had done in books or in expansive essays in the *Edinburgh*, the *Quarterly* and other *Reviews*.[25] Crookes had tried, with *The Quarterly Journal of Science*, to make a scientific review work, but the forces of specialism were too strong. To see 'two cultures' in nineteenth-century Britain is too simple. There were divisions not just between scientists and those brought up in the humanities (overlapping groups at the academic end) but also between different kinds of scientists, with a ranking order. The social history of science is a complicated business, perhaps especially in England where amateurism continued longer than in France and Germany, and where the class system seems to have been maintained with unusual and particular care. Democratic Americans love to rank scientists as quantitatively as possible; the British had, and maybe have, more subtle and probably more offensive ways of doing it.

Australia was very distant from Britain geographically, but not necessarily in other ways; it still seems very British. Banks and Darwin both went there, and for neither of them were social distances a problem. Their visits can be seen as exemplifying the Baconianism so much invoked in nineteenth-century Britain.[26] Bacon's most accessible work, the fragmentary *New Atlantis*, had after all described a small island from which fact-collectors went out and brought back information, to be reasoned upon by those further up the social pyramid. Banks's move to the administration of science and Darwin's into theorising fit this pattern, for one need not remain in the fact-gathering stage throughout life. Visiting Australia was thus a stage in a career: the fauna and flora of Australia, the stars of the southern hemisphere, or the behaviour of wombats in the wild, cannot be observed in London.[27] There is some science that must be done in Australia, just as there was some that had to be done in the provinces: the Silurian and Cambrian rocks had to be studied in Wales,[28] and pendulums swung down mines in Cornwall and in Durham as well as in Tasmania in studying gravitation.

The reasoning upon and publication of these and other pieces of 'Humboldtian' science was done, as Humboldt had done it, in great European centres, where there were scientific societies, museums and libraries. This pattern was extended to the British Association: those in the provinces might furnish, especially in natural history and astronomical observation, material for those in the metropolis to think about. Londoners may well have felt that Australia was no more distant intellectually than the industrial North, and certainly that the science done by gentlemanly visitors and the colonial administrators who supported them was of central importance. Right through the nineteenth century reputations were made on scientific voyages, and important posts might follow. Not only Banks but also Sabine, Hooker and Huxley became President of the Royal Society, and the voyage of HMS *Challenger* was one of the most important bits of big science done in late Victorian Britain. This also involved scientists from a number of countries in its writing up, in a self-conscious repudiation of nationalism.

The problems came, no doubt, where geographical distance was reinforced, in sciences that did not need to be done in Australia and that in the British empire were already being done at home, for example in physics and chemistry in their more abstract branches, specialised disciplines where even in Britain the communities were not large. In such fields one might be worse off in Australia than in Japan,[29] for which there was no metropolitan cultural centre at a distance; and one would certainly be worse off than those in the provinces in Britain, who had a rapid penny postage by the second half of the century and speedy train services making visits to London or to BAAS meetings practicable. Australian physical science could not but be provincial, and those thinking of teaching science in Australian institutions had to bear that in mind. Symbolic visits to the colonies by the BAAS every twenty years or so were no substitute for easy contact. Science in Britain was affected by all sorts of distances in the shifting pattern of class-consciousness in the nineteenth century: the empire presented opportunities to Hooker and Sabine, Rutherford and Bragg, but to others it must have closed doors and made the centre of things feel very far away.

Australians probably also felt that they were treated as second-class citizens in terms of resources available, as colonials to be exploited.

158

50

Australians and Americans use the word 'colonial' in a rather different way from everybody else. Their experience, except for those who are Aborigines or American Indians, has been so very different from that of Indians or Nigerians that it is misleading to use the word 'colonial' for both situations; 'provincial' seems a better way of describing science in the former case. A recent analysis of the British empire in the second half of the nineteenth century[30] confirms the British impression that running an empire was expensive and that those on the periphery did best out of it, an impression that Americans looking back on the years since 1945 might share. The close links between home and colonies that one imagines do not show up in economic terms, and Australian education like Australian defence seems to have been indirectly supported by the taxpayer in Britain; Australians are probably right in thinking theirs has been a lucky country.

Whether jumping out of the British fire into the American frying pan would make life more comfortable for Australian scientists remains to be seen; the Canadian experience here[31] might well be instructive. What seems certain is that the *New Atlantis* model is inevitable, whether one is in Melbourne or in Durham, until the local scientific community can somehow grow large enough to be self-sustaining. Then patriotism is not enough – in nineteenth-century Britain contact with men of science abroad gave the chance to escape from the stultifying effects of social distances – though in the last resort one had to work within the social system in which one found oneself. The British experience seems to be that a multiplicity of international connections helps science grow. For different people at different times and in different disciplines, important contacts in the nineteenth century were not just with France or Germany, but also with Switzerland, Sweden, Belgium, Russia, Italy and the Netherlands. These geographically-distant connections were sometimes easier than socially-distant ones within the same city; for as well as being a period of great class-consciousness, the Victorian era in Britain was also one of momentous rows. Distances of different kinds – geography, language and status – may tyrannise science, or they can be a great resource of variability. Australian science may have suffered because Britain was more important to Australia than the other way round; but to the outsider it looks as though it had quite a flourishing growth.

159

NOTES

[1] H.E. Roscoe and A. Harden, *A New View of the Origin of Dalton's Atomic Theory* (London, 1896), p. 122.

[2] M.J.S. Rudwick, *The Great Devonian Controversy: The Shaping of Scientific Knowledge among Gentlemanly Specialists* (Chicago, 1985); C.A. Russell, *Science and Social Change, 1700–1900* (London, 1983); D.M. Knight, *The Age of Science* (Oxford, 1986, new. ed. 1988).

[3] J. Morrell and A. Thackray, *Gentlemen of Science* (Oxford, 1981); and the associated *Correspondence* (London, 1984); R. MacLeod and P. Collins (eds.). *The Parliament of Science* (London, 1981).

[4] L. Stone, *The Family, Sex and Marriage in England, 1500–1800* (new ed., Harmondsworth, 1982), p. 179. I. Inkster and J. Morrell (eds.), *Metropolis and Province* (London, 1983); I. Inkster (ed.), *The Steam Intellect Societies* (Nottingham, 1985). For earlier London dominance, see P. Earle, *The Making of the English Middle Class: Business, Society and Family Life in London, 1660–1730* (London, 1989). On provincial geology, see H.S. Torrens, 'Arthur Aitken's Mineralogical Survey of Shropshire 1796–1816, and the Contemporary Audience for Geological Publications', *British Journal for the History of Science*, 16 (1983), 111–53.

[5] S. Forgan, 'Context, Image and Function', *British Journal for the History of Science*, 19 (1986), 89–113, looks at the buildings and arrangements of scientific societies.

[6] On British geology at a triumphal period, see Robert A. Stafford's paper in this volume. On Albert Gunther, see A.E. Gunther, *A Century of Zoology at the British Museum* (London, 1975); on an industrialist, S.E. Koss, *Sir John Brunner: Radical Plutocrat, 1842–1919* (Cambridge, 1970). On Mueller, see R.W. Home's paper in this volume; it is interesting that the telescope at the Sydney Observatory came from Germany. See also A.M. Lucas, 'Baron von Mueller: Protégé turned Patron', pp. 133–52 in R.W. Home (ed.), *Australian Science in the Making* (Melbourne, 1988), a very useful collection of papers giving an excellent overview and some close studies.

[7] Two important scientists who remained in Ireland were Lord Rosse and Hamilton; see T.L. Hankins, *Sir William Rowan Hamilton* (Baltimore, 1980).

[8] H.B. Carter, *Sir Joseph Banks* (London, 1988). On Brown, D.J. Mabberley, *Jupiter Botanicus* (Braunschweig, 1985).

[9] H. Davy, *Collected Works*, ed. J. Davy, Vol. 1 (London, 1839–40); H. Mayhew, *The Wonders of Science, or Young Humphry Davy: The Life of a Wonderful Boy written for Boys*, 2nd ed. (London, 1856).

[10] T.H. Levere, 'Dr Thomas Beddoes; Science and Medicine in Politics and Society', *British Journal for the History of Science*, 17 (1984), 187–204; D.A. Stansfield, *Thomas Beddoes* (Dordrecht, 1984).

[11] S. Forgan (ed.), *Science and the Sons of Genius: Studies on Humphry Davy* (London, 1980); my paper on fishing is on pp. 201–30.

[12] D.P. Miller, 'Between Hostile Camps; Sir Humphry Davy's Presidency of the Royal Society, 1820–1827', *British Journal for the History of Science*, 16 (1983), 1–47. H.B. Carter (ed.), *The Sheep and Wool Correspondence of Sir Joseph Banks, 1781–1820* (Sydney, 1979).

52

[13] D. Gooding & F. James (eds.), *Faraday Rediscovered* (London, 1985); my paper on Davy and Faraday is on pp. 33–49.

[14] A.E. Gunther, *Founders of Science at the British Museum, 1753–1900* (Halesworth, 1980).

[15] W.H. Flower, *Essays on Museums* (London, 1898); M. Girouard, *Alfred Waterhouse and the Natural History Museum* (London, 1981). On Owen, see the paper by Elizabeth Newland in this volume and, for a more sympathetic account, Jacob W. Gruber, 'From Myth to Reality: the Case of the Moa', *Archives of Natural History*, 14 (1987), 339–52. On the row over Children's appointment, see A. Gunther, 'President's Anniversary Address', *Proceedings of the Linnean Society*, 112 (1899–1900), 14–24, pp. 19f.

[16] M.B. Hall, *All Scientists Now: The Royal Society in the Nineteenth Century* (Cambridge, 1984); and see the informal writings of the Society's Assistant Secretary, W. White, *Journals* (London, 1898).

[17] Gunther's address (n. 15) describes Swainson's correspondence preserved at the Linnean Society, and is followed (pp. 25–61) by a summary Calendar arranged under correspondents. Incoming letters only are preserved.

[18] D.M. Knight, 'Ramsbottom Lecture: William Swainson, Naturalist, Author and Illustrator', *Archives of Natural History*, 13 (1986), 275–90; S. Natusch and G. Swainson, *William Swainson; the Anatomy of a Nineteenth-century Naturalist* (Wellington, NZ, 2nd printing 1987); this reproduces the Calendar of Swainson's correspondents, and lists other MS sources; G.M. Swainson (ed.), *William Swainson: Diaries, 1808–1818* (Palmerston North, NZ, 1989). On the trade in specimens, see M.A. Taylor and H.S. Torrens, 'Saleswoman to a New Science: Mary Anning and the Fossil Fish *Squaloraja* from the Lias of Lyme Regis', *Proceedings of the Dorset Natural History and Archaeological Society*, 108 (1986), 135–48.

[19] J.F.W. Herschel, *Preliminary Discourse to the Study of Natural Philosophy* (London, 1830, reprint introduced by A. Fine, Chicago, 1987). J.N. Hays, 'The Rise and Fall of Dionysius Lardner', *Annals of Science*, 38 (1981), 527–42. D.M. Knight, *Zoological Illustration* (Folkestone, 1977); A. Ellenius (ed.), *The Natural Sciences and the Arts* (Uppsala, 1985).

[20] R.V. Jensen, 'Return to the Wilberforce-Huxley Debate', *British Journal for the History of Science*, 21 (1988), 161–79; Ruth Barton, ' "An Influential Set of Chaps": the X-Club and Royal Society Politics, 1864–1885', *British Journal for the History of Science*, 23 (1990), 53–81.

[21] See my paper on Faraday in R. Porter (ed.), *Man Masters Nature* (London, 1987), pp. 126–36; J. Tyndall, *Faraday as a Discoverer* (London, 1868).

[22] R.F. Bud and G.K. Roberts, *Science versus Practice* (Manchester, 1984). P. Alter, *The Reluctant Patron: Science and the State in Britain, 1850–1920* (Oxford, 1987).

[23] C.A. Russell, N.G. Coley and G.K. Roberts, *Chemists by Profession* (Milton Keynes, 1977). Crookes' journal *Chemical News* reached both professional and learned chemists, and its Letters to the Editor in the 1870s are most interesting on the distances between them.

[24] S.F. Cannon, *Science in Culture* (New York, 1978), esp. chapters 2 and 3.

[25] A.J. Meadows (ed.), *Development of Science Publishing in Europe* (Amsterdam, 1980); W.H. Brock and A.J. Meadows, *The Lamp of Learning* (London, 1984).

26 R. Yeo, 'An Idol of the Market-place: Baconianism in Nineteenth-century Britain', *History of Science*, 23 (1985), 251–98.

27 A. Moyal, *'A Bright and Savage Land'*: *Scientists in Colonial Australia* (Sydney, 1987).

28 J.A. Secord, *Controversy in Victorian Geology: The Cambrian-Silurian Dispute* (Princeton, 1986); and 'The Geological Survey of Great Britain as a Research School, 1839–1855', *History of Science*, 24 (1986), 223–75; N.A. Rupke, *The Great Chain of History: William Buckland and the English School of Geology, 1814–1849* (Oxford, 1983).

29 R.W. Home, 'Physics in Australia and Japan to 1914', *Annals of Science*, 44 (1987), 215–35; idem, 'The Problem of Intellectual Isolation in Scientific Life: W.H. Bragg and the Australian Scientific Community, 1886–1909', *Historical Records of Australian Science*, 6 (1) (1984), 19–30, and 'First Physicist in Australia: Richard Threlfall at the University of Sydney, 1886–1898', *ibid.*, 6 (3) (1986), 333–57.

30 Patrick K. O'Brien, 'The Costs and Benefits of British Imperialism, 1846–1914', *Past and Present*, 120 (1988), 163–200, and the ensuing debate between O'Brien and P. Kennedy, *ibid.*, 125 (1989), 186–99.

31 Trevor H. Levere, 'The History of Science in Canada', *British Journal for the History of Science*, 21 (1988), 419–25. On the organisation of science in Australia in the last hundred years, see R. MacLeod (ed.), *The Commonwealth of Science: ANZAAS and the Scientific Enterprise in Australasia, 1888–1988* (Melbourne, 1988).

XII

THE APPLICATION OF ENLIGHTENED PHILOSOPHY: BANKS AND THE PHYSICAL SCIENCES

On 30 October 1815 Banks wrote to Humphry Davy on hearing about his newly-invented safety lamp for coal miners[1]:

"Many thanks for your kind letter, which has given me unspeakable pleasure. Much as, by the more brilliant discoveries you have made, the reputation of the Royal Society has been exalted in the scientific world, I am of the opinion that the solid and effective reputation of that body will be more advanced among our cotemporaries [sic] of all ranks by your present discovery, than it has been by all the rest. To have come forward when called upon, because no one else could discover means of defending society from a tremendous scourge of humanity, and to have, by the application of enlightened philosophy, found the means of providing a certain precautionary measure effectual to guard mankind for the future against this alarming and increasing evil, cannot fail to recommend the discoverer to much public gratitude, and to place the Royal Society in a more popular point of view than all the abstruse discoveries beyond the understanding of unlearned people. I shall most certainly direct your paper to be read at the very first day of our meeting. We should have been happy to have seen you here; but I am still happier in the recollection of the excellent fruit which was ripened and perfected by the very means of my disappointment, your early return to London."

This magnificent epistle allows us to raise the uncomfortable questions which all those concerned with Banks and the physical sciences must feel. Polarity was a fashionable scientific concept during his reign[2], and pairs of polar opposites may illuminate Banks's position. We may contrast an explorer to a natural philosopher; an administrator to an active researcher; and a utilitarian to a theorist or explainer. In these categories, Banks was unusually polarized by comparison with most eminent men of science (especially those concerned with the physical sciences) even in his own day.

J.J.Thomson, writing about his work on the electron, wrote of William Crookes, a predecessor in the field of cathode rays, that "In his investigations he was like an explorer in an unknown country, examining everything that seemed of interest, rather than a traveller wishing to reach some particular place, and regarding the intervening country as something to be rushed through as quickly as possible".[3]

Sir Joseph Banks: a global perspective

Instead of recording all sorts of interesting observations, Thomson had a theory to test; and devised apparatus which would do the job quickly and elegantly, going straight to the heart of the matter. Banks was literally an explorer, whose scientific training was based upon the descriptive science of botany, and whose real science had been learned on board *HMS Endeavour* with James Cook. Rigorous simplifying, and testing of theories, had little part in his science; as we can see from the letter to Davy. References to "brilliant discoveries" are offset against "solid and effective reputation", and "abstruse" is the word used to characterize Davy's electrochemical work.

Davy was himself, as with laughing gas, sometimes an explorer; but the work which had exalted the Royal Society in the opinion of the scientific world had been his demonstration that chemical affinity was electrical.[4] This research of 1806 had received a prize from the Institut[5] in Paris: it involved careful experiments to establish what Davy had been sure of, that water was decomposed by an electric current into oxygen and hydrogen only, and that everything else that had been noticed by more exploratory chemists was a side-reaction. To this end, Davy required from the Royal Institution apparatus of silver, gold and agate; building up the idea that fundamental science could only be done by men of genius with splendid equipment.[6]

Priestley had been an explorer among the airs or gases he had isolated;[7] the newer chemistry of Lavoisier[8] and Davy required the capacity to rush through intervening country and reach the objective. Banks was of course pleased at the success of his protegé, but resolute and undistracted aiming at a theoretical goal was not his style. My old physics teacher used to talk of "the less exacting discipline of the more descriptive sciences"; which is perhaps another way of emphasizing different casts of mind within the sciences; and may indicate why those involved in physics and chemistry were not always enthusiastic subjects of Banks's learned empire. Natural history and natural philosophy had distinct approaches to nature, which we may forget when we group them all as science. Rutherford may have said that all science is either physics or stamp collecting; he was probably not the first to feel it. But as well as being a descriptive botanist and explorer, Banks was an administrator:[9] the Royal Society filled his life — his home in Soho Square was open to men of science; and, as in the letter to Davy, invitations to his country houses also went to scientific associates.

Political skill is a great gift. It may well be, and Davy found it so, that eminence in scientific discovery is not the best preparation for high administrative office: wrestling with nature in the laboratory is not quite like dealing with the busy world of men. But we have been accustomed to a world in which an office such as the Presidency of the Royal Society goes to someone eminent for research. It involves promoting institutional rather than individual good, and recognising that one's role may well be encouraging rather than active; which can lead to a rather shattering recognition of being middle-aged, whereas physical science is taken to be a game for the young. Davy was 41 when elected to succeed Banks, with a distinguished career behind him: but Banks had been younger (35) when chosen as PRS, and his importance in science came through his long reign. It is perhaps significant that in the portraits at the Royal Society[10], Banks is shown with the mace, the sign of office; whereas Davy stands proudly beside his lamp, the invention which had promoted him, as he put it, to be a general in the army of science. We expect the office to be the crown of a career made in the laboratory or the field; but Banks was different, and much of his authority depended upon his social position as a landed gentleman — which was not welcome to meritocrats like Davy who later wrote of him: "He was a good-humoured and liberal man, free and various in conversational power, a tolerable botanist, and generally acquainted with natural history. He had not much reading, and no profound information. He was always ready to promote the objects of men of science; but he required to be regarded as a patron, and readily swallowed gross flattery. When he gave anecdotes of his voyages, he was very entertaining and unaffected. A courtier in character, he was a warm friend to a good king [George III]. In his relations to the Royal Society he was too personal, and made his house a circle too like a court".[11] This is no doubt unfair and ungrateful; it is very much the way people in universities talk about their professional administrators.

For Banks, as in the letter, references to the Royal Society came naturally; by 1815, he had been President for more than half his life, and it must have seemed to him as to the unsympathetic that "l'état c'est moi". For him, the interests of science were identical to those of the Society. We are accustomed to seeing scientific discovery as the achievement of a person, or perhaps a team; and the various medals and prizes awarded for science enshrine this idea. We recognize that Rutherford's Cavendish Laboratory was more than the sum of its parts, and Research Rating Exercises make us look hard at today's institutions; but Banks's remarks to Davy are still surprising because the good of the Royal Society is quite so prominent. Davy had been Secretary from 1807 until 1812, but he had since then been abroad for a year and a half, and probably was not thinking of the good of the Royal Society when he undertook his research on the lamp; yet Banks wrote to him as to a team member.

There is no doubt that the solid and effective reputation of the Royal Society, and placing it in a more popular point of view, was one of Banks's overriding concerns; and indeed promoting both science and the public understanding of it remains a major task for his successors in every generation. As President for over forty years (Davy, his successor, was born a few days after Banks's election), Banks undoubtedly brought great dedication to his task and saw the Society through various crises;[12] sometimes with help, from for example Henry Cavendish.[13] He has been heavily and not always fairly criticized by later men of science who were operating in a very different world; but his charm and dedication went with the good administrator's horror of rocking the boat. He did indeed keep the Society afloat and on the course he set. By contrast, active researchers, discoverers, are inveterate boat rockers; keeping the balance, while allowing a little pitching and tossing, is crucial if intellectual institutions are to flourish.

Active researchers in Banks's time were sometimes awkward customers; they might, like Joseph Priestley, be political radicals — and the Royal Society gave Priestley little support in his hour of need[14] after the "Church and King" mob had wrecked his house and laboratory in Birmingham. Or they might be provincials, perhaps also nonconformists, like Dalton who saw nothing for himself in Banks's genteel Royal Society. Its very success as a gentleman's club, and metropolitan society, meant that those included were by no means all active, or even very interested, in science; and it did not by any stretch of the imagination fully represent the British scientific community of the day.[15]

166

Britain is a very small country, and yet science during the Banksian era was subject to a tyranny of distance:[16] in this case, social distance, coupled with metropolitan[17] disdain for the provincial. Banks was sufficiently secure to be able to cross social boundaries, and to get on with the various gardeners and others to whom he was a patron; but the Royal Society was essentially for officers and gentlemen. The socially-mobile Davy,[18] "indebted for his address to the narrowness of his original circumstances", could not but be aware of this; and no doubt that gives an edge to his remarks about Banks. Science in the nineteenth century became a vehicle for those who would rise above the station and its duties into which they had been born, but they could not play a full part in the Royal Society until it lost its *ancien regime* character.[19]

Banks found it hard also to recognize that science might be carried on outside the Royal Society. His opposition to societies devoted to particular sciences, except for the Linnean Society, is well known.[20] The Animal Chemistry Club[21] which brought together physiologists and chemists in what should have been a promising conjunction was allowed only as a subset of the Royal Society; other groups received a Banksian frown, and Davy for example was bullied into withdrawing from the infant Geological Society. This should not be put down to imperialism, *tout court,* for Banks feared that if men of science did not hang together they would hang separately; but it was a failure to read the signs of the times, and aroused fears of decline. Banks, despite his efforts to be fair, in effect represented only one constituency among men of science in Britain.

Decline is always an emotional business. What is clear is that during Banks's reign the leading scientific power was France. This was true right across the board. Able Britons made interesting discoveries, modified French theories, adopted French notations, and did their best to keep their end up; but it was as citizens of a second-class power. 1815, the year of the Safety Lamp and Banks's letter, was the year of Waterloo; and Sadi Carnot, in his pioneer work on thermodynamics, reflected that Britain had won the war because of her industrial power[22] — notably her steam engines. Visitors to defeated France, and especially scientific ones, nevertheless saw a more modern country than their own, with scientific institutions that might profitably be copied.

Davy's safety lamp was not the only one; in response to the same crisis in the industry, George Stephenson had come up with the Geordie lamp, in which access and egress of air was controlled by thin tubes. He had arrived at this lamp by the traditional method of trial and error; whereas Davy had employed the new technique whereby laboratory investigation of the properties of the explosive gas led to the invention of a device — and because he understood the principle, he could do further research on flames and on heterogeneous catalysis. Davy's lamp and his originality were endorsed by the Royal Society, rather as Newton's position had been in the celebrated conflict with Leibniz; Stephenson was outside Banks's empire, and for the Royal Society it was clearly important that credit should go wholly to one of their Fellows.[23] Stephenson later became a famous man through his steam engines; the dispute over safety lamps, partly carried on by supporters of both sides in Thomas Thomson's respectable *Annals of Philosophy*, reveals the social tensions in Regency science and the narrow base of the Royal Society.

Banks's commitment to the Society he administered was a source of strength as well as weakness; but it does mark him off from some of his more creative contemporaries. One scientific society he did back was the Royal Institution;[24] where his investment paid off with Davy, and his illustrious successors. The R.I. was essentially complementary to the Royal Society. They drew upon the same social groups; but whereas Royal Society meetings involved papers being read, usually by the Secretary, to an all-male auditory, the Royal Institution was founded as a centre for attractive lectures. In accordance with their "unprofessional" character, they were open to ladies: science was to be presented as a part of general culture, perhaps[25] to an audience "composed of the gay and idle, who could be tempted to admit instruction only by the prospect of receiving pleasure": the general public, or at least the wealthier part of it.

Moreover, there was in the basement a laboratory; something the Royal Society had always lacked. Here the celebrated demonstration-experiments could be prepared, and research could go on; sometimes, in Davy's case, in public, before a small audience. We think of the Royal Institution in connection with Davy, Thomas Young and Faraday, as a place where fundamental research was carried on at a level the French had to admire. Faraday in particular is remembered for giving up his "professional" chemical analyses done for a fee,[26] in favour of work in electricity and magnetism where the outcome was quite unclear. But this was not what the founders,[27] whose first formal meeting had been in Banks's house, had in mind. They hoped, and Banks saw it in the safety lamp, for fruit; especially on behalf of the landed interest.

Francis Bacon had believed that experiments of light would in the end lead to experiments of fruit, an idea he more clearly expressed as, knowledge is power. In the seventeenth century, there was little direct pay-off from the new science; but by Banks's time mastering nature was a great project, in which all the sciences would have their role. Bacon's *New Atlantis*, a scientific utopia set on a small island, seemed designed for a metropolis set at the hub of a far-flung empire; and the Royal Society might be developed into Salomon's House, the island's seat of power, authority and knowledge, perhaps with Banks as its Prospero. Bacon's cautious, inductive method was also in fashion (in scientific rhetoric if not practice), because it was feared that speculative science, based on systems rather than method, lay behind the dreams of the French revolutionaries of 1789 and succeeding years. Indeed, revolutionary Paris's supremacy in the sciences seemed to indicate that they might be destabilizing. Bacon, contrariwise, had believed that deep knowledge, based upon cautious generalization, would lead to faith in God, and no doubt in the best of constitutions; and in the 1790s this was what people wanted to hear.

At the Royal Institution, the idea had been that artisans would be admitted to the gallery for the lectures; and indeed it had been hoped to do more for them, and to have exhibitions of new machinery. This came to nothing, but Davy on arrival was put to work to study tanning and agriculture. His researches[28] essentially confirmed the best practice, giving it some scientific rationale; and in lectures he urged the importance of science for the economy. His election to the Royal Society followed, in 1803; and in 1805 he was awarded the Copley Medal of the Society for his chemical researches. The Royal Society's highest award thus came to the young Davy for fairly straightforward researches in applied chemistry, in a clear endorsement of useful knowledge.

A year later, he received the French prize for his electrochemical work; this had no obvious utility about it, but cast light on the nature of chemical affinity.

Sir Joseph Banks: a global perspective

Our distinction between pure and applied science was not made in Banks's day; there was science and there were arts, and all those involved in science hoped that sooner or later their work would improve the arts, fine or useful. Indeed, the systematic pursuit of knowledge without reference to the good of one's fellows did not then seem an evident good. It might be self-indulgence in mere curiosity; a rather childish impulse carried on into what should be maturity. We should therefore be careful about seeing Banks's concern with utility as extreme; but in his letter to Davy the emphasis upon usefulness is as striking as the references to the Royal Society and its public image. Defending society from a scourge was the proper business of natural philosophy.

What distinguished Davy's work from that of Stephenson and many others who were involved in the Industrial Revolution was that it was based in the laboratory, and followed upon the discovery of an underlying principle. It was not just rule of thumb, or inspired analogical thinking. The chemist had to be a practical man, and Banks (welcoming a new instrument in 1816) deplored the way chemists wanted to be gentlemen;[29] but effective chemical thought had to be done with the head as well as the hands. Swift's *Voyage to Laputa* had satirized Newton's Royal Society as a collection of learned fools; a sensible man like Banks did not want to preside over such an outfit. Enlightened usefulness was on the other hand a very proper objective for the Royal Society, and prominent among the hopes of its founders in the seventeenth century.

To Sir William Huggins, President a hundred years after Banks at the beginning of our century, it seemed that: "The supreme value of research in pure science for the success and progress of the national industries of a country can no longer be regarded as a question open to debate, since the principle has not only been accepted in theory, but put in practice on a large scale, at a great original cost, in a neighbouring country, with the most complete success.[30] What Huggins saw in Germany, Banks could not have seen in France a hundred years before. British seamen apparently preferred captured French warships because they were better designed than British ones;[31] under Lavoisier French gunpowder had been superior to British; and French bridges were more elegant. But in general, in Banks's day, British industry and British agriculture were well ahead of French, despite the Parisian lead in the sciences.

Paley's *Natural Theology* was first published in 1802, and the appeal of its vision of the world as a watch[32] seems to owe something to the newly developed marine chronometer[33] so vital to the explorer. But while Banks no doubt believed in a Creator, the idea that a major role of science should be finding out about God[34] through his works seems to have been alien to him. It could bring seriousness to the otherwise apparently frivolous inquiry into the workings of nature; and many in Banks's generation were affected by the Evangelical Revival, initiated by the Wesleys. Nevertheless, faced with childlessness and the gout, Banks apparently found an undergoing stomach to bear up against what might ensue not in the consolations of religion, but in hard work and domesticity. It was straightforward usefulness that he sought from the science over which he presided.

The years around 1800 were a time of poor harvests and of dearth; in which the Faraday family for example found themselves very near the breadline, and the eminent Quaker chemist William Allen distributed relief to the poor. Banks's hope was that work such as Davy's on fertilisers and insecticides would lead to agricultural productivity; and this was very close to his own interests in naturalizing animals and plants in regions new to them. Benevolence and scientific usefulness[35] went readily together, as the association of Sir Thomas Barnard and others like him with the Royal Institution shows. Hunger was undoubtedly an evil and also a threat to society; and it seemed that science, especially in the form of organized common sense in which it was so important to Banks (and right through the nineteenth century), might abolish famine and much disease. Davy indeed wrote[36] that "science is nothing more than the refinement of common sense making use of facts already known to acquire new facts": although for him this was not the whole story.

Davy's safety lamp went beyond organized common sense; and it was also far more dramatic than steady improvements in crop yields could ever be. The chronometer, invented by a clockmaker and a not a man of science, was a mechanical device that would save lives by indicating to sailors where they were; but here was something depending upon the latest chemical and physical science, which would directly and obviously save lives. Metropolitan science at last had something to boast about. Banks could feel unspeakable pleasure at this vindication of long-standing hopes.

The picture we get from the letter is of the natural philosopher not closeted in an ivory tower, but ready to come forward when called upon to defend society from a tremendous scourge of humanity. The enlightened chemist is a new St. George, bodly tackling an alarming and increasing evil — the experiments were dangerous, and in the Royal Institution's laboratory, safety precautions usually only seem to have been taken when there had in the past been an explosion. Because he is applying enlightened philosophy, he can (as St. George could not) promise that he has a "certain precautionary measure effectual to guard mankind for the future"; he knows the principles behind the device. Banks's letter notes that the lamp "cannot fail to recommend the discoverer to much public gratitude": the Royal Society's came in the form of the Rumford Medal, which is awarded biennially for the most important discoveries made in heat and light. This seems fair enough, especially because Rumford had distinguished himself in the invention of economical stoves, and in the founding of the Royal Institution; and had intended to encourage[37]: "such practical improvements in the management of heat and light as tend directly and powerfully to increase the enjoyments and comforts of life, especially in the lower and more numerous classes of society".

The public, or rather the powers that be, showed their gratitude in the form of a baronetcy; raising Davy to the same rank as Banks, and a higher recognition than Newton had received — scientific peerages were still a long way off. Nevertheless, to Davy as to Banks, it was an empty honour to a childless man; and as a reward for saving lives in the year of Waterloo, it contrasted with the more generous honours poured upon the soldiers and sailors who had destroyed so many of their fellow-men.[38] Nevertheless, useful science undoubtedly did win gratitude; and in promoting it, Banks did not differ from Rumford (who had received the first of his own medals), or Davy, or others involved in the physical sciences.

This must surely be the key to understanding Banks's view of the physical sciences. One who had sailed with Cook could not doubt the value of astronomy, although he opposed the setting up of a society dedicated to it; and he promoted

170

the Board of Longitude, dedicated to the practical use of astronomy in navigation. In Banks's day, as we can see from the arrangement of the *Encyclopedia Metropolitana*, planned by Coleridge about the time Davy was inventing his lamp, the "pure sciences" were those without an empirical component: — pure mathematics, and logic. All the rest were "mixed sciences", chemistry and mechanics along with manufactures. Science was distinguished from mere practice, based upon rule of thumb and associated with unreasoned resistance to change;[39] and was expected to be useful, sooner or later. Nature was there to be mastered; but those like Banks who had little patience with abstruse investigations would miss some of the pleasures and satisfactions of the physical sciences. Unlike Davy, he would not readily have seen the chemist as "animated by a spark of the divine mind", following a pursuit which exalted the understanding but did not depress the imagination.[40]

Banks thus emerges as an explorer, an administrator of science, and a promoter of useful knowledge. His emphases were a bit different from those of contemporaries devoted to the physical sciences, but not wildly different: Priestley invented soda-water, W.H. Wollaston manufactured platinum apparatus, Davy invented his lamp. It was a younger generation, including those nowadays called the Cambridge Network,[41] who began to move away from the kind of utilitarian emphasis natural to Banks. If science was to be a part of a liberal education, then it must be the theoretical parts which take precedence. Gradually too the lesson of the safety lamp was assimilated, and the model of technology as applied science, so boosted by Huggins, accepted as a reasonable picture of what normally happens. It did fit the new electrical and dye industries, but by no means all the others. Our view of Banks in relation to the physical sciences will depend upon what we require of a President of the Royal Society: he had little empathy with discoverers in these fields, but was prepared to give them a fair measure of the Society's support, and if their work could benefit humanity then he would be enthusiastic about it. Enlightened philosophy was his delight; and that is no bad thing for a man in his position.

Notes

(1) J. Davy, *Fragmentary Remains, Literary and Scientific, of Sir Humphry Davy*, London, 1856, 208.

(2) A. Cunningham and N. Jardine (eds.), *Romanticism and the Sciences*, Cambridge, 1990.

(3) J.J. Thomson, *Recollections and Reflections*, London, 1936, 379.

(4) D.M. Knight, *Humphry Davy: Science and Power*, Oxford, 1992, chap. 5; J.Z. Fullmer, *Sir Humphry Davy's Published Works*, Cambridge, Mass., 1969.

(5) M.P. Crosland, *Science under Control: the French Academy of Sciences, 1795–1914*, Cambridge, 1992, 23ff.

(6) J. Golinski, *Science as Public Culture: Chemistry and Enlightenment in Britain, 1760–1820*, Cambridge, 1992.

(7) On the Chemical Revolution, see W.H. Brock, *Fontana History of Chemistry*, London, 1992.

(8) A. Donovan, *Antoine Lavoisier*, Oxford, 1992.

(9) H. Lyons, *The Royal Society, 1660–1940: a History of its Administration under its Charters*, Cambridge, 1944, chap. 6, esp. 198ff.

(10) N.H. Robinson and E.G. Forbes, *The Royal Society Catalogue of Portraits*, London, 1980, 18, 82.

(11) J. Davy, *Memoirs of the Life of Sir Humphry Davy*, London, 1836, vol. 2, 126.

(12) H.B. Carter, *Sir Joseph Banks*, London, 1988, chap. 9; this biography is invaluable.

(13) R. McCormmach, 'Henry Cavendish on the proper method of rectifying abuses', in *Beyond History of Science* (ed. E. Garber), Bethlehem, Pa., 1990, 35ff.

(14) On Priestley, see the historical papers in Royal Society of Chemistry, *Oxygen and the Conversion of Future Feedstocks*, London, 1983, (special publication, 48).

(15) J. Morrell and A. Thackray, *Gentlemen of Science: the early years of the B.A.A.S.*, Oxford, 1981, chaps. 1–3.

(16) D.M. Knight, 'Tyrannies of distance in British Science', in *International Science and National Scientific Identity*, (eds. R.W. Home and S.G. Kohlstedt), Dordrecht, 1991, 39–53.

(17) I. Inkster and J. Morrell (eds.), *Metropolis and Province*, London, 1983, esp. chap. 1.

(18) Obituary of Davy, *Annual Register* (1829), **71**, 505.

(19) M.B. Hall, *All Scientists Now: the Royal Society in the Nineteenth Century*, Cambridge, 1984.

(20) For a view a hundred years on from Banks, see W. Huggins, *The Royal Society*, London, 1906, chap. 2.

(21) See the papers by N.G. Coley in *Notes and Records of the Royal Society*, (1967), **22**, 173–85, and *Ambix* (1988), **35**, 155–68.

(22) S. Carnot, *Réflexions sur la puissance motrice du feu* (ed. R. Fox), Paris, 1978 [1824], 62.

(23) S. Smiles, *The Lives of George and Robert Stephenson*, reprint, London, 1975, [1874], chap. 6.

(24) There are some MSS relating to Banks's involvement in the early days at the Royal Institution, and I am very grateful to the Librarian for supplying me with xeroxes of them.

(25) Obituary of Davy, *Annual Register* (1829), **71**, 507.

(26) J. Tyndall, 'Faraday as a Discoverer', [1868], reprinted in Royal Institution Library of Science, *Physical Sciences*, Amsterdam, 1970, vol. 2, 116ff.

(27) M. Berman, *Social Change and Scientific Organization: the Royal Institution, 1799-1844*, London, 1978, chap. 1.

Sir Joseph Banks: a global perspective

(28) D. Knight, *Ideas in Chemistry: a History of the Science*, London and New Brunswick, NJ, 1992, chap. 8.

(29) Carter, op. cit. (12), 517.

(30) Huggins, op. cit. (20), 21.

(31) Society of Arts, *Lectures on the Results of the Great Exhibition of 1851*, London, 1852, 547.

(32) J.H. Brooke, *Science and Religion: some historical perspectives*, Cambridge, 1991, chap. 6.

(33) D. Howse, *Nevil Maskelyne: the seaman's astronomer*, Cambridge, 1989, 122-7.

(34) D. Knight, *The Age of Science: the scientific world view in the nineteenth century*, Oxford, 1986, chap. 3.

(35) Berman, op. cit. (27), chap. 1.

(36) Knight, op. cit. (4), 44.

(37) Lyons, op. cit. (9), 219.

(38) Davy, op. cit. (1), 210.

(39) R.F. Bud and G.K. Roberts, *Science versus Practice: Chemistry in Victorian Britain*. Manchester, 1984.

(40) H. Davy, *Collected Works*, London, 1839-40, vol. 9, 361; part of a dialogue concerned with "chemical philosophy".

(41) S.F. Cannon, *Science in Culture*. New York, 1978, chap. 2.

XIII

A note on sumptuous natural histories

In illustrations of natural history, science and fine art inosculate; but when the illustrations are coloured and large, and the paper and binding of the book containing them are worthy of works of fine art, then economics also enters in and we have the eternal triangle. The most sumptuous works of natural history have always been too expensive for the majority of men of science to buy; hence such publications have gone perforce to wealthy amateurs or institutions, and have often played rather a small role in the development of science. The more puritanical biologists and historians of science indeed sometimes seem to feel the deepest suspicion of beautiful illustrations, though it is good to see Martin Rudwick drawing attention to the use of visual as well as verbal language in geology in the early nineteenth century.[1] Outside the range of experts, handsome illustrations and magnificent tomes have earned due respect; so that from the nineteenth century it is Audubon and Gould who are remembered, rather than MacGillivray or Newton, as ornithologists. Certainly illustrations when done by someone who understood the specimen have a much longer useful life than verbal descriptions, as we can see from modern bird books which re-use pictures from Gould or Thorburn. Admirers of the great natural-history illustrators of the past may sometimes forget that their plates had their context, being done for a specific book against a background of zoological or botanical theory, and subject to the constraints involved in any process of reproduction. Zoological and botanical art is also subject to aesthetic fashion, and is no more timeless than any other kind of art.

We live in a great age of colour printing, when the constraints to which our ancestors were subject have been lifted—except that really fine coloured illustrations are still very expensive. An artist's coloured pictures can now be reproduced without the intervention of an engraver or lithographer, and without the vagaries of hand-colouring, using either collotype with its continuous tones, or multi-coloured litho-graphy which involves screening. The 1970s have seen the publication of a number of extremely handsome illustrated works of natural history which are of some interest to the historian of science, and which can be considered sumptuous in that the price of all of them exceeds £100; and at some of these we can now glance. Often they are advertised as an investment, and no doubt they are a better bet than money, as well as much more beautiful, in the current state of things; but they represent serious science too.

The first group of such books contains reproductions of handsome works published in the past. Often, as with Audubon's and Gould's bird pictures, these will have been familiar through reproductions in many recent books; but cheap and small repro-ductions do not do justice to the originals which were monumental in scale. A selection of Audubon's plates from the *Birds of America* has been published in full-size facsimile by Ariel Press in nine-colour collotype, the volumes being 98 × 68 cm

---

[1] M. J. S. Rudwick, 'A visual language for geology ', *History of science*, **14** (1976), 149–195.

174

in size and thus being quite difficult to carry around; their effect is duly magnificent.[2] Similarly, a selection of Mark Catesby's eighteenth-century plates of animals and plants of Carolina, smaller (50 × 35 cm) than Audubon's but still very arresting, has been published;[3] the colours in some of these plates look more gaudy than some originals, but no doubt this is because of the variations in hand colouring in the early editions.

A larger-scale enterprise, and one of more direct value to the historian, is the reprinting of complete works; it is one thing to have reproductions of the best plates, and another to have all of them with the original text. Two splendid orchid books, Lindley's *Sertum orchidaceum* (1838) and Bateman's *Orchidaceae of Mexico and Guatemala* (1837–42), have been reprinted with their full text.[4] This enables one to see that, while both appeared at the same time and were illustrated chiefly by the same artists, Lindley's is more a work of botany while Bateman's is more concerned (as became a wealthy amateur) with their cultivation—though it has delightful descriptions of the whereabouts of the species in the wild state, and charming vignettes, which give it a strong character of its own. It opens to us the world of men of leisure, seriously devoting themselves to the study and cultivation of plants; while Lindley's, with its many oriental orchids, sometimes drawn by Indian artists, can perhaps recall the Empire. One of Bateman's most celebrated vignettes is of a team of lilliputian librarians handling the book with a hoist, for its original dimensions were 53 × 71 cm; in the reprint, which is reduced to match the Lindley, the page size is 35 × 49 cm, which makes it still a volume of princely size although less splendid than the original. Both volumes are beautifully printed, and the reprints make available works which include many first publications of pictures of species, and which are otherwise excessively rare and expensive.

Another even more ambitious complete reprint is of Gould's *Birds of Australia* (1848–69) in eight large volumes (37 × 53 cm) containing nearly 700 plates.[5] Some plates, those of bower-birds in vol. IV, occupy a double-page spread; it is a pity that they have been so bound as to leave a gutter down the middle. In general, the standard of reproduction is very high, and makes one realise what an achievement it was to describe and illustrate the bird-life of Australia only about half a century after the First Fleet had sailed to Botany Bay. One can see why not everybody got on with Gould: vol. VII, plate 71 is signed ' Edward Lear ', but its caption has the names of Gould and Richter only; plate 48 has the same caption, although it is made clear in the text that Elizabeth Gould drew it. Some of the plates are dull; Gould was a ' splitter ', and was engaged in an enormous enterprise, and as he worked through a family, especially when he was depicting species ' collected ' by someone else, his plates sometimes conformed too closely to a formula. The business of ' collecting ' is illustrated in vol. VII, plate 53, where a dying bird wounded by the ornithologist is shown; and Gould's descriptions of the birds and their habitats are always interesting. The large size of his books meant that like Audubon he could usually show birds life-size, and the subscription-list at the beginning of the first volume indicates Gould's market.

As well as reproducing old books in facsimile, modern methods of printing can be used to reproduce illustrations from the past which have never been published. A magnificent example of these is the flower paintings done by Ferdinand Bauer on

[2] J. J. Audubon, *The Birds of America* (2 vols., 1972–73, London, Ariel Press), 40 plates. See also his *Original water-colour paintings* (2 vols., 1966, London); these are reduced in size to 34 × 27 cm, and are done directly from his pictures and not from the printed plates.

[3] M. Catesby, *The natural history of Carolina, Florida and the Bahama Islands* (1974, Savannah, Georgia, Beehive Press), 50 plates.

[4] J. Lindley, *Sertum orchidaceum*, and J. Bateman, *Orchidaceae of Mexico and Guatemala* ([1974], Amsterdam, Theatrum Orbis Terrarum; and New York, Johnson Reprint), 50 and 40 plates.

[5] J. Gould, *Birds of Australia* (8 vols., 1972–76, Melbourne, Landsdowne Press), 681 plates. See also A. McEvey, *John Gould's contribution to British art* (1973, Sydney); and C. E. Jackson. *Bird illustrators* (1975, London).

175

his voyage round Australia with Flinders in 1801–05.[6]   Robert Brown, the botanist of the voyage, taught him to use the microscope; and his drawings, which must be among the finest ever made, show a superb combination of aesthetic sense and of understanding of a plant.   In the book, which measures 63×45 cm, they are mounted on a green background, which sounds a bit arty but does enhance them. When the book came out, the publishers took the risk of showing the plates beside the originals, when it was really hard to tell which was which.   The book has both historical and botanical text, and is a useful contribution to our knowledge of botany in Brown's time as well as being very beautiful.

The Bauer drawings are in the British Museum (Natural History); as are some drawings sent back at the same period by John Reeves from Canton, a selection of which has been recently published.[7]   These were done by Chinese artists.   Some are stock pictures, while others were commissioned by Reeves; some of them depict type specimens.   All those printed are very decorative; both botanical and zoological subjects are illustrated, but some of the drawings are of more scientific usefulness than others.   The text of the book is valuable as casting light on the internal and external history of natural history in the early nineteenth century, and there is also a useful description of the collotype process by which the plates have been exquisitely printed—it has even done white on white effectively.   The book measures 53×40 cm.

At the end of the nineteenth century, Major Henry Jones, who had served in the Army in India but about whom little seems to be known, painted large numbers of birds from skins at the British Museum (Natural History); these he bequeathed to the Zoological Society of London, and a selection from them has been published to mark the 150th anniversary of the Society.[8]   His style is characteristic of his time, and yet clearly his own, with attractive colouring and with backgrounds depicted although the drawings were made from museum specimens; and the attitudes in which they are shown seem realistic.   It is clearly a good thing that this little-known zoological artist should achieve, even if long after his death, the fame he deserves, and that these pictures should get into circulation; even though they do not have the importance in the history of science that those of Gould or Bauer have because they are ' types ' that are valuable for taxonomic purposes.   The book is handsomely printed, but the binding—with a flat spine in the American manner—is only moderately attractive; the book is an oblong shape, the plate size being 46×38 cm.

It should not be supposed that there are no longer artists capable of preparing magnificent illustrations; there are indeed modern examples of sumptuous works on orchids, on a group of Australian birds, and on the flora of a region of Australia, which can stand comparison with those we have already mentioned, despite the weight of their authority.   The *Orchidaceae* of Hunt and Grierson is a delightful volume, with stiff heavy paper and vellum binding;[9] the illustrations, like those in Bateman, are from living plants.   Plates 24 and 32 depict the same orchids as Lindley's plates 20 and 44; the second is more beautifully shown in the twentieth-century volume, while for the first it is hard to opt one way or the other.   Some of the plates in the new volume are comparative, showing different forms of leaves and flowers within the family; and some show hybrids.   The plates are 35×47 cm in size.

[6] W. T. Stearn (ed.), *The Australian flower paintings of Ferdinand Bauer* (1976, London, Basilisk Press), 25 plates.   Further sumptuous examples of the reproduction of manuscript and illustration are L. Baldner, *Vogel-, Fisch-, und Thierbuch, 1666* (2 vols., 1973–74, Stuttgart, Verlag Miller und Schindler), 108 plates, 30×19 cm, the first regional fauna of Europe; and W. Bligh, *Log of H.M.S. Providence, 1791–1793* (1976, Guildford, Genesis Publications), 12 plates + 5 fold-out maps, 22×34 cm, describing the successful transport of bread-fruit trees from Tahiti to the West Indies.

[7] P. J. P. Whitehead and P. I. Edwards, *Chinese natural history drawings* (1974, London, British Museum (Natural History)), 20 plates.

[8] B. Campbell, *The bird paintings of Henry Jones* (1976, London, Folio Fine Editions with the Zoological Society), 24 plates.

[9] P. F. Hunt (illustrated by M. Grierson), *Orchidaceae* (1973, Bourton, Berkshire, The Bourton Press), 40 plates.

Inviting comparison with Gould are the illustrations in Morris's *Birds of prey of Australia*.[10] Gould's illustrations of this group come in his first volume, were by Richter, and are very impressive. Morris's are perhaps a little tauter; he does not go in for much background, favouring the branch of a tree—often dead—rather than the landscape usually found in Gould's plates. On the whole, Morris emerges well from the competition; and he also draws a silhouette, with that of a magpie for comparison, and a full-page field sketch of each bird also—so that one gets a feeling for how it moves, as well as seeing from the main plate how every feather lies. Morris's remarks on classification can be interestingly compared with Gould's, revealing that taxonomy is rather different from stamp-collecting. His plates are about the size of Gould's (36×51 cm), but the birds tend to be drawn smaller.

Finally, we have appearing a work on the *Endemic flora of Tasmania*, under the patronage of the late Lord Talbot de Malahide, with a text by Winifred Curtis and plates by Margaret Stones.[11] It is a pity that the only Tasmanian plant collected by Brown in the Bauer volume, *Brunonia australis*, is not yet illustrated in the *Flora*, so that we cannot compare them directly. The scale of the twentieth-century volume is smaller (29×41 cm), and the plates, particularly in the earlier parts, do not always show dissections; but they are done with verve and accuracy. In the fifth part, a fern which is green when growing, but had gone brown by the time it got to the artist, is shown brown—which seems to show perhaps an excessive devotion to accuracy. Not only is this work like those of the last century in having a noble patron; but also in appearing in parts, so that the plants are not in systematic order, each part containing an assortment of plants from many families. As happened frequently in the past also, the project grew in that further species were discovered while it was in train; altogether it will fill six parts, of which five have already come out. The text, sometimes describing the finding of the type specimens of the plants, is of interest to the historian of botany; Tasmania after all is an island in which Brown and J. D. Hooker botanised, and where Sir John Franklin was, as governor, a notable patron of science.

Sumptuous illustrated works, then, are not simply a good investment but can also be intellectually valuable to the historian of science. It is a pity if he does not know about them, and never sees them, for they can be useful sources—and are sources of aesthetic pleasure.

[10] F. T. Morris, *Birds of prey of Australia* (1973, Melbourne, Landsdowne Press), 24 plates.
[11] W. Curtis (illustrated by M. Stones), *The endemic flora of Tasmania* (1967, in progress, London, Ariel Press).

XIV

Scientific Theory and Visual Language

The idea that the sciences and the arts are not very different is an attractive one. There appears to be no one method appropriate to all sciences at all times, and the pursuit of science seems to require certain unprovable assumptions – which may be a general causal principle, or something more complicated like the paradigms of Thomas Kuhn. Humphry Davy made Wordsworth appreciate the creative and imaginative aspects of chemistry, and induced Coleridge to place the scientist along with the lover, the lunatic and the poet – 'of imagination all compact'.[1] And yet for all its 'academic' training and dogma, there is surely something progressive about science which is lacking in the fine arts. It would be paradoxical to say that our astronomy was not so good as that of Galileo, but not to say that we have no painter to touch Michelangelo; though perhaps we might say that no living astronomer was as great as Galileo, we would feel that science has moved on since his day.

The classics of science are thus quickly or slowly made obsolete, even though they may have a long run;[2] their content is squeezed into textbooks, and their authors become names to which hagiographical anecdotes are attached. The very language in which they are written, being theory-laden, changes so that in a generation or two the very meaning of the text is hard to recover – 'battery' was used in early-nineteenth-century electricity, and 'evolution' in pre-Darwinian biology, but not in the sense in which we use them. Instruments are improved, so that old observations also generally become obsolete; or irrelevant, like very accurate atomic weights upon the advent of Rutherford's atomic model. Some old eclipse-observations are of value to astronomers, and old floras and faunas to ecologists; but in general the pattern in science is of more or less rapid obsolescence.

Scientific illustrations are pictures designed not to stand on their own, but to accompany a text: they are in partnership with prose (or occasionally, as in Erasmus Darwin, verse) intended to convey knowledge old or new. They may be diagrams: showing apparatus with, or in our century without, an elegantly cuffed hand holding it; illustrating mathematical propositions, like the epicycles in Copernicus, or psychological ones, as in works of physiognomy or phrenology; or, like the only illustration in the *Origin of Species*, showing hypothetical divergence and extinction over time. Some are thus more theory-laden than

[1] See my paper "Chemistry and Poetic Imagery", *Chemistry in Britain* (1983), 19: 578–82; and my *The Nature of Science* (London, 1976), Chapter 3. On dogma in science, T.S. Kuhn, *The Structure of Scientific Revolutions* (Chicago, 1970).

[2] A.J. Meadows, ed., *Development of Science Publishing in Europe* (Amsterdam, 1980).

others, but even diagrams of apparatus (especially without hands or supports) make sense only to those who have learned the conventions. They are concise visual languages which must be taught before they can be read, and sooner or later get out of date. In scientific illustration, the artist's intentions can be illuminated by the text; they are pictures with a clear context.

Linnaeus classified on the basis of external characters, and natural history is a science concerned with the outsides of things. Theories of classfication involve tables and diagrams, and so does physiology; but we shall be concerned with natural history, a descriptive science, where theory loading is less obvious, and the visual language a subtle one. We are apt indeed to assume that description can be theory-free, that what is noticed does not depend upon time and place, despite Goethe's reminding us[3] that we see only what we know. Certainly some of the great illustrations of birds, flowers or butterflies are admirable works of art, which have passed the test of time in that they are appreciated long after they were produced, and adorn walls and coffee-tables where they give genuine aesthetic pleasure.[4] In this they are just like portraits and landscapes painted long ago; and yet they were drawn or painted, and generally printed, to accompany a scientific text, and are thus a part of science; the picture may indeed be a type specimen, if its subject has perished. The artist, who may or may not have also written the description, will have been expected to show features useful in diagnosis or deemed important in the organism's life; and while he may, like Edward Lear painting parrots,[5] have done portraits he might well have been expected to generalise a little and paint a member of a species rather than an individual – rather like Zeuxis painting Aphrodite from a number of Crotonian girls. Natural-history illustrations therefore have some theory loading; they are science in visual language as well as art.[6] Sometimes the theory is considerable, as in dinosaur-reconstructions in the last century.

When we look at illustrations from the past, we notice at once that there are conventions characteristic of times and places, just as there are in other kinds of painting. Some of these, such as the way the plant or animal is placed on the page, are conventional. Outside Europe, artists in India, China, and Peru[7] were trained to paint natural history subjects and often did so very beautifully; but something of their native tradition shows through (Fig. 1 A, 1 B). In the seventeenth and eighteenth centuries, the requirement of the natural history artist was that he should help one to recognise an isolated and probably dead specimen. There was

[3] J.W. Goethe, *On Art*, ed. and trans. J. Gage (London, 1980), p. 6. On natural history illustration, see W. Blunt, *The Art of Botanical Illustration* (London, 1950) and my *Zoological Illustration* (Folkestone, 1977); and contrast J. Rawson, *Animals in Art* (London, 1977), where the pictures were not associated with scientific texts.

[4] A. Savile, *The Test of Time* (Oxford, 1982); cf. W. Hazlitt "Why the arts are not progressive", 1814, in his *Selected Essays*, ed. G. Keynes (London, 1948) pp. 603–9.

[5] S. Hyman, *Edward Lear's Birds* (London, 1980); and on types, W. Pater, *The Renaissance* [1893], ed. D.L. Hill, (Berkeley, 1980), p. 51.

[6] M.J.S. Rudwick, "A Visual Language for Geology", *History of Science* (1976), 14: 149–95;

N. Rupke, "The apocalyptic denominator in English culture in the early nineteenth century", in M. Pollock, ed., *Common Denominators in Art and Science* (Aberdeen, 1983) pp. 30–41; A. Desmond, *Archetypes and Ancestors* (London, 1982), chapter 4, on dinosaurs.

[7] M. Archer, *Natural History Drawings in the India Office Library* (London, 1962); R. Desmond, *The India Museum, 1801–1879* (London, 1982); W. Roxburgh, *Icones Roxburghianae* (Calcutta, 1964 – in progress); P.J.P. Whitehead and P.L. Edwards, *Chinese Natural History Drawings... from the Reeves Collection* (London, 1974); J.C. Mutis, *Flora de La real Expedicion Botanica del Nuevo Reino de Granada* (Madrid, 1954 – in progress).

Fig. 1A. Purple three-toed kingfisher, drawn and lithographed in India. From T.C. Jerdon, *Illustrations of Indian Ornithology,* Madras, 1847, pl. XXV. Quarto. Some plates in this book were prepared by European artists and printers, and have an interestingly different look.

Fig. 1B. *Miconia prasina*, from a painting by Vincente Sánchez, an American Indian. Plate 37 in J.C. Mutis, *Flora de la real expedicion botanica del Nuevo Reino de Granada*. Madrid, vol. XXX, 1976. Some flowers are diseased (showing little sticks) – here the artist has painted what he saw, rather than idealised: the expedition went in the eighteenth century (Mutis corresponded with Linnaeus) but has been slowly published.

a famous maxim that what is hit is history and what is missed is mystery; the zoologist had to be quick on the draw, however poorly this may fit with our conservationist impulses.

Occasionally this alliance between hunting and natural history is actually demonstrated in a painting, as when Gould shows[8] a seabird bleeding from his shot and being succoured by its mate – which no doubt he 'collected' with the other barrel. More often it can be inferred, as when Willughby showed some birds as gibbetted corpses, or more usually when a bird or animal is shown stiffly posed either standing on nothing or on a standard studio mossy stump. Before the coming of binoculars and cameras with fast film, catching animal motions was very difficult; and does not seem anyway to have been wanted. Natural history has come some way since Gould's time; for example, no nineteenth-century artist could have knowingly drawn a bird defending its territory. But in some fields, like entomology, illustrations tend still to show what the dead insect looks like, and to be formal rather than naturalistic. This formality was sometimes contrived, as when Wilkes arranged moths to form a pleasing pattern; and this also had some connection with techniques of reproduction (Fig. 2 A, 2 B). An engraver generally came between the artist and his printed work, and even with the cheaper technique of lithography most plates were prepared by somebody other than the artist. Because copper plates were very expensive, and needed to be printed on a different press and on different paper from type, the author who was not aiming at the wealthy had to economise on plates and therefore depict several species at once. He might aim at an agreeable overall plate, perhaps with an attractive border; or he might leave it higgledy-piggledy, in which case it may serve as illustration but cannot be claimed as art.[9] If a plate is being treated as a whole, the arrangement may like that of Wilkes be based chiefly on aesthetic grounds; but often there will be some theory behind it, and even in lithographs or woodcuts the artist may prefer to put two species on one plate. He will show what he believes to be related species, for when two or three are depicted side by side it is easier to know which has just been seen in the flesh; or he will try to show animals which might be seen together – these will not usually be close species, though they may be. Here theory, systematic or ecological, will necessarily begin to creep in.

The text may well cast light on the theory, but it is notorious that illustrations may surpass the text, or an excellent text have wretched or unappealing plates[10] – the author and illustrator may not be in step, or if they are one person he will be better at one job. Sometimes a plate will be re-used with new text over a

[8] J. Gould, *Birds of Australia* (London, 1848), vol. 7, plate LIII; F. Willughby, *Ornithology*, trans. J. Ray (London, 1678), plate XX.

[9] For coherent plates, see J. Barbut, *The Genera Vermium* (London, 1783), and for disorderly ones, W. Borlase, *The Natural History of Cornwall* (Oxford, 1758) – both authors were keen Linneans. G.D.R. Bridson, "The treatment of plates in bibliographical description", *Journal of the Society for Bibliography of Natural History* (1976), 7 : 469–88.

[10] W. Yarrell, *British Birds* (London, 1837–43) was more successful than W. MacGillivray, *A History of British Birds* (London, 1837–52): though the text was less original, the woodcut

illustrations were more attractive. E. Tyson, *Orang-Outang* (London, 1699); G. Cuvier, *The Animal Kingdom*, ed. and trans. E. Griffiths, C. Hamilton Smith and E. Pigeon (London, 1827–35), vol. 1, p. 252; N. Grew, *The Anatomy of Plants* (London, 1682) plates XXII ff; R. Thornton, *The Sexual System of Carolus Linnaeus* (London 1798–1807), plate of "Transverse sections through wood," publ. May 1st 1798, (see R. Thornton, *The Temple of Flora*, ed. G. Grigson and H. Buchanan (London, 1951) pp. 13f for bibliographical remarks); H. Davy *Elements of Agricultural Chemistry* (London, 1813), plates VIII-X.

182

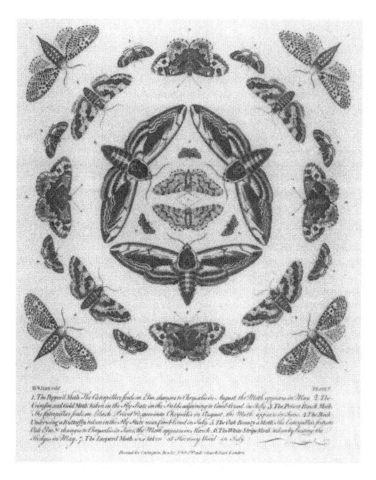

Fig. 2A. Moths handsomely arranged, with details of when and where collected. Engraving, from B. Wilkes, *The British Aurelian,* [1742] ed. R.S. Wilkinson, Faringdon, Oxon, 1982, pl. V. Quarto.

surprisingly long period. This may be because the plate is a part of a science which has made little progress: thus the study of the anthropoid apes was amazingly static so that plates from Edward Tyson's *Orang-Outang* of 1699 were used in the standard English version of Cuvier's *Animal Kingdom* of 1827–35. Chimpanzees were hard to come by, and there was nothing better than Tyson. Similarly Nehemiah Grew's sections through various kinds of trees, from the *Anatomy of Plants* of 1682 were used by Robert Thornton and by Davy in the early nineteenth century. Here, microscopes had not yet been greatly improved, and plant anatomy had been little studied – it was not a major concern to Thornton or Davy – and the elderly illustrations were therefore still as good as new.

In more recent years, there have been a number of bird books using illustrations over half a century old. T.H. Huxley used Tyson's pictures in his *Man's Place in*

111

Fig. 2B. *Sepia,* from J. Barbut, *The Genera Vermium of Linnaeus,* London 1783; engraved from
  Barbut's drawing by J. Newton. The Preface refers to "the immortal Linnaeus" and his
  "infinite judgement".

Nature of 1863, but only as historical documents; and old illustrations continue to
be used in that way. But Gould's bird plates, and others by Thorburn (see p. 148),
Lodge and Finch-Davies have all in recent years been reissued with new texts;[11]
the former two had a text which went with them, while the latter did not. The

[11] J. Gould, *Birds of Europe,* text by A. Rutgers (London, 1966); G.C. Sauer, *John Gould, the bird man* (London, 1982). A. Thorburn, *Birds,* text by J. Fisher (London, 1967); G.E. Lodge, *Unpublished Bird Paintings,* text by C.A. Fleming (London, 1983); C.G. Finch-Davies, *The Birds of Southern Africa,* text by A. Kemp (Johannesburg, 1982); J. Sepp, *Butterflies and Moths,* text by S. McNeill (London, 1978).

184

Fig. 3. Diamond firetail finch, "L'Acalanthe". Engraving, from L.J.P. Vieillot, *Songbirds of the Torrid Zone*, [1805–9], ed. P.R. St. Clair, trans. E. Jentoft, Kent, Ohio, 1979, pl. XXXII. Quarto.

same thing has happened with the butterfly and moth plates of Jan Christian Sepp, which first appeared over two hundred years ago. In all these cases the belief is that the modern naturalist will find the old pictures helpful – by implication, more helpful than anything which could be done today, and not simply more beautiful. So in these plates, the test of time seems to have been passed with flying colours: they score a double first in science and art.

Yet on the whole, plates older than about a century are prized if at all for their aesthetic merit (or perhaps their curiosity) rather than their use in natural history; and it may be worth examining why this might be so. We can see a sequence of

Fig. 4. Diamond firetail finch, "Spotted gross-beak". Engraving, from J.W. Lewin, *The Birds of New South Wales*, [1838], ed. A. McEvey, Melbourne, 1978. pl. XI. Quarto. The 1813 edition of this was the first illustrated book and the first natural history book known to have been published in Australia, where the engravings were done also.

plates of the same species if we begin with Vieillot's plate of the Diamond Firetail finch from Australia (*Emblema guttata*);[12] where the bird, although apparently one studied in captivity rather than described from a corpse, is shown in a decorative but unnatural attitude perched on a studio stump (Fig. 3, 4, 5). This plate was

[12] F.T. Morris, *Finches of Australia* (Melbourne, 1976), plate XXVII.

186

Fig. 5. Diamond firetail finch, "Spotted-sided finch". Lithograph, from J. Gould, *The Birds of Australia*, London, 1848, vol. 3, pl. LXXXVI. The artist was H.C. Richter. Folio. Note, despite Linnaeus, the instability of names, Latin as well as English.

done in the first decade of the nineteenth century; only a few years later we find Lewin's plate of the same bird, this time portrayed on some Australian vegetation; but here we have a pair displayed, more naturalistic in pose and stuffing, by one who had observed them wild. Gould's plate shows us a pair, on some

Fig. 6. Young Salmon, or grilse. Engraving from W. Jardine, *British Salmonidae*, [1839–41], ed. A. Wheeler, London, 1979. pl. II. Elephant folio. The uncoloured background shows the River Tweed.

grasses which form part of their diet; so that here in place of vague local colour we are getting definite information. As well as seeing how the bird actually comports itself, we learn other things about its life. Although Gould was thoroughly pre-Darwinian, these concerns are much more modern than an interest in birds either simply for classification or for caging. Some contemporaries of Gould gave what we would think of as redundant information: illustrations of fish are hard to bring to life – it is elegance rather than anthropomorphism that the artist must seek[13] – and Victorian illustrators often embellished their fish plates with a background drawing or painting of the river from which the specimen had been caught. This can, like the blue sky lightly washed in behind birds in the 1838 plates of Lewin (Fig. 4) or the lightly-sketched backgrounds of Lear and G.H. Ford, help to get the tone of the plate right (Fig. 6, 7). If overdone – and some artists showing fossils depicted their matrix too lovingly – background detail can distract, and while making an attractive picture diminish its scientific usefulness. The freer technique of lithography allowed such lightly-sketched backgrounds more easily than engraving, and could give a more 'lifelike' plate.

With butterflies and moths, some nineteenth-century plates similarly give more information than is now customary. Thus they try to show the insect at all the stages of its life, its food-plant(s), and also close species; which makes an attractive if crowded plate (Fig. 8). It also makes one which gives hostages to fortune: the

[13] On depicting fish, see A.C.L.G. Günther, *An Introduction to the Study of Fishes* (Edinburgh, 1880) chapter 1. M. Twyman, *Lithography, 1800–1850* (London, 1970); C.E. Jackson, *Bird Illustrators* (London, 1975) and *Wood Engravings of Birds* (London, 1978).

Fig. 7. Rhinoceros, from A. Smith, *Illustrations of the Zoology of South Africa*, London, 1840, pl. 1 of Mammalia. Quarto. Lithograph, lightly tinted but with sketchy background, by G.H. Ford, later best-known for his pictures of reptiles and fish; this rhinoceros owes nothing to Dürer.

caterpillars may not feed exclusively or even mainly on the plant shown after all, the imago shown in a natural attitude may be harder to discriminate than if it was spread out when dead and photographed, and the "close species" may be put some way apart in more modern taxonomy. Serious butterfly books of recent years[14] therefore tend to revert to the earlier practice of showing in effect a cabinet of specimens, with additional plates to show larvae, food plants, and perhaps dissection. These, which became necessary for diagnoses,[15] are needed for invertebrates and sometimes for reptiles and fish; their incorporation into a plate diminishes the effect of naturalism, but can be done to make a handsome whole (Fig. 9A, 9B, 9C.)

Egerton in 1852 could refer to a fossil fish which he had lithographed as a 'pretty little specimen',[16] and this sort of anthropomorphic language – like the 'advertisement language' used by Darwin of insects and orchids – was characteristic of its time and place. It went with a serious concern with how the organism

[14] J.D. Bradley, W.G. Tremewan and A. Smith, *British Tortricoid Moths* (London, 1973–9); J. Heath, ed., *The Moths and Butterflies of Great Britain and Ireland* (London, 1976 – in progress).

[15] W.T. Stearn, ed., *The Australian Flower Paintings of Ferdinand Bauer* (London, 1976); see my "Note on Sumptuous Natural Histories", *Annals of Science* (1972), 34 : 311–4.

[16] P. de M. Grey Egerton, *Memoirs of the Geological Survey, decade VI* (London, 1852) p. 2; in this volume plate III was engraved, but all the others were lithographed. C. Darwin, *On the Various Contrivances by which British and Foreign Orchids are Fertilised by Insects* (London, 1862); M. Allen, *Darwin and his Flowers* (London, 1977); R.D. Fitzgerald, *Australian Orchids* (Sydney, 1882–94), plates II and CV. L. Agassiz, *Contributions to the Natural History of the United States*, vol. 2 (Boston, 1857) plates XXVI and XXVII.

Fig. 8. Large and Small Tortoiseshell butterflies, and Camberwell Beauty; on elm, stinging-nettles, and willow. Lithograph, from J.O. Westwood, *The Butterflies of Great Britain*, London, 1854, pl. VII. Quarto. The caterpillars and crysalids are also shown, and the plates highly coloured.

must have lived; and perhaps with a taxonomy taking note of ecology rather than with the arrangement of bird skins in patterns on the parlour floor. One way of giving interest to plates of flowers was to put in some butterflies or other insects; this was done in both Chinese and European flower pictures, where the decorative insects have as a rule no scientific function and do not always belong to definite species at all. Illustrations of insects, whether seen as pretty or as

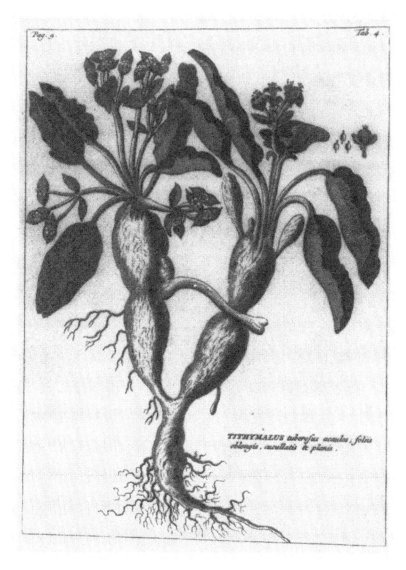

Fig. 9A. *Tihymalus* in J. Burmann, *Rariorum africanarum plantarum*, Amsterdam, 1738, pl. 4. These early engravings of the Cape flora show some magnified flower-parts.

pestiferous, often included their food plant; and conversely with study of plant fertilization we find the fertilising insect sometimes shown in the picture. Darwin's studies of orchids indicated to him that all the curious forms of the flowers are adaptations to bring about efficient insect pollination, often by only one species; to show that species of insect with the flower, as Fitzgerald did in his work on Australian orchids, was therefore to provide an evolutionary theory-

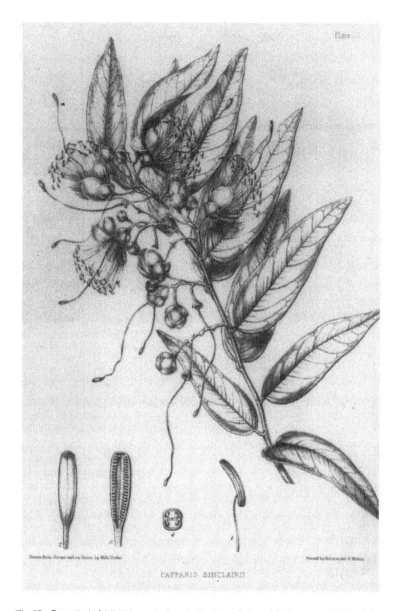

CAPPARIS SINCLAIRII

Fig. 9B. *Capparis sinclairii.* Lithograph, from G. Bentham, *Botany of the Voyage of H.M.S. Sulphur,* London, 1844, pl. XXVII. Folio. The artist was Miss Drake, of whom not much is known except that she did many botanical illustrations; the plate shows dissections. The book was one of a series of publications sponsored by the Admiralty, and supported by government grants, which included Darwin's *Zoology of the Voyage of H.M.S. Beagle.*

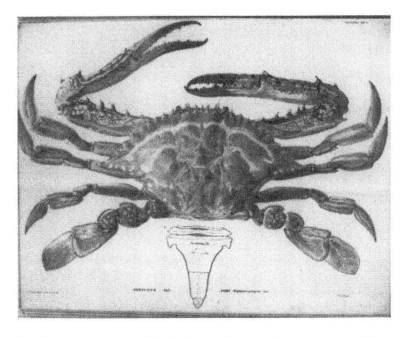

Fig. 9C. *Portunus pelagicus*, from P.F. Siebold, *Fauna, Crustacea* (ed. W. de Haan), Leyden, 1850, pl. 9; an autolithograph by A.S. Mulder, including dissection.

loading of a tactful kind, and to get more information into a picture. One can see preoccupation with the range of variation within a species in Louis Agassiz's illustrations of turtles, published two years before the *Origin of Species* – but Agassiz refused to accept evolution. (Fig. 10).

The most extreme cases of anthropomorphic illustration showed animals in exciting scenes; in bird books it is always agreeable to see the pair with young, and it may be instructive to see them with prey – though here there is a risk of theory or myth loading, as in Buller's plate of the Kea of New Zealand shown attacking a lamb[17] – or perhaps to show two males fighting or dancing for a female. Joseph Wolf, the greatest of zoological illustrators of the mid-nineteenth century (see p. 150), did a series of pictures of animals in dramatic scenes: a crocodile fighting a tiger (Fig. 11), a gazelle dislodging a leopard from its back and so on. Here the actors are not quite like the animals with human faces of the medieval bestiary but the pictures are meant to display fear and bloodthirstiness – because Wolf was so good, they are not caricatures; they are the zoological equivalent of the anecdotal or moralising pictures or the programme music of the same period, which in the hands of a genius can produce real works of art. Lear, the bird-painter and nonsense-writer, had earlier managed to give his birds strong individual charac-

[17] W.L. Buller, *Birds of New Zealand* [1888] ed., E. G. Turbott (London, 1967); B. Rowland, *Animals with Human Faces* (London, 1974); on Lear, see note 5. Wolf expounded Goethe's maxim to read "we see distinctly only what we know thoroughly", see his *Life*, by A.H. Palmer (London, 1895).

Fig. 10. *Ptychemys rugosa*, from L. Agassiz, *Contributions to the Natural history of the U.S.A.*, Boston, 1857, II, pl. 27; this coloured lithograph showed the variation within a species.

ter, and seems to have favoured species like parrots and pelicans and toucans (painted from life when possible) which lent themselves to this kind of portraiture.

In Gould's books illustrations by Lear and Wolf particularly jump out at one; by leaving his backgrounds sketchy Lear was able to focus on the bird, which in the wild is often hard to see in its shadowy or leafy environment. Money being no

Fig. 11. Crocodile catching a tiger. Engraving, from J. Wolf, *The Life and Habits of Wild Animals*, text by D.G. Elliott, London, 1874, pl. III. Folio. The engraving was by J.W. and Edward Whymper, the famous mountaineer; who were eminent as translators of drawings into printed illustrations.

object, he also avoided another difficulty by generally illustrating only one species. Systematic illustrations showing two or more species (which Gould was praised for judiciously using in some books) may be overtaken by events: thus one of Lodge's plates of New Zealand birds, just published after some fifty years, shows three specimens which were then in the same genus but are now in three different genera and would not be put on one plate by an artist working now.[18] This is the easiest way in which an illustration can become obsolete; but it is perhaps a special case of a picture becoming useless or imperfect for science if it does not include diagnostic characters. Because taxonomy has more or less steadily moved away from the Linnean concentration on externals, illustration has become less crucial than it was – fine discrimination may depend on the bones of the ear of a bird (which can be illustrated, but may not generate great art) or chemical investigation of egg-white proteins, which does not lead to pictures at all. The characteristic illustration in these circumstances might be that of the field guide, not intended to show how every feather lies but to assist rapid identification in the open.

While then various illustrations from the past continue to give pleasure, their usefulness tends to diminish with the passage of time because scientific language and the concerns associated with it changes, whether it be visual or ordinary language. Particularly to the layman, the visual language of past science – provided it is not the remote past – may be more intelligible than that of current science, as older paintings may be easier than those of the avant-garde; and it may still have some scientific usefulness. If it is the work of a great artist, it may pass time's test and live on, passing into 'art' if it is no longer 'science', rather than being a casualty of progress.

[18] Lodge, see note 11, plate XXVI. On taxonomy, see my *Ordering the World* (London, 1981); A. Wheeler and J.H. Price, ed., *History in the Service of Systematics* (London, 1981).

XV

WILLIAM SWAINSON: TYPES, CIRCLES, AND AFFINITIES

William Swainson was the pioneer in England of hand-coloured litho-graphed illustrations in natural history. He was also a "declinist" in the early 1830s, believing that science was far better done in France than at home; and he was the author of a series of zoological works in Lardner's *Cabinet Cyclope-dia* in which he developed the Quinary System of taxonomy invented by William MacLeay. His correspondence is preserved at the Linnean Society, and shows him to have been a contentious person who fell out in the end with most of those to whom he wrote; and the annotated pattern-plates for his *Zoological Illustrations*, done to assist the colourists, are also there.[1] The correspondence illustrates various networks, and patronage; and shows how far natural history was a trade in the 1820s and 1830s. But we shall be concerned with Swainson's intellectual history, and in particular with his taxonomic system: trying to see what it was that made what most have seen as an eccentric arrangement seem to him a satisfactory arrangement of the order and variety we see in the world.

Swainson was born in Liverpool in 1789; one of his *Cabinet Cyclopedia* volumes contains, after a section on stuffed animals, biographies of zoologists, Swainson's own entry being the longest.[2] His father worked in the Customs Office, and Swainson was to follow him there; but he had developed a passion for natural history and travel, and joined the Commissary of the Army, being sent in 1807 to Malta. There he met Rafinesque-Schmaltz, and made some collections of Mediterranean fauna; and after the Napoleonic Wars ended he chose to go on half-pay. He had always believed that his family had gone down in the world, having once been landed; and in his military life, where he had risen to a high position, he was touchy and awkward: "bred up with somewhat of aristocratic notions, and accustomed, when on service, to *command* rather than to *obey*, I had a rooted dislike of all commercial affairs, and would rather have

1. On the Swainson Correspondence, see A. Günther, "The President's Anniversary Address", *Proc. Linn. Soc.*, 1899–1900, 112, pp. 14–61. On his illustrations, see my Ramsbottom Lecture of 30 March 1984, forthcoming in *Archives of Natural History*, and on the general background, my *Ordering the World*, London, 1981.

2. W. Swainson, *Taxidermy, with the Biography of Zoologists*, London, 1840; the engraved frontispiece is a portrait of Swainson, whose autobiography occupies pp. 338–52 and includes bibliography; quotation from p. 347.

J.D. North and J.J. Roche (eds.), The Light of Nature. ISBN 90-247-3165-8.
© *1985, Martinus Nijhoff Publishers, Dordrecht. Printed in the Netherlands.*
Reproduced by kind permission of Kluwer Academic Publishers.

gone once more on active duty than have sat behind a desk''. He also, like Priestley, suffered from a speech-impediment: it may be that this had some connection with his controversial style in print and in correspondence. Certainly, he was a man who found it hard to keep a friend; his correspondence is full of rows.

In 1816, he set out for Brazil, where he made some collections and on his return was elected FRS. He got to know William Leach at the British Museum, who encouraged him to use lithography rather than engraving for zoological illustrations, perceiving him to be an excellent draughtsman; and in 1822 Swainson hoped to become Leach's successor. In the event, the post went to J.G. Children, and Swainson, henceforth a disappointed man, found that he inherited very little from his father and had to support himself by his pen and his pencil. After working on an encyclopedia for Longman, he contracted in 1833 to produce fourteen volumes of three hundred pages each, illustrated with wood-engravings and with engraved title-pages, to be produced at the rate of one every three months: in 1834 he duly received £200 for the first of them, the *Preliminary Discourse*.[3] Lardner was known as the Tyrant, and he expected books at three-month intervals; even if they consisted of material from the aborted encyclopedia, this was unrealistic, and Swainson soon got hopelessly behind. In 1840, having failed again to get a post at the British Museum, he emigrated to New Zealand, where he died in 1855.

In his pioneering *Zoological Illustrations*, which began to come out in 1820,[4] Swainson in prefaces to the various volumes put down two of his preoccupations. In the first volume, he referred to his countrymen's "prejudiced adherence to the strict Linnean system, [which] has been the primary cause why Zoology has been more neglected with us, than on the Continent". There is a footnote of praise for W.S. MacLeay's *Horae Entomologicae* of 1819–21: "a work which for acutness of reasoning and profound research, has never been equalled either in this, or perhaps in any other country". Despite this, the next volume was dedicated to the purchaser of the Linnean Collections and doughty defender of his system, J.E. Smith, the persecutor of S.F. Gray whose *British Plants* of 1821 was organised on the natural method of Jussieu. But in volume three we find it remarked that "Paris has become the zoological university of Europe; and that the principles which have emanated from it, are now con--sidered the only true ones by which nature is to be studied".

3. Swainson Correspondence, under "Lardner" and "Longman". See J.N. Hays, "The Rise and Fall of Dionysius Lardner", *Annals of Science*, 1981, 38, pp. 527–42. On Davy and Children, see H. Hartley, *Humphry Davy*, London, 1966, pp. 88, 99; and P.B. Miller, "Between hostile camps", *BJHS*, 1983, 16, pp. 1–47, esp. 42ff.
4. W. Swainson, *Zoological Illustrations*, 3 vols., London, 1820–3; i, vi, and iii, x. On classifying invertebrates, see M. Winsor, *Starfish, Jellyfish and the Order of Life*, New Haven, 1976, dealing with Cuvier's Radiata and mentioning the Quinarians. There is a brief account of Swainson, by N.F. McMillan, in *DSB*, but the fuller biography in *DNB* remains useful.

Admiration for the French, and for MacLeay, and a certain concern for patronage coupled with a certain tendency to bite the feeding hand, were to characterise Swainson all through his active period in the 1820s and 1830s. MacLeay[5] was the son of an entomologist who was Secretary of the Linnean Society; and he was a francophile who in 1815 had gone to France as a member of a commission appointed to clear up war claims, and had met Cuvier, Latreille and Geoffroy St Hilaire. Later he persuaded Darwin to undertake the editing of the *Zoology* of the *Beagle* voyage, and in Australia (where he had gone in 1839 after being appointed President of his section of the British Association) he encouraged the young T.H. Huxley during the voyage of *HMS Rattlesnake.* Darwin's work on barnacles was partly directed against MacLeay's *Horeae Entomologicae*, which was already a very rare work.

For MacLeay, natural groups properly discerned revealed neither the bushy system of Cuvier, nor the dichotomous system of Fleming, but rather an arrangement of circles. Extremes met. Moreover, at every level one might hope to see the pattern repeated, as the various circles took their places in fives. Swainson was commissioned to write up and illustrate the birds for the volumes *Fauna Boreali-Americana*[6] which John Richardson was editing, based upon the collections made on the various expecitions to the Canadian Arctic in the 1820s; notably the overland journeys to the Arctic Ocean of John Franklin and Richardson himself. This book was subsidised by the government to the tune of £1,000 to pay for the illustrations, which were hand-coloured lithographs; it was thus an official publication of an official expedition. In his "introductory observations on the natural system", Swainson again praised MacLeay for promoting the "tendency to raise Zoology to the rank of a demonstrative science". The groups in his system were definite; arrangement was no longer a matter of choice or convenience. Moreover, the scheme all hung together; it was up to opponents to propose "some other way, more calculated to show the harmonious combinations of nature". He noted that the circles might be incomplete, either because of extinction or through the non-discovery of certain types.

Also in 1831 Swanson published in Loudon's *Magazine of Natural History* a defence of "certain French naturalists".[7] This is a document in the "decline of science" debate, but it also marked a furious row between Swainson and

5. On MacLeay, see *DNB*; he is omitted from *DSB*.

6. W. Swainson and J. Richardson, *Fauna Boreali-Americana: the Birds*, London, 1831, p. xliv. This was the first book to be thus subsidised.

7. *Magazine of Natural History*, 4, 1831, pp. 97–108; controversy continued on pp. 199–206, 206–7, 316, 319–37, 455ff., 554, 481–6, 487f., 559f.; then in 1832, vol. 5, pp. [109]f., [208]ff.; the bracketed pages, and some of the others, were appendices published at the expense of the authors. In view of Swainson's relations with Vigors and the Zoological Society, there is an irony in the fact that the collections used in *Fauna Boreali-Americana* were divided between that Society and the University of Edinburgh.

86

N.A. Vigors, Secretary of the Zoological Society and until then a supporter of Swainson and of the system of MacLeay as applied in ornithology. In many fields, nineteenth-century science seems characterised by leading men not being on speaking terms, but the apostles of a new paradigm should stand united. Swainson's tone was condescending: anyone declaring that his contemporaries are merely pigmies or in a decline faces an argument *ad hominem*, and one cannot help reflecting how well Swainson's doctrines stand up to his own criteria.

The *casus belli* was an international argument over the naming of the Malay Tapir – not a very important affair. Swainson produced four evidences of decline: denial of the "greatest and most acknowledged truths by bold and specious reasoners"; "the zealous adoption by some, and the unqualified rejection by others, of theories or systems *which neither party understood*"; "substitution of flowery and sententious oratory for the result of deep and patient research"; and "a spirit of dissention and of invective, against all who thought differently from ourselves". Swainson's forte was not the seeing of beams in his own eyes. The real occasion for the assault upon Vigors was that Swainson had not been given the freedom to use the Zoological Society's collections when writing up the birds for the *Fauna Boreali-Americana*, having resigned from the society pleading poverty, though he was able to subscribe to Gould's *Birds of the Himalayas* (1831) and Lear's *Parrots* (1832). He had therefore gone off to Paris in dudgeon, and had been welcomed at the Museum – the centre of the "zoological university of Europe" – and had been given a room in which to develop his heterodoxy. A visit to France had given him a vision of how things might be done in England.

What is curious is that despite his spirited defence of the French, he was by no means a convert to French views. He saw the work of Cuvier and of Latreille as fragmentary,[8] while MacLeay's work offered the possibility of seeing nature as a whole. It was not the placing of a bird or mollusc which really interested Swainson, but the links, relationships and general principles. This brings us to the other strand which seemed to compose the rope with which Swainson could tie up zoology or else hang himself: belief in God as the guarantor of order. As he wrote in the *Fauna Boreali-Americana*: "Let us . . . adore that God, who, to increase our faith in his word, enables us to discern, however dimly, in *earthly* things, the shadows of such as are heavenly". This was written in the context of a description of the threefold divisions in nature, such as the animal, vegetable or mineral distinction. Faced then with the prospect of writing a series of semi-popular works, Swainson was in much the same position as Mendeleev later: he had to devise some way of classifying which would make his masses of material intelligible. It might also help by, as it was later put, 'investing with a

8. W. Swainson, *Preliminary Discourse*, London, 1834, p. 197; *Fauna Boreali-Americana*, London, 1831, p. lvi; A. Günther, *op. cit.*, note 1, p. 23.

cloak of originality his treatises on those classes of animals with which he was not well-acquainted".

The first volume in the series was a *Preliminary Discourse*, for which the model was John Herschel's *Preliminary Discourse*.[9] This has become a classic work, and was so regarded in its own time: it describes a scientific method appropriate to what we would call physics, with more emphasis upon induction than one might expect, with Wells' *Essay on Dew* being given some prominence. The book hangs together as an extended essay, and its teachings became accepted as thoroughly sound. Swainson's *Preliminary Discourse* was rather different. The first part is an exposition of the quinary system. What is odd is that the conclusions were presented here, while the more detailed evidence was to be set out in the other volumes: one is reminded of Darwin's evolutionary writings, where the *Origin of Species*, an "abstract" without notes or bibliography, preceded the treatises on variation under domestication, on orchids, and on sexual selection. The *Preliminary Discourse* then changes tack, and the second half is a contribution to the "decline" debate. Instead of ever being seen as sound and reputable, the book seemed eccentric and propagandizing. Swainson himself wrote later[10]:

> I verily believe, that, had I expressed my convictions in a more subdued tone, many of those who now differ from me would have adopted these views, – at least in a general way; but I am always so delighted with detecting a new link of relation, or in bringing an isolated fact to bear upon general principles, that my enthusiasm sometimes overcomes my judgement. I forget, in fact, that no one, unacquainted with the other instances of a similar nature, – all converging to the same point, – can possibly attach the same importance to a *single* instance, that I do myself.

The reader may be permitted to smile at the "at least in a general way", but may also feel some respect for the speculative and synthetic imagination Swainson displays.

The first surprise about the book is that opposite the title page we find a quotation from William Jones of Nayland. In passing, one might add that this and other volumes on natural history, including J.S. Henslow's on botany, have three title pages, one engraved and two typographic; one is headed *The Cabinet Cyclopedia*, and the other *The Cabinet of Natural History*. In all volumes that I have seen, the former is at the front and the latter at the back. This is a curiosity, but Jones is more than that. He formed an important link between the nonjurors of 1689 and the Tractarians, and was highly regarded in high church cir-

---

9. J.F.W. Herschel, *Preliminary Discourse*, 1830/2; there is doubt about how to cite this because the engraved and typographic title pages differ; reprint, intr. M. Partridge, New York, 1966. On Herschel's importance, see S.F. Cannon, *Science in Culture*, New York, 1978.

10. W. Swainson, *op. cit.*, note 2, p. 349.

cles in his lifetime (1726–1800). He was also a follower of John Hutchinson, author of *Moses' Principia*, 1724–7, who saw nature as a network of symbols and opposed Newtonian theory on the grounds that a void was impossible.[11]

To put Jones of Nayland where Herschel had put Bacon was thus to make a statement about one's religious and scientific allegiance: while generally gentlemen, Jones' followers were not "gentlemen of science". The quotation, from 1784, goes:[12]

> The world cannot show us a more exalted character than that of a truly religious philosopher, who delights to turn all things to the glory of God; who, in the objects of his sight, derives improvement to his mind; and, in the glass of things temporal, sees the image of things eternal.

A follower of Jones would be expected to emphasise symbolic and analogical relations, even to see how the crysallises of stinging caterpillars point head downwards towards Hell; a notion of Swainson's which A.R. Wallace mildly characterised as "rather fanciful", correcting this as "going too far beyond common sense".[13] For Swainson, a satisfactory explanation or taxonomic scheme had to include a symbolic dimension and take the whole scheme of nature into account.

Such an outlook was perhaps uncommon among leading men of science in the 1830s, but it was shared by William Kirby, the leading British entomologist of the day and the author of the *Bridgewater Treatise* on the History, Habits, and Instincts of Animals, 1835.[14] We should not allow ourselves to see liberal-minded divines from Cambridge making all the running in natural theology of significance at this time. Kirby saw close affinities between the views of Henry More and those of Humphry Davy on the relations of matter and spirit; but while filled with interest in the interpretation of the more obscure passages of the Scriptures, Kirby never embraced the system of MacLeay and Swainson.

On the engraved title page of his *Preliminary Discourse*, where Herschel had

11. On Jones see *DNB*; on Hutchinsonians, C.B. Wilde, "Matter and Spirit as Natural Symbols in eighteenth-century British natural philosophy", *BJHS*, 1982, 15, pp. 99–131. Jones' writings are among those recommended in the interesting reading lists which accompanied various of the *Tracts for the Times*, Oxford, 1833–40.

12. J. Morrell and A. Thackray, *Gentlemen of Science*, Oxford, 1981; W. Swainson, *Preliminary Discourse*, London, 1834, facing engraved title page.

13. W. Swainson, *The Geography and Classification of Animals*, London, 1835; Wallace's annotated copy, at the Linnean Society, p. 248. He appears to have read the book twice, dating it 1st September 1842, annotating it in pencil and then amending his notes in ink.

14. W. Kirby, *On the Power, Wisdom, and Goodness of God, as manifested in the Creation of Animals*, new ed., ed. T.R. Jones, London, 1853, ii, pp. 185f.; Swainson corresponded with Kirby, and quoted from Kirby and W. Spence, *Introduction to Entomology*, London, 1815–26 (a classic work) in *op. cit.*, note 12, pp. 115f.

put a bust of Bacon, Swainson had one of Aristotle, whom he considered to be the true discoverer of the law of correlation attributed to Cuvier: Aristotle was there in spite of a letter from Lardner advising no bust, and especially not one of Aristotle or Ray − Swainson did not mind standing up to his publisher.[15] He believed that the systems of Cuvier and Latreille were not "built upon philosophic principles" because they exhibited nature not as a whole but in pieces. One of that "harmony of plan" which must "necessarily form part of the system of nature" was exhibited by them, in that they used different characters in the categorizing of different groups. This kind of opportunism could not bring out the natural system, and while undoubtedly "even their errors are the errors of genius" natural history is progressive and must move on from them.

For Swainson,[16] Galileo's dictum must be modified: "the book of nature is a book of symbols". Its key was the idea of representation, a system of "general analogy", in which one object or group represents another at the same or a different level. This Swainson considered a "truth universally admitted"; what lay behind it was "one universal and consistent plan". What was needed was first the dismissal of prejudices, for most taxonomists had been much too narrow in their views: there was but one "natural mode", "the true series or chain of being" in which every organism held an unambiguous place. This place could generally be determined from external characters: only when external form failed to provide a clear answer should one turn to the insides of animals. This meant that whereas MacLeay had dismissed conchology, writing about shells, as beneath science, Swainson's *Malacology* was in fact based on shells rather than on their inhabitants, and it is shells rather than molluscs which appear in his *Zoological Illustrations.*

Emphasis on externals had been a feature of the systems of Linnaeus and of Werner for the same reason as Swainson gave[17]: so that students should not be "embarrassed . . . with unnecessary difficulties". But in a natural system one might expect a follower of Aristotle to take all characters into account, rather than just those most visible, even if he is writing for "the generality of mankind" who cannot follow all Cuvier's ideas about organisation. "It would be absurd", wrote Swainson, "to suppose that the internal construction of an animal is not deserving of great attention. This study, in fact, constitutes, of itself, a distinct branch of physical science; useful, indeed, to the zoologist, as the means of assisting and guiding his studies, but by no means so essential as is

15. Swainson Correspondence, under "Lardner", 12th September 1833; W. Swainson, *op. cit.*, note 12, pp. 197−9, 377.

16. *Ibid.*, pp. 116, 114, 115, 152ff., 156, 171. W. Swainson, *A Treatise on Malacology; or the Natural Classification of Shells and Shellfish*, London, 1840, pp. 8, 11.

17. W. Swainson, *op. cit.*, note 12, p. 167, 170, 171; *The Natural History and Classification of Quadrupeds*, new ed., London, 1845, pp. 143ff.

90

generally supposed. Whenever external peculiarities are sufficient to supply us with clear definitions, we require no other." This can be, and no doubt was in part, an excuse for ignorance; certainly it was going against the tide in nineteenth-century zoology. It led Swainson to place the "Tasmanian Wolf" with the wolves, breaking up the marsupials of Cuvier; but to one who saw harmony and design everywhere, the idea that the inside and outside of an animal might tell a different story would be unacceptable.

Swainson's key terms were *affinity* and *analogy*. Affinity is really opposed to "kindred" and does not mean family resemblance, but a link through marriage – the union of opposites. It is indeed in this sense that it was used by chemists, and provided the metaphor for Goethe's novel, *Elective Affinities*; but naturalists throughout the nineteenth century used it as though it meant family resemblance.[18] For Swainson, the affinities arranged themselves in circles; they were the "links of a chain", and their degrees were few and narrow. In each group, there would be found to be three circles; this was not a piece of *a priori* reasoning from the orthodox Trinitarianism of Jones, but the result of sophisticated induction as recommended by Herschel! The first circle in any grouping is called the Typical one; the next the Subtypical; and the third, arranged below the others, is called the Aberrant. This one can itself be divided into three small ones; thus five and three are the magic numbers in the system.

As Wallace again remarked[19] in his copy of Swainson's *Geography and Classification of Animals*,

But does not the assumption of any particular number in the first instance lead to a forming in a manner of groups, and placing in a manner otherwise than, without such a theory, would appear to be natural. There appears to be not the slightest reason for believing a priori that all groups of animals are divided into the same number of types of form or divisions though the proofs brought forward in subsequent parts of this volume are certainly most extraordinary.

And later he added:

among those naturalists who are much opposed to Mr Swainson's views – the primary divisions of the animal kingdom are acknowledged to be 5 – . . . and the remaining divisions are still so much disputed & all at present open to so much objection that Mr. S.'s may be as good as any.

Swainson did think that all groups were divided into the same number of types, because of his beliefs about God and the Creation being harmonious; Wallace

18. T.H. Huxley, in R. Owen, *Life of Richard Owen*, London, 1894, ii, p. 302 note. W. Swainson, *op. cit.*, note 12, pp. 186, 205ff.
19. W. Swainson, *op. cit.*, note 13, pp. 222f. and endpapers; p. 5.

in contrast wrote mockingly of Prichard (quoted by Swainson) "To what ridiculous theories will men of science be led by attempting to reconcile science to Scripture!"

Swainson was not doing any ordinary reconciliation, but was using his beliefs to support the search for analogies and representation throughout nature. His contemporaries looked for representative species in different regions, and wrote of the analogy of nature; but Swainson took these notions further than they did. He saw a web of analogies linking all creatures together[20]:

> Nature seems to delight in showing us glimpses of that beautiful and consistent plan upon which she has worked, by giving us a few instances of symbolical or analogical representations, so striking and unanswerable in themselves, that they are perceived and acknowledged by all. What for instance could be more perfect than the analogy between the Bengal tiger and the African zebra?

Both animals, for those slower than Swainson to perceive perfect analogies, are striped and cannot be tamed; an analogy is the "type or *emblem* of other animals with which it has no positive connection, or consanguity". Particular circles were determined by affinities, but parallels between circles were matters of analogy, which could determine for example the number and even the nature of the animals which should form a new circle.

This was particularly important for placing man. Linnaeus had put the orang-utan in our genus, as *Homo sylvestris*, but speculative evolutionists had since then made man's place in nature controversial, along with the date of his first appearance on earth. In a draft classification of the mammalia of about 1829, now at the Linnean Society, MacLeay[21] put man among the primates and organised the races of mankind into five circles: two civilised, the Caucasian and Mongolian, and three savage or aberrant, the American, Negro and the Malay/Polynesian. With references to Blumenbach and Lawrence, he wrote that there was but one genus and one species of man, with these constant varieties. Elsewhere he wrote that "the ourang outang & the chimpanzee most certainly approximate to man but a number of species are wanting to complete the chain of affinity which is probably linked (?) by animals of a rather lower grade than man". It is curious to find references to the Great Chain of Being common among devotees of circular arrangements.

Swainson was altogether more circumspect about mankind.[22] He thought he

20. W. Swainson, *op. cit.*, note 12, p. 184, 187.

21. W.S. MacLeay, packets at the Linnean Society with title "Draft classification of mammalia on quinary principles".

22. W. Swainson, *Quadrupeds, op. cit.*, n. 17, pp. 74, 75ff. Swainson's system was popularised in [R. Chambers], *Vestiges of the Natural History of Creation*, London, 1844, pp. 236ff.; but his arguments for the separation of man were scouted in this evolutionary text.

92

had enough analogies to fix the number of animals in a circle, and that the circles of the apes and monkeys were full. It was right that there should be no room for man among them, for man was not an animal but a spiritual being. He saw continuities in nature, such as from the fish to the frog to the tortoise to the penguin and then to quadrupeds!, but saw a break before mankind was reached. Man was the lord and governor only of the animal creation, not part and parcel of it: a view now confirmed not merely "from an enlarged view of his moral and physical qualities" but also now from a purely scientific analysis, relating only to his form. There is no place in the circles for a sole representative of a mammiferous order: we need no longer therefore worry about the "degrading and humiliating" theory that we are animals.

In grouping animals therefore we have numbers as well as general analogies to guide us.[23] As we work upwards, "bearing in mind, that the greater the degree of harmony and unity we can produce in our arrangement, the greater is the probability of our discovering the order of nature", so we "try the strength of the law thence assumed, upon a more extensive scale". Swainson was trying to be the Laurent of zoology, bringing in a sophisticated inductive procedure rather than straightforward fact-collecting and generalising. He duly found that all levels of his system paralleled each other. Even the main divisions of the animal and vegetable creations could be made parallel, but it was within zoology that he really tried to make the system work.

Thus he sets out tables like Table 1[24]:

Table 1.

Orders of quadrupeds	Typical characters	Orders of birds
I *Typical Group* QUADRUMANA	Pre-eminently organised for grasping	INSESSORES
II *Sub-typical Group* FERAE	Claws retractile, carnivorous	RAPTORES
III *Aberrant Group*		
CETACEA	Pre-eminently aquatic, feet very short	NATATORES
GLIRES [RODENTS]	Muzzle lengthened and pointed	GRALLATORES
UNGULATA	Crests or other processes upon the head	RASORES

These are at the same level, but we can compare the circles of the old-world monkeys with the quadrupeds (Table 2):

23. W. Swainson, *op. cit.*, note 12, pp. 220f.; *op. cit.*, note 22, p. 42.
24. W. Swainson, *op. cit.*, note 22, pp. 54, 72.

Table 2.

	Typical analogical characters	
SIMEA	Grave, intelligent, inoffensive: typical of the	QUADRUMANA
CERCOPITHECUS	Mischevous, malicious	FERAE
PAPIO	Head very large, little or no tail	CETACEA
MACUCUS	Tail comparatively long, hare-lipped	GLIRES
INUUS	Head conspicuously crested	UNGULATA

This becomes a game any number can play, and makes treatises on zoology more fun than one might have expected. Thus we are informed[25] that small eyes tend to go with long noses, in humans as in elephants, pigs and humming birds; that rhinoceroses, elephants, oxen, deer, hornbills, crested birds, and beetles all in a sense have horns; and that in the animal world we may find types of evil and of good (which would hardly be news to the reader of Aesop or of the Bible). Like Mendeleev, Swainson left gaps for undiscovered creatures, and filled others with extinct ones.

Swainson was a good enough zoologist to recognise differences between ornamental and useful tails (the horse and peacock, and the cebid monkey and woodpecker) and more seriously to recognise gradations through a class;[26] for instance, in parrots' feet from ground to tree-dwelling species. His illustrations of details of birds and bird-families were indeed used by Alfred Newton half a century later in his *Dictionary of Birds*, 1893–7. But it is the analogies that leave one rather breathless, along with the dismissal of the "verbose technicality and minute investigation" involved in the use of internal characters, and even of metamorphosis among invertebrates; though here we should note that Swainson had high praise for J.V. Thompson's work on the development of the crab.

The use of analogy was deliberate. Davy had spoken of the importance of analogy in science,[27] which in his own work led to his recognising iodine as an element analogous to chlorine; and Swainson quoted from Dugald Stewart in support of the power of analogical reasoning. Swainson believed that "every thing in this world is evidently intended to be the means of moral and intellectual improvement, to a creature made capable of perceiving in it this use. This perfect analogy between the moral and the natural world, no Christian in these days will even think of questioning, much less of disputing". This was of course a field in which Bishop Butler had made everyone aware of the value of anal-

---

25. W. Swainson, *op. cit.*, note 12, pp. 252f., 261.

26. *Ibid.*, p. 256, 269, 247, 276, 345, 360 note, 437.

27. H. Davy, *Collected Works*, London, 1839–40, viii, pp. 167, 179, 285f., 317; see my *Transcendental Part of Chemistry*, Folkestone, 1978, chap. 5. W. Swainson, *op. cit.*, note 12, pp. 282ff.

94

ogy, but Swainson refers to Hampden rather than directly to Butler. Swainson believed that

> Analogy, or *symbolical representation*, is, therefore, the most universal law of nature, because it embraces and extends its influence over the natural, the moral, and the spiritual world; a property which no other law, yet discovered, is known to possess. Hence, we may infer that, in its more restricted application to natural history, it is equally paramount; and that to this science, it is what the law of gravitation is known to be to astronomy.

Swainson thus saw himself not as the Laurent but as the Newton of natural history; analogy was not merely a device of method but the key to nature.

Had Swainson, like Newton, been justified in supposing that the numerical law which he had discovered was a reliable guide, then he could have used the sort of analogies we find in his books. Newton's laws worked; and this pragmatic test is after all what in the last resort we have to apply, for different people appreciate different harmonies. Swainson's was a static and balanced world; his system did give him a key to zoology which allowed him to write a whole series of books which hung together, many of them about creatures of which he knew little. To those who did not fall in with the paradigm of symbols, types and harmonies, the basis of the whole system would never seem convincing, and the analogies look singly, and even *en masse*, like rather amateurish trifling; Swainson's friend P.M. Roget[28] in his *Bridgewater Treatise* was prepared to write about analogy, unity of design, and a "definite *type* or ideal standard", but to recognise something hypothetical about Cuvier's unity of composition, and to make little use of analogies. Swainson rated philosophical naturalists far above illustrators, and before simply saying that he should have known his place (as Vigors did), we might reflect that there are ways in which he was a mirror of his age, and that audacity merits some respect.

28. P.M. Roget, *On Animal and Vegetable Physiology*, London, 1834, ii, pp. 625ff. Swainson, *op. cit.*, note 2, gives one page to Audubon as an "Animal Painter" and eight to Wilson as "The Ornithologist of America". For Vigors, see note 7, esp. pp. 320ff.

XVI

William Swainson: naturalist, author and illustrator

Anyone who proclaims the decline of science exposes himself to the charge of failing to advance it himself; and it is an irony that three notable 'declinists' of the 1830s[1] failed to give good examples of 'Modernity'. David Brewster never came to terms with the wave theory of light; J. F. W. Johnston of Durham never opened his copies of the *Philosophical Transactions*; and William Swainson became notorious for adhering to the 'Quinary System' of classification. The message of decline had been proclaimed by Charles Babbage (1830) and a year later Swainson (1831) took it up in Loudon's *Magazine of Natural History*, with a paper in defence of 'certain French naturalists'.

It was a standard complaint among declinists that things were ordered better in France. Swainson was born in 1789, and was thus too young to have experienced that blissful dawn followed by horror as the Revolution failed to usher in a glorious day. For him and his contemporaries, Paris was the scientific capital of the world; and in all the sciences, the 'paradigms' were French. Men of science in France were not regarded as following an eccentric hobby, but were honoured and respected— even enriched. Babbage as a mathematician had emphasized the lowly and provincial state of natural philosophy in Britain, and Swainson resolved to expose the poor state of natural history.

The Malay Tapir, belonging to a genus hitherto only known in the New World, had been described by Farquhar (1816) in the *Researches* of the Asiatick Society of Bengal and illustrated by Thomas Horsfield (1824)[2]. A specimen was in the menagerie of the Governor-General of India at Barrackpore, where it was seen by French visitors; and Desmarest, Lesson and Frederic Cuvier had implied that it was a French discovery. For this they were rebuked by Bennett (1830) in his little book describing the newly founded Zoological Gardens in Regent's Park. Swainson did not care to see these great men ill-treated by pigmies of his own nation and he leaped to their defence. His essay turned out to be particularly an assault on his former friend, N. A. Vigors, who was accused of publishing a private letter from MacLeay; Vigors, Swainson and Macleay being the upholders of the Quinary System, though sometimes disagreeing over details of its application.

Swainson (1831) wrote that science could be seen to be in decline if one found a denial of the 'greatest and most acknowledged truths by bold and specious reasoners'; 'the zealous adoption by some, and the unqualified rejection by others, of theories or systems *which neither party understood*'; and the 'substitution of flowery and

Reproduced by permission of the Society for the History of Natural History; first published in the *Archives of Natural History* (1986) 13: 275–290

210

sententious oratory for the result of deep and patient research'. He added to these 'a spirit of dissention and of invective, against all who thought differently from ourselves'. It was unfortunate for Swainson that all these charges could readily be turned against him.

Vigors (1831) replied in the devastating way open to those of higher social status[3]: 'What, Sir! do we live in an era and in a country which will tolerate such an outrage upon all honourable feelings and principles?... In the circle of society in which Mr Swainson appears to revolve, such insinuations may perhaps be little regarded; their shafts may most probably fall blunted from the coarse and callous feelings of the individuals of his caste: but among gentleman and men of honour the case is different'. Swainson had declared himself too poor to pay his subscription to the Zoological Society; and yet as Vigors remarked he was 'a person, deriving, according to public rumour, considerable emoluments from zoology'. Certainly, he was a subscriber to Gould's *Birds of the Himalayan Mountains* (1830–33)[4] and to Lear's *Parrots* (1832).

Vigors's conclusion was 'on the head of the aggressor let the odium rest', but two correspondents wrote to Loudon expressing their distaste for controversy; which rumbled on, but later mostly in supplementary pages paid for by the antagonists—a device which modern editors might note! We may agree with Loudon's correspondents that the matter was unedifying, and that neither party emerges with enhanced reputation: what is striking is how it illustrates that parting of friends which seems so characteristic of nineteenth-century science, an epoch of monumental rows and of scientific parricide. Swainson seems to have been particularly unable to keep a friend; his correspondence[5] shows a series of fallings-out and attempts, sometimes successful, to clear the air. By 1831 Swainson was already a disappointed man: it is an irony that a lecture in memory of John Ramsbottom, whose fruitful life was spent at the British Museum (Natural History), should be devoted to a man whose unattained ambition was a post at the British Museum.

At the back of his book on Taxidermy,[6] Swainson presents as it were a cabinet of naturalists; that is a series of biographies of eminent naturalists, dead or alive. Modesty was never Swainson's problem and it should not surprise us that he gives himself much the longest entry, including a bibliography. He was born in Liverpool, where his father worked in the Custom House, and where he acquired an interest in natural history and in travel. He started in his father's footsteps, but soon got himself transferred to the Commissary Department of the Army and was sent to Malta and then to Sicily. There he met Rafinesque, with whom he later corresponded, and who increased his enthusiasm for natural history. At the end of the Napoleonic Wars, he left the army on half pay, by now accustomed to command rather than obey, and got himself on an expedition to Brazil. Here, he ran into war and had to make collections only on the coast; but on his return, with the effrontery needed in a society based upon patronage, he introduced himself to Joseph Banks, William Hooker and William Leach of the British Museum.

Leach became his friend and, perceiving that Swainson was an excellent draughtsman, persuaded him to take up the new technique of lithography. In 1820 he began to bring out his *Zoological Illustrations*, of birds, insects and shells, in monthly parts; and in 1821 came his *Exotic Conchology*. In 1821 Leach, for whom Swainson

211

in the *Taxidermy* book wrote a pleasant anecdotal account, had to retire and Swainson applied for the job, collecting testimonials from eminent naturalists. In the event, it went in 1822 to J. G. Children, a friend of Humphry Davy and a mineralogist rather than zoologist; but somebody who was the social equal of the Trustees, as Gunther (1980) has pointed out, and whose career as a zoologist was if undistinguished certainly not disgraceful. It seems to have been Davy, as President of the Royal Society, who got Children appointed, but Swainson blamed the Archbishop of Canterbury for what he saw as injustice,[7] and he plied his Liverpool friend Thomas Traill with material for an assault upon the British Museum which discredited the national institution.

When in 1826 Swainson's father died, he found that he did not inherit as much as he had hoped, and he therefore decided to take to professional authorship. As well as his illustrated works, he had begun at the end of 1827 on a project of an encyclopedia of zoology for Longman's, for which Lindley was to provide the botanical part. Swainson intended to organize his contributions using the Quinary System, but Lindley would not admit it into the botanical articles, adding however that he saw 'no necessity or propriety in combating a system which as far as Botany is concerned can scarcely be said to have an existence'[8]. In the event this project was absorbed into the *Cabinet Cyclopedia* of Dionysius Lardner, known to contributors as 'the Tyrant' and later to become notorious when he ran off with an army-officer's wife (Hays, 1981).

Lardner's volumes, published by Longman, were one of the fruits of the new printing technology, cheaper paper and cased bindings which brought the price of books down so sharply about 1830; by which time there was also a large class of readers. They were not organized as an encyclopedia, but more like a library of books of uniform size on all topics. The most famous volume was the *Preliminary Discourse* on natural philosophy by John Herschel,[9] the declinists' candidate for the Presidency of the Royal Society in 1830 when he was narrowly defeated by the Duke of Sussex. Herschel's volume is one of the classics of scientific method of the nineteenth century, emphasizing both Baconian inductive procedure and Newtonian mathematical techniques as extended by the French. Darwin admired the book, but it has little to say to those working in natural history. Like most philosophy of science, it is based on the assumption that real science means physics. Other branches of science should be pruned and nourished so as to grow into something looking like physics.

In 1833 Lardner concluded arrangements with Swainson[10] to write fourteen volumes of 300 pages each on zoology, beginning with a *Preliminary Discourse* on natural history. The natural history volumes of the Cabinet Cyclopedia have two typographic title pages: they also have an engraved one, as do all the Cyclopedia volumes. One title page has *The Cabinet Cyclopedia* at the head; the other (which I have never seen anywhere except at the back of the book) has *The Cabinet of Natural History*. Darwin's patron, J. S. Henslow (1836), undertook the volume on botany; Swainson was in good company. Swainson provided half-page outlines of each volume and a prospectus, and he undertook to produce them at the rate of one every three months. Lardner was not called the Tyrant for nothing, but to expect this rate of writing, at a time when Swainson was also producing fine illustrated works, seems unreasonable and Swainson duly fell hopelessly into arrears and in the end fell out with Lardner.

212

212

278   WILLIAM SWAINSON: NATURALIST, AUTHOR AND ILLUSTRATOR

Despite Herschel's book having Bacon's bust on its engraved title page, Lardner wrote to Swainson that he was doubtful about busts, preferring vignettes, and that anyway Aristotle or Ray would not do. One's admiration for Swainson is therefore increased when one finds a bust of Aristotle there on the title page; presumably for Swainson this made a visual point about his book and Herschel's. In the summer of 1834 Swainson duly received £200 for this book. It was followed, though not as fast as Lardner would have liked, by a string of others. In setting this timetable and expecting this vast scope, Lardner and Swainson must have hoped that the materials collected for the abortive encyclopedia could be readily reworked into a series of treatises.

The whole set was an exercise in the Quinarian System. Swainson (1840a: 348) wrote that he could have just followed Cuvier, but (despite his defence of the French in 1831) had decided not to. In his Presidential Address to the Linnean Society, describing the recently-acquired Swainson Correspondence, Albert Günther (1900: 23) remarked that 'we cannot overlook the fact that the quinarian method[11] served him well, investing with a cloak of originality his treatises on those classes of animals with which he was not well-acquainted', adding that his fishes were 'disastrous'.

Swainson's *Preliminary Discourse* (1834) is a curious mixture, unlike Herschel's coherent volume. Much of it is concerned with arrangement, with the interpretation of the book of nature which for Swainson is a book of symbols: as in other contemporary discussions, types and representatives appear prominently even if one could not say that Swainson himself represents the zoologists of his era. The second half is concerned with the decline of science. It is very interesting on the formality of meetings of many societies, where discussion was not permitted; it calls for more official patronage for science, making unfavourable comparisons with foreign countries; remarks on the need to subsidize zoological literature because illustration is necessary and expensive; points out how unfair the copyright laws, requiring deposit of copies in great libraries, are to those producing hand-coloured works in very small editions; and asks for more public honours for men of science. Swainson himself had been the recipient of patronage and subsidy when he had done (again, behind the deadline) the plates and descriptions of birds for John Richardson's *Fauna Boreali-Americana* (1831), for which the government had set a precedent of providing £1,000 as a grant towards publication; but he never became Sir William.

Swainson believed that zoology was at a stage where its general laws were only just emerging (1834: 376). Everywhere in nature he saw networks of relationships, of affinities and analogies. The attraction of the quinary system was that on every level the groups paralleled each other, and that the analogies became a real expression of harmonies rather than of accidental resemblances (1834: 182). Like most zoologists of the day, but unlike the more accurate chemists, Swainson used the term affinity as though it were synonymous with kindred; really it implies relation by marriage, and thus (as in chemistry and in Goethe's novel *Elective Affinities*) the attraction of opposites. Swainson also believed, with many contemporaries for this was the epoch of the Bridgewater Treatises, that zoology could well find a place in university courses as a part of natural religion, and that 'the Christian philosopher ... may observe the close analogy which exists between the revealed character of God, and the material creatures of His creation' (1834: 372, 117 i.e. note).

The main attraction of the quinary system was that in every group it seemed to show the Coleridgean phenomenon of extremes meeting: as one arranged the creatures in a group, a circle seemed to be the arrangement which emerged if one dismissed prejudice and relied upon facts and inference. To Swainson (1834: 152ff, 207f, 214f), it was his opponents who were blinded by preconceptions, he being the only one in step. Three circles then united themselves (and this called forth the note about the revealed character of God) forming a larger one. The circles were drawn in a pattern with two on top and one underneath: the top left is the 'typical' forms, the top right the 'subtypical', and the bottom one the 'aberrant'. This last circle is found to divide itself into three, so we get our five—two large upper circles and three smaller ones below drawn inside a big one.

The scheme made sense if one believed that zoology could thus become a demonstrative science, rather like physics, and also if one could say (Richardson and Swainson 1831: xliv, lvi) 'Let us ... adore that God, who, to increase our faith in his word, enables us to discern, however dimly, in *earthly* things, the shadows of such as are *heavenly*'. It was an advantage that, the classes being limited in size, there was no room for man among the monkeys: man, the great amphibium, did not for Swainson—and for many of his contemporaries—belong in the animal kingdom (Swainson, 1845: 6ff, 74f, 46f, 54). At different levels, we may compare the vertebrates with the whole animal kingdom:

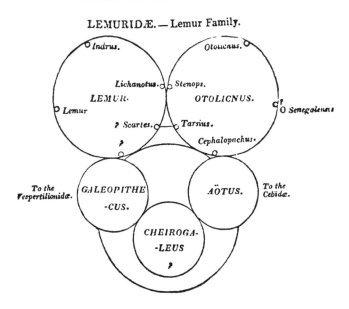

Figure 1. Pattern of circles, from SWAINSON, W., 1845. *The natural history and classification of quadrupeds*, London (p 89). Despite Swainson's insistence on the circular nature of affinities, he makes more use of tables—perhaps to help the printer.

Mammalia	Vertebrata	(highest)
Birds	Annulosa (insects)	(flight)
Fish	Radiata	(Aquatic)
Amphibia	Acrita (polyps)	(variable shape, lowly)
Reptiles	Testacea	(creepy-crawlies)

While at the same level, we can compare mammals and birds:

Quadrumana	Insessores	(grasping)
Carnivores	Raptores	
Ungulates	Rasores	(domestic)
Glires (rodents)	Grallatores	(run about, having long muzzles or bills)
Cetacea	Natatores	

The modern reader may not feel that the parallels are compelling, or be struck with the observation that long noses and small eyes go together in nature or in man, or agree (Swainson, 1834: 252f, 184) that nothing 'can be more perfect than the analogy between the Bengal tiger and the African zebra', both being striped and impossible to tame.

In the Quinary System the search for analogies became as important as that for affinities, and the affinities were by Swainson (1834: 171f, 246, 174, 197f, 345, 360, 437) determined from external characters[21]. Chemistry he believed should be avoided where possible and, because of the harmonious adjustment of things, internal investigation was generally a waste of time. So might field observation be, for analogies of nature would indicate the habits of a bird to one who had only seen its skin. Among lower groups, indeed, recourse may be necessary to dissection, and Swainson praised J. V. Thompson's *Zoological Researches* (1828–34); but in general 'whenever external peculiarities are sufficient to supply us with clear definitions, we require no other'. This method allowed him to break up the marsupials[12], putting the Tasmanian Wolf among the dogs, which was unfortunate in one who accused Cuvier and Latreille of proposing mixed and scrappy systems not 'built upon philosophic principles'. Swainson (1840a: 349) did admit in his autobiography that 'I am always so delighted with detecting either a new link of relation, or in bringing an isolated fact to bear upon general principles, that my enthusiasm sometimes overcomes my judgement'.

Swainson's adherence (1840b: 5, 8ff) to 'external and more obvious characters' curiously led to his malacology being in fact conchology; that is, a system of classifying shells rather than the creatures which live in them. The shape of the shell was after all more intelligible to 'the generality of readers' than the nervous system, and often the shells were more diversified than the animals. So we are left in these books with a web of relationships, into which extinct creatures like the mastodon can be fitted, and which had certain gaps (like Mendeleev's later chemical classification) where undiscovered animals might be predicted[13].

To write these volumes Swainson assembled considerable collections. Some of the Arctic birds described in Richardson and Swainson (1831) went to him, and he also built up a collection of specimens and plates; his correspondence shows how far natural history about 1830 was a trade or profession as well as a branch of learning. In his *Animals in Menageries*, 1838b, which differs from Bennett's most notably in having at the end 'two centuries and a quarter of new or little known birds' with bills and tails illustrated, we find an advertisement for the sale of the Swainsonian Museum[14]. The Quadrupeds and Birds were already finished with, but the shells and insects were still in use and the purchaser would have to wait a little. Swainson was willing to accept purchase money by instalments, but he does not give a price. In his autobiography, he says that the 'greater part' of the collections will accompany him to New Zealand; Günther reports that after being offered to Lord Derby, to the British Museum, (and to Prince Lucien Bonaparte), the specimens went to Cambridge and only the drawings to New Zealand.

As a naturalist and author Swainson enjoyed only moderate success, his books in the *Cabinet Cyclopedia* series not being taken seriously by working zoologists. During the 1830s he became increasingly isolated, and after the death of his wife he resolved to emigrate with his family to New Zealand, where he died in 1855. Out there he did no more writing, though he tried his hand at some botanizing in Australia (Galloway, 1978) and did some attractive drawings of primeval trees dwarfing mere humans, now at the Linnean Society. But one reader of his *Geography and Classification of Animals* (1835) was A. R. Wallace, whose copy is now at the Linnean Society and who seems to have read it twice, annotating it the first time in pencil and the second time in ink, and dating it 1 September 1842[15]. Wallace made the crucial point: 'But does not the assumption of any particular number in the first instance lead to a forcing in a manner of groups, and placing in a manner otherwise than, without such a theory, would appear to be natural'. Nevertheless, on the endpaper he wrote that there was such disagreement over taxonomic systems that 'Mr. S.'s may be as good as any'.

Certainly Swainson tried to see zoology whole, and to see reason behind it, and he perhaps deserves some credit for his boldness even though as things turned out he was no Wallace or Darwin. What is striking is that as an illustrator Swainson was exceedingly able: had he stuck to his last, we might feel, he might have acquired a reputation like that of Lear. Of Lear it could be written (admittedly by himself) 'How pleasant to know Mr Lear'; nobody would have written that of Swainson, but his talents were undoubted. For him, the problem seems to have been partly of status: among his little biographies (Swainson, 1840a) are Audubon and Wilson, the former getting one page as an illustrator and the latter eight pages as an ornithologist. It was Swainson who helped Audubon to achieve recognition on his visit to Europe. Audubon had hoped that Swainson would help him with *Ornithological Biography*, but Swainson refused because his name would not have been on the title page and the two men's friendship cooled. Swainson saw himself as on a different level as a philosophical naturalist, but we can now look at him as an illustrator.

In Alfred Newton's *Dictionary of Birds* (1893–97) there is high praise for Swainson's illustrations (coupled with heavy criticism of his taxonomy), and many of the illustrations in the *Dictionary* are taken from Swainson; this is a tribute indeed because they were over half a century old[16]. These were not Swainson's plates, but his pictures of details of birds, often comparative (bills, feet, tails and other external characters), which had been engraved in the *Fauna Boreali-Americana* and in other publications of Swainson's. Swainson allowed himself to write implausible things about affinity and analogy, but when drawing he could depict characters without prejudice or distortion. The natural-history artist's work cannot be entirely free of aesthetic fashion and will also emphasize taxonomic characteristics felt important at the time; as with other art, its passing the test of time is probably the best argument for its excellence.

In the 'Advertisement' to his *Exotic Conchology* (1821–22) and reprinted in Hanley's edition of 1841, Swainson set out the principles he intended to follow in his fine illustrations. In Swainsonian manner, the style is contentious and self-promoting, but the content is none the less interesting for that. He began with the remark that 'while the perfection to which the Fine Arts have attained in this country, is so great, as to be obvious in the embellishments of the minor pamphlets which daily issue from the press, the delineations of Zoological subjects in general remain uninfluenced by this universal improvement; and with a few exceptions, present lamentable deficiencies in design, drawing, perspective and the most common principles of light and shade; any one of which would not be tolerated, even in the frontispiece to the most humble of our periodical publications'.

Zoological illustration, then, should follow current aesthetic canons, but that was not all of Swainson's message. He believed that 'Natural History in this country, until very lately, has been little pursued, and still less estimated'. This had meant that because producing illustrated works was expensive and sales small, it was easier to copy plates from old books: and 'thus, we have reiterated copies of the misshapen figures given by the writers of the last century, perpetuated with all their glaring faults, in the hot-pressed volumes of the present day'. Pictures may pass the test of time for the wrong reasons, and attractive images may indeed survive, like Dürer's rhinoceros, into times that ought to know better.

Swainson believed that part of the problem was the low status of the zoological artist, who was widely supposed to need only the slightest knowledge of drawing: that if a bird was 'painfuly copied in the exact position it stands in the Museum', and a shell 'has its due proportion of colour, everything is done'. People forget 'that, in Birds particularly, every family has a decided peculiarity of form and habit, and that all originally possessed the gracefulness of life and action, which does not remain with the preserved skin; and, that, to delineate a shell with a proper degree of accuracy, as complete a knowledge of design, colouring, and *chiaro-scuro*, is requisite, as in painting a cabinet picture of still life'. Swainson praised Barraband's parrots and Ferrusac's shells as models of how things ought to be done, and gave qualified approval of Martyn's *Universal Conchologist*.

Swainson concluded his manifesto with the 'wish of inducing others, more competent than himself, to rescue this branch of the Fine Arts from the neglect it has hitherto experienced, and at the same time to stimulate as much as possible the increasing taste now manifested (particularly among the gentler sex) for a pursuit, which, while it brings together some of the most lovely of nature's productions, has furnished material for deep and philosophic inquiry, not only into the nature and economy of organized beings, but also into the formation of the planet which we inhabit'. Swainson declared that, having brought together in this book a selection of handsome shells, some new and some 'scattered in the works of the last age' and imperfectly represented there, 'at the termination of the work the whole will present a scientific arrangement'; but in the event this was left for Hanley to do, adopting the quinary system in deference to the illustrator who was by then in the Antipodes.

217

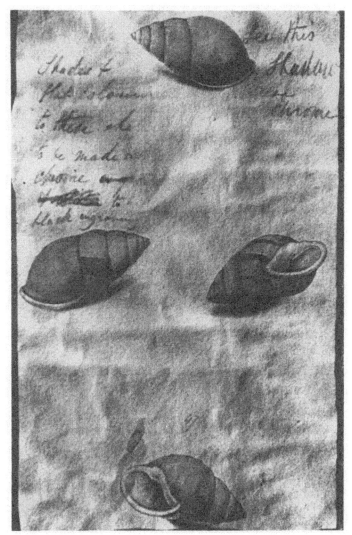

Figure 2. Pattern-plate for plate 47, showing Swainson's attention to colour and shading, and referring to his lithographed lines as 'engraving'.

His criticism of Martyn was that the etching of the plates was so light, leaving so much to 'those who perform the mechanical process of colouring', that scarcely two copies will be found alike. They had also been done in (opaque) body-colouring, which looked gaudy; whereas Swainson would use only (translucent) water-colour, and 'engraving in a decided manner all such delicate characters as belong to the

218

Figure 3. Pattern-plate, where Swainson is calling for colouring 'to hide the stripes of engraving' which again is curious in a lithograph. With autolithographs like these, pattern plates are presumably about the nearest we can get to 'originals'.

species, instead of leaving them to be put in by the uncertain ability of those who colour'. Since the plates were autolithographs, it is curious that Swainson calls them here 'engraving', which he must have used simply to mean anything printed; he did the same in manuscript annotations.

Figure 4. Pattern-plate for 78, showing Swainson critical of his own colouring, and insisting on getting shadows and tones right.

We are very fortunate that the pattern plates for the *Zoological Illustrations*[12] (which Swainson brought out at the same time as the Conchology) have come to light recently. Mr Howard Swann of Wheldon and Wesley kindly let me use them and they are now at the Linnean Society. These are prints from the stone, coloured and annotated by Swainson as examples for his colourists, in order that they should not fall into the errors of Martyn's. This book was the first in Britain to use hand-

220

Figure 5. Pattern-plate 60, showing the life-cycle of a moth, with the colourists' attention being called to contrast between the caterpillar and its background.

coloured lithographs, thus beginning a tradition of a century of the finest illustrated works. The book is a large octavo size and was issued in parts. These, containing five illustrations, came out monthly (in principle) and cost 4/6, with an extra 2/6 for title page and index at the end of the series. The plates are of birds, shells and insects; two insect plates were engraved by Curtis, but all the rest were autolithographs.

Figure 6. Pattern-plate 38, where particular attention is given to getting the background right.

Swainson's annotations to the pattern plates are full of exhortations about getting colour and shadows right, about usually having a grey tint behind a bird, and sometimes about details of background. He did not attempt like Audubon to introduce drama into his plates; sometimes he shows the life cycle of an insect. Many of the birds have little background and are like those of Barraband perched stiffly but decoratively on a studio mossy stump. Occasionally the composition is balanced with some foliage, and in a few cases there is a full background; occasionally bird plates show a pair, and some contain a detail, such as a drawing of the bill, at the bottom. There is in general no scale to the pictures, but dimensions are given in the descriptions, and some toucans do have a scale. The shells are shown uninhabited and casting shadows.

Some of the pattern plates are uncoloured, and there one can see more clearly than through the water-colour that Swainson's technique was to use many short lines so that the plates do look like engravings. No doubt this is often a feature of pioneers of new techniques; those who bought Swainson's book would hardly be aware that it marked a great departure in printing method. Swainson never adopted the flowing line and large format of Lear, and can therefore be said not to have really exploited lithography; but no doubt its cheapness in comparison with engraving helped in the success of the *Zoological Illustrations*[18].

Some of Swainson's pictures were engraved for Jardine's *Naturalists' Library*; he did the plates for *West African Birds* and for the *Flycatchers*, which makes a handsome little volume and was organized on the quinarian line. There are letters[19] from Lizars, who did the engraving, remonstrating with Swainson about his charges, soothing him after a row with Jardine, and remarking of the *Flycatchers:* 'in so far as the *mechanique* of getting up is concerned I think may challenge comparison with any work ever published and I have no doubt but that the learned in natural science together with all lovers of nature's self will again appreciate your labours'. As far as illustrations are concerned, we can agree with him; Swainson's are almost always a source of pleasure.

While then to know Swainson might not have been especially pleasant, and while his systematic work shows the weaknesses of the pre-professional era in which he worked, we can perhaps sympathize with him even as a naturalist and an author if we are prepared to meet a bold generalizer and determined controversialist. In his writings we can see the problems connected with words such as types, affinities, representatives and analogies which keep coming up in contemporary works, where they were usually taken less seriously. As an illustrator, we can see him as a conservative innovator who succeeded in making zoological illustrations into a branch of fine art in Britain, as he thought it already was in France.

NOTES

[1] On the 'decline of science' and the early years of the British Association for the Advancement of Science see Morrell and Thackray (1981).

[2] The plate of the tapir is reproduced in Knight (1972: 99).

[3] Bennett had replied at pp 199–206; further controversy continued on pp 455f, 481–, 486, 559f, and in 1832, **5**: in appendices, [109ff] and [208ff].

[4] Sauer (1982) has a number of references to Swainson subscribing to Gould's publications.

[5] Günther (1900) includes a catalogue of the letters now preserved in the Linnean Society of London.

[6] Swainson (1840a): in his collected biographies of zoologists. Swainson's life occupies pp 338–352 and his portrait is on the engraved frontispiece; see also McMillan (1976).

[7] Günther (1900) and Miller (1983: 42ff).

[8] Correspondence, Linnean Society, 15 November 1827.

[9] Herschel (1830) On its influence, see Cannon (1978) chapter on 'The Cambridge network'.

[10] Günther (1900: 42). Swainson like Priestley had a speech impediment, but in neither case did this slow their pens.

[11] On the quinary system as applied to Cuvier's Radiata, see Winsor (1976): and for more on Swainson's version, see my paper 'William Swainson: types, circles, and affinities', published in North and Roche (1985).

[12] On the thylacine, see Swainson (1845: 143ff).

[13] Swainson (1845: 193f). Mendeleev's system of chemical taxonomy was also worked out for a relatively elementary textbook.

[14] Swainson (1838b). See McMillan, N. F. (1980). The drawings included some by John Abbot, which survived and are being published: Abbot (1983-); Günther (1900: 23).

[15] Swainson (1835): Wallace's quoted comments come at pp 222f and endpaper.

[16] See my paper on 'Scientific theory in visual language', published in *Ellenius* (1985), on the test of time. There is a useful discussion of bird painting at this period (omitting Swainson) in Parker (1984). On some of the types from Franklin's expedition, illustrated in Richardson and Swainson (1831), see Hood (1974: 168, 175, 178).

[17] Swainson (1820-23; 1829-33): each series formed three volumes. Prices were the same for the first and second series; see *Magazine of Natural History* 1829 2: 51. Jackson (1975) remarks on the conservatism of Swainson's technique.

[18] Twyman, (1970); and see Knight (1977, chapter 2) and Rudwick (1976).

[19] Günther (1900: 44)

REFERENCES

ABBOT, J., 1983-   *Insects of Georgia*, intr. Parkinson, P. G. Wellington, New Zealand.

BABBAGE, C., 1830 *Reflections on the Decline of Science in England*. London.

BENNETT, E. T., 1830 *The Gardens and Menagerie of the Zoological Society delineated*. London.

CANNON, S. F., 1978 *Science in Culture*. New York.

ELLENIUS, A., 1985 *The Natural Sciences and the Arts*. Uppsala.

FARQUHAR, W., 1816 On a new species of tapir. *Asiatick Researches* 14: 417.

GALLOWAY, D. J., 1978 The botanical researches of William Swainson F.R.S. in Australia, 1841-1855. *Journal of the Society for the Bibliography of Natural History*, 8: 369-379.

GÜNTHER, A., 1900 The President's anniversary address, *Proceedings of the Linnean Society* 112: 14-61.

GUNTHER A. E., 1980 *The Founders of Science at the British Museum*. Halesworth.

HAYS, J. N., 1981 The rise and fall of Dionysius Lardner. *Annals of Science* 38: 527-542.

HENSLOW, J. S., 1836 *Description and Physiological Botany*. London.

HERSCHEL, J. F. W., 1830 *Preliminary Discourse*. London.

HOOD, R., 1974 *To the Arctic by canoe, 1819-1821*. (Ed. C. S. Houston). Montreal and London.

HORSFIELD, T., 1824 *Zoological Researches in Java*. London.

JACKSON, C. E., 1975 *Bird Illustrators: some artists in early lithography*, London.

KNIGHT, D. M., 1972 *Natural Science Books in English, 1600-1900*. London.

KNIGHT, D. M., 1977 *Zoological Illustration*. Folkestone.

KNIGHT, D. M., 1985 Scientific theory in visual language. *Ellenius*: 106-124.

McMILLAN, N. F., 1976 William Swainson. In *Dictionary of Scientific Biography*, New York, 13: 167f.

McMILLAN, N. F., 1980 William Swainson (1789-1855) and his shell collections, *Journal of the Society for the Bibliography of Natural History*, 9: 427-434.

MILLER, P., 1983 Between hostile camps. *British Journal for the History of Science* 16: 1-47.

MORRELL, A. & THACKRAY, A., 1981 *Gentlemen of Science*. Oxford.

NEWTON, A., 1893-97. *A Dictionary of Birds*. London.

NORTH, J. D. & ROCHE, J. J., 1985 *The Light of Nature*. Dordrecht.

290   WILLIAM SWAINSON: NATURALIST, AUTHOR AND ILLUSTRATOR

PARKER, R. D., 1984 *Birds of Ireland*. (Ed. M. Anglesea). Belfast.

RICHARDSON, J. & SWAINSON, W., 1831 *Fauna Boreali-Americana*. London.

RUDWICK, M., 1976 A visual language for geology. *History of Science* 14: 149–95.

SAUER, G., 1982 *John Gould, the bird man*. Melbourne and London.

SWAINSON, W., 1820–23 *Zoological Illustrations*. London.

SWAINSON, W., 1821–22 *Exotic Conchology*. London; Reprint, intr. N. F. McMillan, Ed. E. T. Abbott, 1968, Princeton.

SWAINSON, W., 1829–33 *Zoological Illustrations*. 2nd series. London.

SWAINSON, W., 1831 A defence of 'Certain French naturalists'. *Magazine of Natural History* 4: 97–108.

SWAINSON, W., 1834 *Preliminary Discourse*. London.

SWAINSON, W., 1835 *The Geography and Classification of Animals*. London.

SWAINSON, W., 1837 *Birds of Western Africa*. Edinburgh.

SWAINSON, W., 1838a *Flycatchers*. Edinburgh.

SWAINSON, W., 1838b *Animals in Menageries*. London.

SWAINSON, W., 1840a *Taxidermy, with the biography of zoologists*. London.

SWAINSON, W., 1840b *A Treatise on Malacology, or Shells and Shellfish*. London.

SWAINSON, W., 1845 *The Natural history and classification of quadrupeds*. New edn. London.

THOMPSON, J. V., 1828–34 *Zoological Researches*, Cork; facsimile, Ed. A. Wheeler, 1968, London.

TWYMAN, M., 1970 *Lithography, 1800–1850*. London.

VIGORS, N. A., 1831 Reply to Art. 1 No. XVIII of this Magazine. *Magazine of Natural History* 4: 319–337.

WINSOR, M., 1976 *Starfish, Jellyfish and the Order of Life*. New Haven.

XVII

Ordering the World

We are apt to be so impressed with Newton's mathematical way[1] that we forget that the eighteenth century belonged to Linnaeus and his disciples as much as to the astronomers and the natural philosophers. Chemists indeed in Hegel's day looked for a Kepler or a Newton of chemistry, and tried the mantle on themselves[2]: and the assumption has often been made by philosophers of science and by men of science that there is a 'natural history stage' through which every discipline must pass, like an awkward teenager, on its way to maturity and respectability. To Rutherford the saying is attributed that all science is either physics or stamp-collecting: such a sentiment is in accord with Galileo's remark that mathematics is the language of the book of nature; and generally with the view of the great men of the Scientific Revolution of the seventeenth century, that real science is a matter of explanation through experiment and mathematics. Description and classification was mere preliminary work, suitable for under-labourers who would provide a foundation upon which natural philosophers could build their splendid edifices.

Despite the low esteem of mathematicians, natural history in the seventeenth and eighteenth centuries continued to flourish[3]; and the classification of animals, plants and minerals, and also of knowledge, of sciences and of languages, occupied some of the greatest intellects of the day — who certainly supposed that they were doing something less arbitrary than stamp-collecting, though we may sympathise with Rutherford when we look at the 'cabinets of curiosities' assembled by the virtuosi. It is certainly possible to classify arbitrarily, as a child might order a stamp-collection by size or colour, or into those that have birds

---

1 H. Guerlac, 'Newton's mathematical way', *British Journal for the History of Science* (henceforward *BJHS*), 17 (1984) 61–4.

2 A. Thackray, *Atoms and Powers*, Cambridge, Mass., 1970; T. H. Levere, *Affinity and Matter*, Oxford, 1971; and my *Transcendental Part of Chemistry*, Folkestone, 1978.

3 K. Thomas, *Man and the Natural World*, London, 1983; D. E. Allen, *The Naturalist in Britain*, London, 1976; D. G. Charlton, *New Images of the Natural in France*, Cambridge, 1984, esp. Ch. 4.

on them and those that do not. Frank Buckland about a hundred years ago took a monkey and a tortoise on a train [4], and was told that he must pay for the monkey because it was a dog, but not for the tortoise because it was an insect: but nobody has ever claimed that British railways are wholly rational. The great question facing the taxonomists of the eighteenth and nineteenth centuries was whether it was possible to achieve a natural classification: reflecting the way things really were ordered as Newton's laws were supposed to reflect the real plan of the solar system.

Classification by railway officials is clearly socially determined; and indeed Durkheim [5] has suggested that in a deeper sense all classifications reflect the structure of the society in which they are made. This clearly fits some classifications, like that of the mentally ill. In his book *Mystical Bedlam* [6] a study of the case-notes of an English clergyman-psychiatrist of the early seventeenth century, McDonald argues that his eclectic method of treatment, with prayers, purges, amulets, astrological horoscopes, and sometimes consultations with an archangel, fits with a world view in which the insane were viewed with sorrowful wonder. They were to be helped in every possible way, spiritual and physical. But because Puritan sectarians were always ready to have prayer meetings over them, and Roman Catholics were quick with bell book and candle to exorcise them, spiritual help seemed to the establishment in church and state to be dangerously unorthodox. By the end of the seventeenth century, they had handed over the insane to the medical profession, classifying them as ill and needing merely physical remedies; and it took quaker pietists at the end of the eighteenth century to bring humanity back into lunatic asylums, in a struggle that still goes on.

It is not surprising that mental health should be a matter of social determination; but there do seem also to be areas of more austere science where social attitudes lie behind taxonomy. Benjamin Stillingfleet, an eminent naturalist and dilettante of the eighteenth century wrote [7]:

Each shell, each crawling insect holds a rank
Important in the plan of Him, who formed
This scale of beings: holds a rank, which lost
Would break the chain, and leave behind a gap
Which nature's self would rue.

4 F. T. Buckland, *Curiosities of Natural History*, 3rd ed., London, 1858, p. 297.
5 E. Durkheim & M. Mauss, *Primitive Classification*, ed. & tr. R. Needham, 2nd ed., London, 1969.
6 M. McDonald, *Mystical Bedlam*, Cambridge, 1981.
7 G. Montagu, *Testacea Britannica*, 1803; on Stillingfleet and Montagu, see *Dictionary of National Biography*.

And this poem was engraved on the title-page of George Montagu's *Testacea Britannica* of 1803; it is a classic statement of the notion that there is a Great Chain of Being. This idea, involving fixed species each with its role to play, seems to chime happily with the ideology of the *ancien regime*.

In Montagu's day the barnacles were still classified with the shellfish, and it was Charles Darwin who brought order into the *Cirripedia*[8]. He also brought in successfully an evolutionary way of looking at species, enshrining ideas then (and now again) popular in the English-speaking world such as 'the struggle for existence' and 'the survival of the fittest', coming from Malthus and from Spencer writing about human society. Indeed John Greene[9] has pointed out that all those who independently hit upon the idea of natural selection were Anglo-Saxons; and this theory can be seen as an expression of liberal ideology.

On the other hand, such comments seem both highly reductive and also inappropriate: it may have helped to make these scientific ideas popular that they fitted so well with political ones, and have helped in the diffusion of metaphors derived from science; but we know that there was more to the Great Chain of Being than everybody knowing his place, and more to Darwinism than 'Victorian values'. Science may be cultural behaviour, but important science, like important art, can transcend its culture. Classification and description, rather than being a mere portal of science or a reflection of the values of a time and place, are a form of theorising. The levels and hierarchies of the taxonomist risk being speculative or over-cautious; like any other science, they ought to be testable, and sometimes (like an excellent hypothesis)[10] they may lead to predictions which are verified.

From the time of Ray and Tournefort (and indeed ultimately from Aristotle's biological works) the distinction was made between artificial systems and the natural method[11]. It was believed that there were real groupings in nature, and that it was the business of the natural historian to find them out; just as the natural philosopher sought the real laws of nature. This gap was not even very wide, for one way of seeing Galileo's work would be to say that he perceived that the Earth and the planets formed a natural group, rather than the planets and the Sun: in England his work was popularised by John Wilkins in a *Dis-*

---

8 C. Darwin, *A Monograph on the sub-class Cirripedia*, 2 vols., London, 1851-4. The life history of the barnacle and other crustaceans was worked out by J. V. Thompson, *Zoological Researches*, 1828-34, reprinted London, 1968.

9 J. C. Greene, *Science, Ideology, and World View*, Berkeley, 1981, esp. ch. 4.

10 Boyle, in M. A. Stewart (ed.), *Selected Philosophical Papers of Robert Boyle*, Manchester, 1979, p. 119.

11 See my *Ordering the World*, London, 1981.

course concerning a new planet (1640), the planet being the Earth, and the taxonomy thus being stressed above the mathematics.

In the natural method the classifier took as many characteristics into account as he could; this meant that like other branches of science taxonomy is provisional, for eighteenth century ornithologists could not for example analyse egg-whites as their successors do in putting birds of paradise rather near starlings [12]. As well as being provisional, the natural method is laborious and requires skill; it was best learned by working for some time with an expert. It could not be picked up rapidly by a ship's surgeon just before setting off on a long voyage, or by somebody living remote from a university or a metropolis. There was therefore something to be said for an artificial system, in which there were clear boundaries between classes, which were based on one character. Such classes might offend the intuition of the expert, and one could not expect that they would lead to predictions; their level could only be that of a good hypothesis, giving shape to what was already known. The judgement on an artificial system must be pragmatic, like that which conservatives preferred for astronomical hypotheses in the sixteenth century; if a new one works better, then the old one should be dropped.

Boundaries and levels are clearest where numbers can be brought in; numbers must be odd or even, prime or not. The Linnean system for botany introduced numbers, because it was based on counting the sexual parts of the flower [13]. Each plant was thus indubitably placed in its major group, for example among the Hexandria trigynia. In Linnaeus' original tables of 1735, there are no plants in the classes Hexandria tetragynia and Hexandria pentagynia; in an artificial system there is no reason to predict that there might be, but the use of numbers does make clear the existence of gaps in a classification system; and later naturalists and chemists were to make something of this: Lacépède with fish, William Swainson with mammals, and Mendeleev most successfully with chemical elements. In mineralogy and in zoology, where numbers did not seem to separate groups clearly, and where emphasis on external characters (at the expense of physiological and chemical characteristics) was least effective, the Linnean system was never as successful. It was fortunate that among plants the number of sexual parts does seem to be

---

12 E. T. Gilliard, *Birds of Paradise and Bower Birds*, New York, 1969; W. T. Cooper & J. M. Forshaw, *Birds of Paradise and Bower Birds*, Sydney, 1977.

13 Recent reprints of Linnaeus include *Systema Naturae*, 1735, Nieuwkoop, 1964; and vol. 1 of the 1758 ed., London, 1956; *Species Plantarum*, 2 vols, 1753, London, 1957–9. On Linnaeus and his collections, see H. Stafleu, *Linnaeus and the Linnaeans*, Utrecht, 1971, and W. Blunt, *The Compleat Naturalist*, London, 1971.

generally constant within groups which agree in other characters, so that most Linnean groupings did not look outrageously arbitrary.

In the opening decades of the nineteenth century there were many natural historians content to work with the Linnean system in botany, especially in Britain where Linnaeus' collections had come by purchase after his death. Similarly there were many chemists prepared to use Dalton's atomic theory to construct recipes, but not to believe in atoms [14]. Instrumentalism in science could go well with the high esteem in which the name of Bacon was held; while attachment to systems was believed to be a French disease, liable to lead to dogmatism in science (as for example over the status of chlorine) and to political instability and revolution. It is ironic that it was in France that the opposition to an artificial system developed, and that a workable natural method was published by Jussieu in the fateful year of 1789 — which also saw the appearance of White's genial *Natural History of Selborne*.

Jussieu's contemporary Adanson had proposed a numerical taxonomy [15], in which all characters would be compared with equal weight being assigned to each. This is sometimes tried in our own day, using computers, in placing ambiguous organisms; but we are generally disposed to feel that some criteria are more important than others — among mammals, for example, method of reproduction rather than length of ears. Jussieu's work on plants, and then that of Cuvier on mammals, involved such prejudgements; but all science seems to involve some assumptions of this kind. The natural method became practicable when at the Museum in Paris there were collections of specimens sufficient for comparisons to be made, and for students to learn how to do it — the cabinet of curiosities was transformed into a teaching collection. Museums became the great centres of activity in natural history in the early nineteenth century, and different states competed to build up great collections by exploration, purchase, and barter [16].

Among organisms it had long been recognised that there were both analogies and homologies; the question was how far these were really due to affinities. The language of natural history was full of terms ta-

---

14 See my *Atoms and Elements*, London, 1967; and A. J. Rocke, *Chemical Atomism in the Nineteenth Century*, Columbus, Ohio, 1984.
15 J. H. M. Lawrence, *Adanson*, 2 vols., Pittsburg 1963; see also on Lamarck, R. W. Burckhardt, *The Spirit of System*, Cambridge, Mass., 1977, L. J. Jordanova, *Lamarck*, Oxford, 1984. See also on invertebrates, M. P. Winsor, *Starfish, Jellyfish, and the Order of Live*, New Haven, 1976; on fossils, N. A. Rupke, *The Great Chain of History*, Oxford, 1983 and A. Desmond, *Archetypes and Ancestors*, London, 1982.
16 See for example A. E. Gunther, *Founders of Science at the British Museum*, Halesworth, 1980.

ken from human society, but until the nineteenth century these terms were seen as mere metaphor rather than as any kind of explanation: putting animals in the same genus or family did not imply for the naturalist that they were really relatives. Analogies, in a distinction going back to Aristotle, were organs which had the same function in different creatures but differed in structure, like the wings of insects and birds. Homologous parts had the same structure, but might differ in function, like the wings of birds and the forelimbs of reptiles. For the taxonomist, it was generally agreed that homologies were the most significant. They indicated that two creatures ought to be placed close to each other; while grouping of all creatures that can fly gives a strange kind of system. Homologies must be given more weight than analogies; but it proved not always easy to tell which was which.

Thus Cuvier firmly separated the marsupials from the rest of the mammals, because of their method of reproduction; which also affects their skeleton, their offspring being so minute at birth that they do not need a big opening in the pelvis [17]. William Swainson in the 1830s, on the other hand, spent some pages in arguing that the Tasmanian 'wolf' should be placed with ordinary wolves, because it agreed with them in general form and in its teeth and jaws and feet — the criteria on which Linnaeus had based his system of animals. Where one put an animal depended on one's perception of what were fundamental features, expressed in homologies, and merely analogical ones. Swainson also in fact placed more emphasis on analogies than many contemporaries, because he had a numerical system: everything was to be organised in circles, in patterns of three because three signified the Holy Trinity. His system was called the Quinary System, because one of the three circles on every level was itself divided into three, making two big circles and three small ones each time; and whereas on each level it was the homologies which were important, up and down the hierarchy it was the analogies which linked the species.

Swainson's system thus had vertical and horizontal relationships (even if some examples were doubtful), and also had numbers built into it, through its circles, each of which could only hold a number of species. Swainson was particularly gratified to find that there was no room

---

17 G. Cuvier, *The Animal Kingdom*, ed. & tr. E. Griffith et al., 16 vols., London, 1827–35, vol. 16, p. viii. W. Swainson, *Preliminary Discourse*, London, 1834, p. 116; *Natural History and Classification of Quadrupeds*, new ed., London, 1845, pp. 7 ff., 143 ff. See my papers: William Swainson; naturalist, artist and illustrator', forthcoming in *Archives of Natural History*, and 'William Swainson: types, circles and affinity' forthcoming in J. D. North & J. J. Roche (ed.), *The Light of Nature*, Dordrecht, 1985. On illustration, see A. Ellenius (ed.), *The Natural Sciences and the Arts*, Acta Universitatis Upsaliensis, 22, 1985.

for man in the circles of the apes, for to him man was not a member of the animal kingdom but marked a transition to the level of spiritual beings. It was unfortunate that, like other eleborate systems based on numbers, his system did not work as more animals were found and had to be fitted in: his Coleridgean perception that extremes meet in circles, rather than nature being a linear ladder or scale, was lost along with his triadic pattern. Three and five are not, it seems, the magic numbers he had hoped they were.

Vertical and horizontal relationships and numbers were the outstanding features of the classification of chemical elements by Mendeleev from 1869[18]. Agreement on atomic weights reached after the Karlsruhe Conference of 1860 meant that where previously some different groups (like chlorine, bromine and iodine) had been remarked, now it was possible to see a whole pattern. When the elements were put in order of increasing atomic weight, then in a regular (or periodic) way similar elements recurred. In a table, this could be readily seen as vertical, horizontal and diagonal relationships. Some of Mendeleev's precursors had been mainly concerned with playing with numbers, looking for ratios between atomic weights; but his insistence on the natural method meant that for him chemical properties were paramount, and that when numbers seemed to put elements in the wrong places, as with tellurium and iodine, they were disregarded.

Mendeleev left gaps for undiscovered elements because his table revealed where they must be: unlike plants and animals, elements did come in families of definite size for the most part, though there were some perplexities with those near lanthanum which were hard to characterise. When his predictions (which were very detailed because he could extrapolate from neighbours all around the gap) turned out to be accurate, his periodic table became a feature of every chemical lecture room.

The natural classification of organisms was explained by Darwin as a consequence of evolution by natural selection; what had been metaphors when words like 'family' were used now became theory. But this explanation made no real difference to the business of classifying. Animals and plants do not carry their family trees about with them; the taxonomist must still look at the various characteristics of the organism, perhaps thinking of some new ones, and determine where to place it; then the evolutionist can work out its descent. Classifying is still an

18 Reprints of Mendeleev's and other papers are collected in my *Classical Scientific Papers – Chemistry, series 2*, London, 1970; see also J. W. Van Spronsen, *The Periodic System of Chemical Elements*, Amsterdam, 1969; B. Bensaude-Vincent, 'L'éther, element chimique', *BJHS*, 15 (1982) 183–8, and her forthcoming paper on Mendeleev in *BJHS*.

inescapable task. In chemistry, the periodic table was also in the first flush of Darwinian excitement given an evolutionary significance, though Mendeleev himself was reluctant to look for an explanation. Chemical evolution turned out to be less testable and fruitful than the Darwinian kind; but at the beginning of our century, with Rutherford and Bohr, the periodic table was given an explanation in terms of electronic orbits. This not only accounted for the way elements could be arranged, but also allowed sense to be made of the metals following lanthanum and actinium, groups which because they had no analogues had not been easy to fit in before. But here again, the explanation makes little difference in much of organic chemistry, which has not been thereby suddenly reduced to physics [19]: the properties of heavy metals may be implicit in the Schrödinger wave equation, but they have to be determined in the laboratory and are readily displayed in the periodic table.

Because of these explanations in terms of wider theories, we are prone to believe that these classifications of organisms and of elements are natural. That there may be other systems which will suit the gardener or the hunter or the jeweller, there is no doubt; but all these will be artificial, to be judged simply by their convenience. It looks as though in some areas we have got beyond this, and found out how things really are ordered. This belief is strengthened when we find that in other cultures classifications of creatures are not unlike ours; and recent work has been done on the zoology of the Kalam of New Guinea, indicating that more than two thirds of their divisions of birds correspond to valid taxa in our system.

This confidence that there are natural categories rather than simply imposed ones, or that our ordering is a matter of discovery rather than invention, may be tested when we look at some human artifacts. As Vico noted [20], we can understand history because we have made it, whereas we have not made nature. We were not surprised to find that classification of mental illness was a feature of different societies; but there are also things like languages and sciences, made by us, for which a natural method of classification has been sought rather than simply an artificial system. With languages the story starts with William Jones on the Indo-European languages in the late eighteenth century, and

19 On reduction, see B. Pippard, 'Beyond the Range of Physics', *Proceedings of the Royal Institution*, 56 (1984) 283–8. On the frontiers of physics in the early nineteenth century, see R. Home, 'Poisson's Memoirs on Electricity', *BJHS*, 16 (1983) 239–59; of geology, R. Porter, *The Making of Geology*, Cambridge, 1977. For New Guinea see I. S. Majnep and R. Bulmer, *Birds of my Kalam Country*, Oxford, 1977.
20 G. Vico, *The New Science*, ed., tr., abr. T. G. Bergin & M. H. Fisch, Ithaca, NY, 1970, p. xxxvi.

continues with Wilhelm von Humboldt in the early years of the nineteenth[21]. At just the time that Darwin was propagating his theory, Max Müller in London and at Oxford was applying a theory of evolution of languages to explain the various families which had been found[22].

The analogies between these families, their languages and dialects on the one hand, and the genera, species and varieties in zoology and botany on the other, were very striking. Müller saw everywhere the struggle for life going on in language; particularly where there were synonyms, one of which came to displace the other. He saw two processes going on: dialectic regeneration, where a language gets new vigour from incorporating dialect usages and words; and phonetic decay, where words are shortened or otherwise made easier to pronounce in the course of usage. Development clearly does happen in languages, and it is evidently the Darwinian kind of development which is not necessarily progress: academies are sometimes after all dedicated to stopping it. The metaphors from human family relationships which had worked well for Darwin seemed also to pay off in linguistics; in both these fields, metaphor became theory, and it came to seem that the classification or ordering was natural[23].

The classification of sciences goes back at least as far, and the position of different disciplines in the hierarchy is often not merely of intellectual but also of economic importance. Theology had been queen of the sciences; but from the seventeenth century a more republican vision of the various branches of knowledge became the norm. Early biological compilations were based on alphabetical ordering, but gradually this had been replaced by artificial or natural arrangements which were more convenient or seemed to reflect truth. In the same way the early encyclopedias were alphabetical; and in the Anglo-Saxon world at least the encyclopedia organised by themes or ideas has never really caught on[24]. The great example was the *Encyclopedia Metropolitana*, based on a plan by the poet, critic and philosopher S. T. Coleridge, and published between 1817 and 1845. This began with a series of book-length treatises, some of which later appeared as separate books and became standard works: the first were on the 'pure sciences' of logic and mathematics,

21 P. R. Sweet, *Wilhelm von Humboldt*, 2 vols., Columbus, Ohio, 1978–80.

22 M. Müller, *Lectures on the Science of Language*, 3rd ed., London, 1862.

23 On language, see R. Rappoport, 'Borrowed Words', *BJHS*, 15 (1982) 27–44; and my 'Chemistry and Poetic Imagery', *Chemistry in Britain*, 19 (1983) 578–82.

24 *Encyclopedia Metropolitana*, 29 vols., London, 1817–45. On Coleridge and science, see T. H. Levere, *Poetry realized in Nature*, Cambridge, 1981; on encyclopedias, S. P. Walsh, *Anglo-American General Encyclopedias, 1703–1967*, New York, 1968. Scientific dictionaries and encyclopedias have not perhaps been enough studied.

which have no empirical component; then came a series of volumes on the 'mixed sciences', part analytic and part synthetic; then on 'history and biography'. These volumes were then followed by a series of 'miscellaneous and lexicographical' ones, and by illustrations and index — which is essential in such an arrangement.

There were two great problems with this work: one was that so much (about half the work) got into the 'miscellaneous and lexicographical' volumes, where it was simply ordered alphabetically; so that the philosophical arrangement only worked for parts of knowledge. The other was that revision of the text seemed much harder when it was in the form of treatises than for brief entries; and the various authors were not all speedy in writing their pieces, so that by the time all was published parts were obsolete. The implicit hierarchy, with *a priori* knowledge at the top and the contingencies of history towards the bottom, represents a view of knowledge which is a curious mixture: the terminology of 'pure' and 'mixed' sciences is Baconian, but we associate Bacon with empiricism — for Coleridge, steeped in neo-Platonism, Bacon was a Platonist.

Coleridge was one of those who introduced German thought into Britain in the early decades of the nineteenth century, transmuting it in the process. William Whewell[25], who held various important posts at Cambridge and was a major figure in many scientific societies, wrote a *Philosophy of the Inductive Sciences* in 1840, opposing the empiricist and utilitarian prejudices of his countrymen, usually characterised as Baconian, and owing something to German critical philosophy. Though Whewell was very unhappy at the prestige given to purely deductive reasoning, believing that pure mathematics was likely to lead to atheism (as with Laplace), he recognised in all real science the necessity of some basic Idea in accordance with which empirical data must be organised. The different sciences could then be arranged according to their characteristic Idea.

Pure mathematics did come at the top of his list, characterised by Space, Time, Number, Sign and Limit. Then came mechanics, and chemistry, as we go through Motion and Polarity to Substance; and then biology, psychology, and ultimately natural theology, as we move through the Ideas of Irritabilty and Emotion to First Cause. The most important sciences are thus at the bottom of the page. A feature of his system was that Whewell was stoutly opposed to any kind of reductionism: because each science had its appropriate Idea, any attempt to reduce it to another could only yield a muddle. When we look at the list,

---

25 W. Whewell, *Philosophy of the Inductive Sciences*, 2nd ed., 2 vols., London, 1840, vol 2, p. 117.

we find some strange neighbours: Formal and Physical Astronomy are some way apart, as are Mineralogy and Geology, and Biology and the Distribution of Plants and Animals. This is because they have different governing ideas: Geology, and Distribution, are sciences based for Whewell on Historical Causation, whereas Mineralogy is concerned with Likeness, and Biology with Organization. Comparative Anatomy, based on Natural Affinity, is distinct from both Biology and Geology.

Whewell's table thus tells something about how the sciences looked in his day, and as such it is very valuable to the historian; but it could hardly be claimed that it passes the test of time like a great work of art — it survives only as a curiosity. It is amusing that like biologists of the day, but unlike chemists, Whewell used 'affinity', which means relationship by marriage and thus implies dissimilarity, when he should have used 'kindred'; as a clergyman forbidden to marry those connected by kindred or affinity, he should have known which was which! The real problem is that his classification is linear; it describes a hierarchy, and each science can only have two neighbours. Progress in fact took place on the frontiers of sciences placed inaccesibly far apart in nineteenth-century systems.

The same criticiam can be made of the contemporary system of Ampère [26], published in 1834 when he was already famous for his electrical researches. His was a bifurcating system; the table summarising it at the end was accompanied by a mnemonic rhyme in Latin, but despite this aid few seem to have learned it. The system begins with a division into two kingdoms, the cosmological and the noological sciences; these are divided into two sub-kingdoms, each of which gives two branches. The next levels are sciences of the first, second, and third order. In each kingdom, there are eighty-four numbered sciences of the first and third orders; those of the second order are not numbered. Once again, this produces some strange juxtapositions and separations; psychology is in a different kingdom from physiology; the whole science of chemistry is a third-order science, a division of general physics, while zoology is a sub-branch including eight third-order sciences. This system is a tree rather than a great chain; but at the level of third-order sciences it has the defects of a chain, and the levels do not seem to have been carefully enough worked out: Ampère in the last resort seems to have gone for symmetry rather than for a natural method. Like Whewell's, his system is of historical interest because it shows how an eminent man of science saw the hierarchy of his day; but it could hardly claim to be timeless.

It takes a bold man to suppose that there will be no really new scien-

26 A. M. Ampère, *Essai sur la Philosophie des Sciences*, Paris, 1834, table at the back of the book.

ces, and that his system can embrace what is to be found as well as what is known. But this is after all what Linnaeus and Mendeleev had hoped, and in modified forms their systems do survive. It could be that if sciences were to be classified in a two-dimensional fashion like chemical elements, or in a three-dimensional one, then it would be possible in bringing out more relationships to achieve something like a natural system. And yet this would be a paradox, for while a natural system is appropriate to nature, sciences (where there seems to be flux, and only artificial systems) and languages (where there is something more settled) are more like artefacts. There is a good deal to be said for regarding science simply as what people called scientists do; we regard as great scientists those who extend our view of what constitutes science, as great novelists extend our vision of what can be done in fiction. It seems to be merely conventional that in Britain Geology counts as a science, while Archaeology and Geography do not [27].

This brings us back to where we started, with the idea that classifying was an expression of relations within a society. It is anyway a fundamental preoccupation; it has the same dangers and rewards as other kinds of theorising; and like all theorising it is provisional. We ought never to be too confident that we have got it right. Certainly we can no longer see classifying as a mere threshold of science, when physicists try classifying fundamental particles; and find themselves in California using the eightfold way from Buddhism as Swainson's system had used the Christian Trinity. And in view of the success of some classifications, as of some theories, one is inclined to believe that there is something more there than a social construction; or that social construction has produced a masterpiece which will outlast its society, like a great temple. We do order the world; but there may be an element of discovery about it.

27 On the social history of scientific frontiers in Hegel's day, see J. B. Morrell and A. Thackray, *Gentlemen of Science*, Oxford, 1981; and the associated *Correspondence*, London, 1984; and I. Inkster and J. B. Morrell, *Metropolis and Province*, London, 1983. On problems in the history of science, see my Bericht, 'The History of Science in Britain: a Personal View', *Zeitschrift für allgemeine Wissenschaftstheorie*, 15 (1984) 343–53.

XVIII

PICTURES, DIAGRAMS AND SYMBOLS:
VISUAL LANGUAGE
IN NINETEENTH-CENTURY CHEMISTRY

The language of alchemy was resonant, working on different levels and ready to be exploited by poets such as George Herbert. That of modern chemistry is very different. Students are taught to use the passive voice and abstract nouns so as to give an impersonal authority to their scientific writing; and the very vocabulary of chemistry has been worked out since the late eighteenth century to be precise and exact. Metaphors are to be avoided, and texts should have a single and plain meaning. Chemical writing is therefore very different from literary prose, or even from what we ordinarily write in letters. But like alchemical description, it depends heavily upon illustrations and diagrams to get its meaning across.

In natural history and in medicine, we find scientific illustrations that are at the same time works of art.[1] A sense of wonder and a feeling for beauty are conveyed in a visual language commensurable with scientific and humanistic culture. This was more the case before 1900 than it is now, as diagrams have come to replace some of the pictures; but there is still no lack of scientific illustrations that can move us aesthetically. Moreover, the artist can sometimes express in

---

[1] ALLEN ELLENIUS (ed.), *The Natural Sciences and the Arts. Acta Universitis Upsaliensis*, XXII, Uppsala, 1985; KENNETH CLARKE, *Animals and Men*, London, 1977; WILFRED BLUNT, *The Art of Botanical Illustration*, London, 1950; DAVID KNIGHT, *Zoological Illustration*, Folkestone, 1977, and *The Age of Science*, Oxford, 1986, ch. 7.

visual language[2] ideas that could be expressed in words much less elegantly and concisely, if at all.

In the physical sciences things are a bit different, but we must be careful not to undervalue visual language there too;[3] it may be less beautiful, but it is every bit as important. In the course of a slightly elongated nineteenth century, from 1789 to 1914, chemistry came to fill a central position in the spectrum of sciences.[4] It had descriptive and taxonomic aspects; it was essentially an experimental discipline; and increasingly it moved towards deductive, explanatory theories. It also became ever more specialized: first into organic and inorganic branches, to which physical chemistry was then added; and then into further ramifications. New techniques and ideas had constantly to be got across. Chemistry was the first science in which laboratory teaching became the norm, and it became the service science with which everybody had to be familiar. Its history illuminates all sorts of questions of general social and philosophical interest; and closely involved non-verbal communication. There are endless examples, and I have chiefly used those from Britain – not the centre of things chemically – simply because they were accessible.

In 1789, when Lavoisier's *Traité* was published, there was beginning to be something like a community of professional men of science in France. Elsewhere this was not the case, though in Germany (where chemistry has always been particularly strong) a new journal does seem to have produced and promoted a self-conscious group of chemists[5] in a way that publications can do.[6] Chemistry was widely seen no longer as a craft, but as the fundamental science, offering dynamical explanations much more profound than those advanced in mechanics: as Priestley wrote:[7]

[2] MARTIN RUDWICK, "A Visual Language for Geology", *History of Science*, XIV, 1976, 49-195.

[3] EDWARD R. TUFTE, *The Visual Display of Quantitative Information*, Cheshire, Conn., 1983; KEN BAYNES and FRANCIS PUGH, *The Art of the Engineer*, London, 1981; URSULA SIEBOLD, "Meteorology in Turner's Paintings", *Interdisciplinary Science Review*, XVI, 1990, 77-86.

[4] DAVID KNIGHT, *Ideas in Chemistry*, London, 1992.

[5] KARL HUFBAUER, *The Formation of the German Chemical Community (1720-1795)*, Berkeky, 1982.

[6] See PETER DEAR (ed.), *The Literary Structure of Scientific Argument*, Philadelphia, 1991, papers by THOMAS BROMAN, LYNN NYHART and LISSA ROBERTS.

Hitherto philosophy has been chiefly conversant about the more sensible properties of bodies; electricity, together with chymistry, and the doctrine of light and colours, seems to be giving us an inlet into their internal structure, on which all their sensible properties depend. By pursuing this new light, therefore, the bounds of natural science may possibly be extended, beyond what we can now form an idea of. New worlds may be opened to our view, and the glory of the great Sir Isaac Newton himself, and all his contemporaries, be eclipsed, by a new set of philosophers, in quite a new field of speculation.

Priestley's intellectual heir, who proved the truth of his prophecy, was Humphry Davy.

We are fortunate with Davy[8] because we have a good idea of what he looked like. Paintings of scientists are (or aim to be) works of art; and in the early nineteenth century they also gave clues about what the subject did. The doings of chemists might seem very fascinating, but few had clear ideas about them; and Davy's lectures at the Royal Institution, for example, were directed at an audience ignorant of the natural sciences. The engraving (Fig. 1.), from a portrait by Howard[9] of 1802 when Davy was 23 and already famous, shows his notebooks, and on the mantlepiece some apparatus, probably for electrochemistry. We can also see that he is wearing breeches like a gentleman and not trousers, which at that date indicated sympathies with the sans-culottes: he had about three years before been a left-winger, supporting materialism and determinism, in the company of Thomas Beddoes and Samuel Taylor Coleridge; but at the Royal Institution he argued for the unequal division of property as the foundation of progress.

Twenty years later, as President of the Royal Society, he was painted by Thomas Lawrence (Fig. 2.),[10] who like Davy had risen from humble origins, and was now President of the Royal Academy; Davy's piercing eyes on which contemporaries remarked are again in evidence, but here the pose is almost domineering and the apparatus

---

[7] JOSEPH PRIESTLEY, *The History and Present State of Electricity* [1775], reprint ed. ROBERT SCHOFIELD, New York, 1966, vol. 1, p. XIV.

[8] DAVID KNIGHT, *Humphry Davy, Science and Power*, Oxford, 1992.

[9] Reproduced in HAROLD HARTLEY, *Humphry Davy*, London, 1966, pl. 1.

[10] JOHN AYRTON PARIS, *The Life of Sir Humphry Davy*, London, 1830, frontispiece.

Fig. 1. - Davy in 1802, from a portrait by HOWARD.

is the safety lamp, that wonderful example of applied chemistry which had crowned his research career. Lawrence wrote a paper for the Royal Society with William Hyde Wollaston[11] the chemist on why the eyes in portraits follow one around, involving overlays and

---

[11] WILLIAM HYDE WOLLASTON and THOMAS LAWRENCE, in *Philosophical Transactions of the Royal Society*, CXIV, 1824, 247-256.

Fig. 2. - Davy in 1822, engraving from portrait by LAWRENCE.

flaps; here the effect is certainly strong.[12] In the two portraits we see the young savant, and the formidable President; with indications not only of his status but also of what chemistry was like. Davy worked with Thomas Wedgwood in the earliest experiments on photography: by the middle of the century, photographs (Fig. 3.) replaced engravings from portraits and were often sent to friends and, perhaps with offprints, to fellow-scientists; and were reproduced in obituaries[13] – as of Graham, founder-president of the Chemical Society, and Cannizzaro. These photographs are rather stiff, and give no clue that the subject is a chemist.

Davy's fame rested on lectures often involving his latest discoveries, as we see in the famous Gillray cartoon (Fig. 4.) showing a demonstration of laughing gas; we can see more formally what a scientific lecture theatre was like from an engraving of the 1820s (Fig. 5.).[14] It is steeply raked, and has a large table for demonstrating experiments; the spectators in the front row were very near the orator; the occasion was literally theatrical, a performance by a savant. Chemistry was not only useful; it could be rational entertainment, and we can see how from the pictures we have. When Davy visited Dublin, there was, we are told, a black market in tickets; and not only in the British Isles but throughout Europe, people flocked to hear lectures on chemistry, and to see demonstrations of experiments – preferably exciting and surprising ones.

For most of Davy's hearers, chemistry was a spectator sport; but for those who wished to participate, the laboratory was just emerging as a room separate from the kitchen or the cellar.[15] Some chemistry was industrial even at the beginning of the century, as in a striking picture of a gas works (Fig. 6.), the last we shall have with claims to aesthetic attention.[16] We also find plates of laboratories; Mackenzie

---

[12] On abstraction in subjects, see RICHARD WOLLHEIM, *Painting as an Art*, Princeton, 1987, ch. 3.

[13] THOMAS GRAHAM, *Chemical and Physical Researches*, London, 1876, frontispiece; Stanislao Cannizzaro's obituary, *Journal of the Chemical Society, Transactions*, CI, 1912, 1677.

[14] CHARLER PARTINGTON, *Natural and Experimental Philosophy*, 1828, vol. 2, frontispiece.

[15] SOPHIE FORGAN, "Context Image and Function", *British Journal for the History of Science*, XIX, 1986, 89-113.

[16] COLIN MACKENZIE, *One Thousand Experiments in Chemistry*, London 1822, frontispiece; and plate of "experimentalist's laboratory".

Fig. 3. - Canizzaro, a photograph which accompanied his obituary, 1912.

Fig. 4. - Cartoon by GILLRAY of a Royal Institution lecture: Davy is holding the bellows, and Rumford standing by the door.

INTERIOR OF THE THEATRE OF
THE LONDON INSTITUTION.

Fig. 5. - Lecture theatre at the Surrey Institution, from PARTINGTON's *Natural and Experimental Philosophy*, 1828.

PICTURES, DIAGRAMS AND SYMBOLS

Fig. 6. - A gas works; the heroic or hellish aspect of chemistry, from MACKENZIE'S
One Thousand Experiments in Chemistry, 1822.

shows a large and commodious room with light from the side and
overhead, while W. T. Brande (Fig. 7.) illustrates the Royal
Institution's splendid (though less well-lit) laboratory,[17] where Davy
had sometimes apparently worked in public. Gas lighting was in
Davy's time just coming in, but was smelly and smoky until mantles
were introduced fifty years later. By the middle of the century, in
the wake of Liebig, institutions were offering laboratory teaching
even to undergraduates: a picture (Fig. 8.) of 1846 shows this in
University College, London,[18] where Graham was the professor; we

[17] This and other plates are reproduced in FRANCIS KLINGENDER's valuable *Art and the
Indutrial Revolution*, ed. ARTHUR ELTON, London, 1968.
[18] WILLIAM BRANDE, *A Manual of Chemistry*, London, 1830 ed., frontispiece.

— 329 —

Fig. 7. - The Royal Institution's laboratory, from BRANDE'S *Manual of Chemistry*, 1830.

Fig. 8. - Undergraduate teaching laboratory, University College London, 1846.

can see the benches, and the costumes of the chemists – still at that date, male.[19] Mere descriptions would tell us much less. By 1869 there were enough professional chemists for them to publish comic verse[20] for each other; they were no longer addressing outsiders in their publications, as a rule, and their illustrations could be less pictorial. Laboratories, like professors, were now usually photographed rather than drawn; though of course there are architect's drawings for new ones, in an age where institutions sometimes competed through their scientific institutes.[21]

Illustrations of laboratories reveal some apparatus, giving us an idea of its size, shape, and composition. Since the end of the nineteenth century, such illustrations have been stylized and linear:[22] there were some such in the early nineteenth century,[23] notably when a simplified woodcut was being printed with the text; but most even of these were shaded to give a three-dimensional effect, as for the American J. R. Coxe's paper published in England[24] during the Anglo-American War of 1812. Plates were generally engraved or etched and printed separately, as in Jane Marcet's little book written for girls,[25] and we may see disembodied hands indicating how things were held (Fig. 9.); the crucial thing was to show what things looked like, and it could not be taken for granted that readers were familiar with apparatus and needed only a sketchy diagram. Chemists took a pride in making things throughout the nineteenth century. In Ure's *Dictionary of Chemistry*,[26] a plate (Fig. 10.) shows both

[19] NEGLEY HARTE & JOHN NORTH, *The World of University College, London, 1828-1978*, London, 1978, p. 58.

[20] MARY CREESE, "British Women of the 19th & 20th centuries who contributed to research in the chemical sciences", *British Journal for the History of Science*, XXI, 1991, 275-305.

[21] British Lions (B.A.A.S.), *Exeter Change*, London, 1869, pp. 18f.

[22] SOPHIE FORGAN, "The Architecture of Science and the Idea of a University", *Studies in History and Philosophy of Science*, XX, 1989, 405-434.

[23] W. R. BOUSEFIELD, in *Journal of the Chemical Society, Transactions*, CI, 1912, p. 1452.

[24] SMITHSON TENNANT, in *Philosophical Transactions of the Royal Society*, CIV, 1814, 588.

[25] JOHN REDMAN COXE, in *Annals of Philosophy*, I, 1813, 69.

[26] [JANE MARCET], *Conversations on Chemistry*, London, 1828 ed., vol. 1, pl. facing p. 232.

248

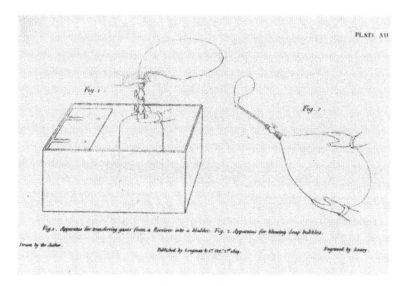

Fig. 9. - Apparatus and how to hold it, from Jane Marcet, 1828.

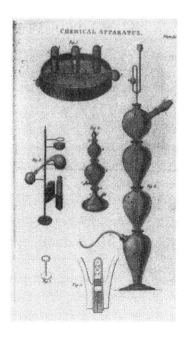

Fig. 10. - Apparatus: pictures and diagrams, from URE's *Dictionary of Chemistry*, 1828.

three-dimensional pictures and some diagrammatic sections, as contemporary botanical plates did. Plates were expensive to make, and wherever possible were reused; otherwise, they were often copied (sometimes reversed in the process) – we keep meeting recycled illustrations.

As late as 1884, we find realistic pictures in a journal[27] showing bricks and wood, but to our taste only revealing the outside, the appearance of things. Photography does not seem to have been used for illustrating chemical apparatus, but it was used for recording such things as spectra which could not have been adequately described verbally, and which were extremely important for chemists;[28] a sketch might have been misleading, whereas the camera was believed not to lie. In this example an important theoretical point depended upon the spectra, for Crookes believed that the "rare earth" or lanthanide metals were not fully separated and distinct species, indicating some kind of inorganic Darwinism – the development of heavy elements from simpler ones.

An illustration, as of Davy's safety lamp,[29] may show stages in the development of a device; and a series of them may show the evolution of theory. Thus crystals are more effectively described in a simplified and diagrammatic, rather than a realistic, illustration; and we see in plates also how explanations in visual language were attempted. Abbé Haüy in 1793 propounded his idea that crystals were built up (Fig. 11.) of unit cells or integrant molecules, like bricks of the appropriate shape; in 1813 Wollaston[30] tried to go deeper and demonstrate how spheroidal atoms could yield different crystal forms (Fig. 12.); while J. F. Daniell returned to an exposition of Haüy in his textbook[31] of 1839, where illustration and text are intimately allied. In the same year, F. W. H. Miller[32] approached

[27] ANDREW URE, *Dictionary of Chemistry*, London, 1828, pl. III.

[28] WILLIAM H. PERKIN, in *Journal of the Chemical Society Transactions*, XXXXV, 1884, 427.

[29] WILLIAM CROOKES, in *British Association Report*, 1886, p. 570.

[30] HUMPHRY DAVY, in *Philosophical Transactions of the Royal Society*, XVI, 1816, pl. 1.

[31] RÉNÉ JUST HAÜY, in *Annales de Chimie*, XVII, 1793, pl. 1.

[32] W. H. WOLLASTON, in *Philosophical Transactions of the Royal Society*, CIII, 1813, pl. following p. 63. The spheroidal balls are preserved in the Science Museum, London.

250

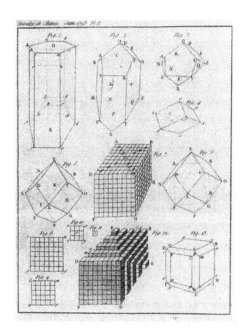

Fig. 11. - Haüy's theory of crystals, 1793.

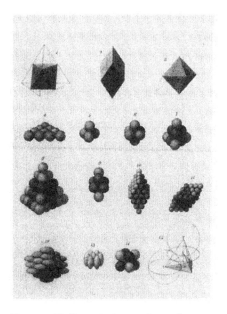

Fig. 12. - Wollaston's theory of crystals, 1813.

PICTURES, DIAGRAMS AND SYMBOLS

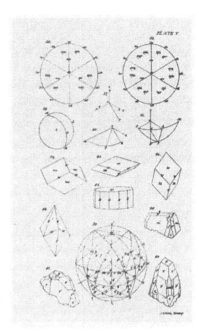

Fig. 13. - Miller on the geometry of
crystals, 1839.

(Fig. 13.) the study of crystals geometrically, proposing the system of indices still in use rather than a "model". It is hard to see how any of this work could have been carried on in purely verbal terms; and yet we have come a long way from realistic pictures, accessible to the outsider.

Tables have long been a feature of chemical publication,[33] and Lavoisier's Table of Simple Substances was very well-known: it finally got rid of the four ancient elements, and of the modern *phlogiston*, and incorporated his new names like *oxygen* and *hydrogen*. His chemistry was not immediately welcomed in England,[34] where there was a strongly empirical native tradition, and where Priestley

---

[33] J. FREDERIC DANIEL, *An Introduction to the Study of Chemical Philosophy*, London, 839, pp. 80-81.

[34] WILLIAM HALLOWES MILLER, *Crystallography*, London, 1839, pl. v.

and others had worked within the framework of phlogiston. At Guy's Hospital in London, chemical lectures were delivered[35] and accompanied by a textbook;[36] some copies at least were interleaved so that students could make notes from the lectures; a table rather like Lavoisier's was included, but incorporating Davy's discoveries – potassium and "chloric gas" are on the list. Medicine was chemistry's nurse as it grew into a developed science.

With the atomic theory came atomic weights, new and important constants; and Thomas Thomson in his textbook was the first to publish Dalton's theory and to appreciate its usefulness. His Table[37] of Atomic Weights (Fig. 14.) is surprising first because it has two different bases, and second because it includes many compounds; the figures are strictly speaking "equivalent weights", and chemists used the term "atom" for many years without intending any claim about indivisibility. In 1869 Mendeleev[38] published his first Periodic Table; later versions turned it through 90°: it gave a structure to the list of atomic weights, turning inorganic chemistry into a taxonomic science, where relationships (vertical, horizontal and diagonal) could be perceived from the Table far more economically than they could ever have been before from the prose of chemistry texts. The Table was a wonderful device for storing and retrieving information; and new elements corresponding to "gaps" were also predicted from it: hitherto their discovery had been a matter of inspired analogical thinking, or good fortune and opportunism, often with a new device or technique.

Chemists did not only record their results in tables, going back at least to Boyle; but also came to use graphical representations of data, especially as physical methods came into chemistry in the second half

---

[35] Lissa Roberts, "Setting the Table: the Disciplinary Development of 18th-century Chemistry as read through the Changing Structure of its Tables", in Peter Dear, *Literary Structure*, pp. 99-132.

[36] David Knight, "Chemistry and the Romantic Reaction to Science" in William Shea (ed.), *Revolutions in Science: their Meaning and Relevance*, Canton, MA, 1989, pp. 49-69.

[37] Noel Coley, "Medical Chemistry at Guy's Hospital", *Ambix*, XXXV, 1988 155-168; Hermione de Almeida, *Romantic Medicine and John Keats*, Oxford, 1991, ch. 2.

[38] William Babington, Alexander Marcet & William Allen. *Syllabus of a Course of Chemical Lectures*, London, 1811, table at the end.

TABLE II.

EXHIBITING THE ATOMIC WEIGHTS OF BODIES.

	[illegible]	
	Oxygen = 1.	[illegible]
Hydrogen	0·125	1
Carbon	0·75	6
Lithium	0·75	6
Carbo-hydrogen	0·875	7
Oxygen	1	8
Boron	1	8
Silicon	1	8
Carburetted hydrogen	1	8
Water	1·125	9
Silicet of hydrogen	1·125	9
Aluminum	1·25	10
Magnesium	1·5	12
Bicarburet of hydrogen	1·625	13
Azote	1·75	14
Carbonic oxide	1·75	14
Olefiant gas	1·75	14
Lithia	1·75	14
Silicet of carbon	2	16
Phosphorus	2	16
Sulphur	2	16
Silica	2·125	17
Deutoxide of hydrogen	2·125	17
Ammonia	2·125	17
Sulphide of hydrogen	2·1875	17·5
Sesqui-hydret of phosphorus	2·25	18
Fluorine	2·25	18
Glucinum	2·25	18
Ammonium?	2·25	18
Alumina	2·375	19
Hydro-fluoric acid	2·5	20
Calcium	2·5	20
Magnesia	2·625	21
Trito-carbo-hydrogen		

Fig. 14. - Thomson's Table of Atomic Weights, 1831.

of the nineteenth century[39] to express continuously varying quantities. But a diagram could also show how something happened, as in William Henry's textbook[40] where we have (Fig. 15.) not only some pictures of apparatus and crystals, but also a reaction mechanism set out in words with arrows and lines, accompanied by a rather early chemical equation – his formula for hydrogen peroxide being HO^2 rather than the H_2O_2 to which we are accustomed.

Eighteenth-century tables of "affinity" could be used to predict the way reactions would go, and there were in our period attempts to quantify affinities, though this was generally pretty arbitrary. The work of Volta and Davy in electrochemistry led to a new hope of

---

39 THOMAS THOMSON, *System of Chemistry*, London, 1831 ed, table II.
40 DMITRI I. MENDELEEV, *Zeitschrift fur Chemie*, XXII, 1869, 405.

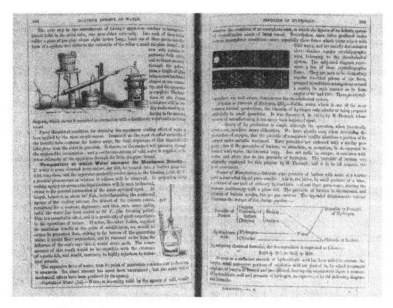

Fig. 15. - Henry's apparatus and reaction mechanism, c. 1853.

quantification, as in John Bostock's diagram of 1814;[41] but in 1840 Poggendorff's table of electromotive forces was again qualitative and comparative.[42] Tables here summarize a great deal of information, but also include an element of theory. The electrochemical series was and still is a useful way of tabulating the elements.[43]

We have met atomic weights and even an equation; although since the seventeenth century, atomism had been broadly accepted among men of science, it had been more like a world view than a theory as far as chemists were concerned: it was indefinite, as Lavoisier complained. His argument that atomic speculations must

[41] WILLIAM RAMSAY & S. YOUNG, in *Journal of the Chemical Society Transactions*, IXL, 1886, p. 806.

[42] WILLIAM HENRY, *Treatise on Chemistry*, ed. JOHN SCOFFERN, [*The Circle of the Sciences*, vol. 6], London, [c. 1853], pp. 304f.

[43] JOHN BOSTOCK, in *Annals of Philosophy*, III, 1814, 4.

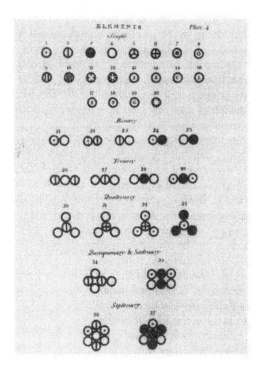

Fig. 16. - Dalton's atomic symbols, 1808.

be metaphysical, detached chemistry from the mechanical natural philosophy of the day; and most chemical thinking ran on separate lines for many years,[44] while prominent philosophers opposed its reduction to mechanics.[45] But Dalton's idea that the atoms of each simple substance or element had a characteristic weight made sense of the laws of chemical combination; and to make his point, and to

---

[44] JOHANN POGGENDORFF, in *Philosophical Magazine*, XVI, 1840, 495.

[45] DAVID KNIGHT, *Transcendental Part of Chemistry*, Folkestone, 1978; ALAN ROCKE, *Chemical Atomism in the Nineteenth Century*, Columbus, 1984; MARY-JO NYE, *The Question of the Atom*, Los Angeles, 1984.

make his theory more useful as well as accessible, he introduced[46] symbols for his atoms (Fig. 16.). The two-dimensional structures of compounds are based on the idea that like atoms repel each other; some of the symbols look rather like those used in alchemy.[47] His theory was received with considerable scepticism; and when in 1826 the Royal Society honoured him with a medal, it was for the laws of chemical combination and not for the atomic speculations which had guided him.

His symbols never caught on, but in 1814 Berzelius' rival plan was published[48] in Thomson's journal. His idea was that the initial letter(s) of the Latin name of an element should stand for an atom of it. Dalton's were still a bit like pictures of the little spheres he had imagined, whereas Berzelius' were pure symbols. There were some protests from mathematicians, to whom chemical equations looked like curious algebra. One great problem was to decide the relative weights, and in a paper by the elderly Berzelius in 1840[49] we see (Fig. 17.) some rather curious formulae: with a superscript dot standing for oxygen; "atom" used for compounds (to mean what was later called "molecule"); underlined atoms counting differently; and some curious formulae, for instance for acetic acid, by our standards. The formulae do involve the idea that the structure of compounds is important, but the symbols cannot express this clearly. Because chemists differed over whether water, for example, was HO or H_2O, their formulae for more complicated compounds came out differently. Whereas the layman could feel with Dalton's symbols some understanding of what was going on, Berzelius' were at best much more recondite and indeed forbidding; they also seemed, in the current state of chemistry, confusing. Symbols brought out the uncertainties more clearly than words would have done.

In the middle of the nineteenth century, Auguste Laurent[50]

---

[46] WILLIAM WHEWELL, *Philosophy of the Inductive Sciences*, London, 1847, vol. 1, pp. 383, 387; MENACHEM FISCH *William Whewell, Philosopher of Science*, Oxford, 1991, M. FISCH & SIMON SCHAFFER (ed.), *William Whewell: a Composite Portrait*, Oxford, 1991.

[47] JOHN DALTON, *New System of Chemical Philosophy*, Manchester, 1808, vol. 1, pl. 4.

[48] FRED GETTINGS, *Dictionary of Occult, Hermetic and Alchemical Sigils*, London, 1981.

[49] JÖNS JACOB BERZELIUS, in *Annals of Philosophy*, III, 1814, 52.

[50] JÖNS JACOB BERZELIUS, in *Philosophical Magazine*, XVI, 1840, 8.

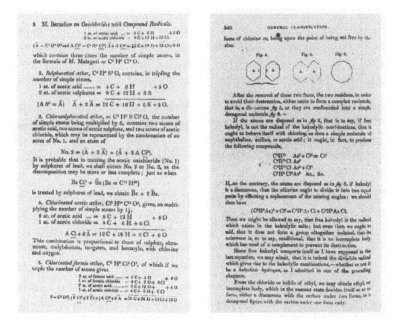

Fig. 17. - Berzelius' symbols and formulae, 1840.

Fig. 18. - Structures, from LAURENT'S *Chemical Method*, 1855.

changed the course of chemical theorizing by proposing that chemists should follow a hypothetico-deductive approach. Dalton had said much the same, but the prevailing notion in his day had been that the science must be inductive:[51] only a secure base in facts, with cautious generalizations from them, could avoid the hazards of "systems" and their dangerous consequences, notably materialism and Naturphilosophie. Laurent argued that no amount of data from analyses could ever entail a particular formula; he did not despair and go for conventionalism, but argued that chemists should propose formulae, work out their consequences, and test them. In his book we find (Fig. 18.) some modern-looking structures in these hexagons

---

[51] AUGUSTE LAURENT, *Chemical Method*, tr. William Odling, London, 1855, p. 340.

(which are not benzene rings), but we should not be misled by anachronism: Laurent was not a predecessor of Kekulé in this detail, but in his general approach. Following the Karlsruhe Conference of 1860,[52] and especially Cannizzaro's paper read there, chemists made this leap towards a new and bolder method, and agreed atomic weights and molecular formulae came in. This made it possible to write much clearer textbooks; and Mendeleev indeed hit upon his Periodic Table while writing a textbook. Symbols and tables made sense of vast quantities of data.

Dalton's structural formulae had been two-dimensional: crystallographers had thought in three dimensions, but the first ball and wire atomic models, called "glyptic formulae", seem to have appeared in the 1860s. They augmented the thinking that could be done with pencil and paper, but led to structures that were essentially flat. J. H. van't Hoff (Fig. 19.) made chemists appreciate that molecules had three-dimensional structure:[53] something again best expressed in visual language. His text actually recommends a particular brand of atomic model: to visualize structure without toys of this kind is probably impossible, and they and their representations (in the form of formulae showing valence bonds) are of crucial importance.

Physical evidence, first from spectroscopy, began to be increasingly important important in chemistry in the second half of the nineteenth century. The idea of energy and its conservation brought classical physics into being, and made it seem the fundamental science; and in the 1890s a chemical question was answered using physical evidence alone. William Ramsay was trying to decide how many atoms there were in the molecule of argon. Since it was a "noble", or inert, gas, forming no compounds, there was no chemical evidence bearing upon the question: and a train of deductions from the kinetic theory[54] of gases was required to determine the answer, using specific heat measurements. His diagram was not what was expected in a chemistry book; but the gap

[52] DAVID KNIGHT, *The Age of Science*, Oxford, 1986, pp. 16 ff.

[53] BERNARDETTE BENSAUDE-VINCENT, "Karlsruhe, septembre 1860: l'atome en congrès", *Relations internationales*, LXII, 1990, 149-169.

[54] JACOBUS VAN'T HOFF, *Atoms in Space*, London, 1898, p. 8.

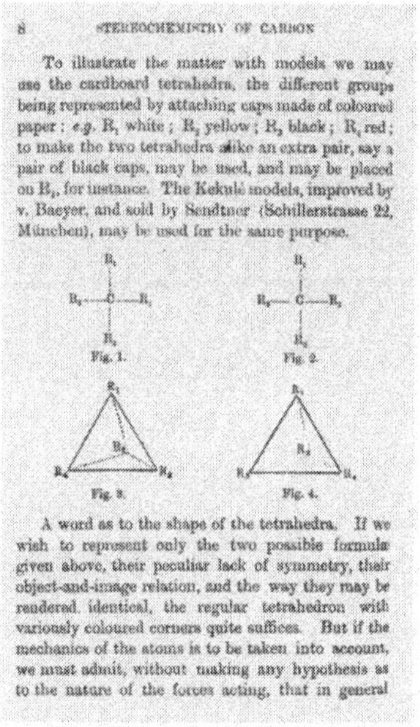

8 STEREOCHEMISTRY OF CARBON

To illustrate the matter with models we may use the cardboard tetrahedra, the different groups being represented by attaching caps made of coloured paper: *e.g.* R_1 white; R_2 yellow; R_3 black; R_4 red; to make the two tetrahedra alike an extra pair, say a pair of black caps, may be used, and may be placed on R_4, for instance. The Kekulé models, improved by v. Baeyer, and sold by Sendtner (Schillerstrasse 22, München), may be used for the same purpose.

Fig. 1. Fig. 2.

Fig. 3. Fig. 4.

A word as to the shape of the tetrahedra. If we wish to represent only the two possible formulæ given above, their peculiar lack of symmetry, their object-and-image relation, and the way they may be rendered identical, the regular tetrahedron with variously coloured corners quite suffices. But if the mechanics of the atoms is to be taken into account, we must admit, without making any hypothesis as to the nature of the forces acting, that in general

Fig. 19. - Van't Hoff's three-dimensional arrangements of atoms, 1898.

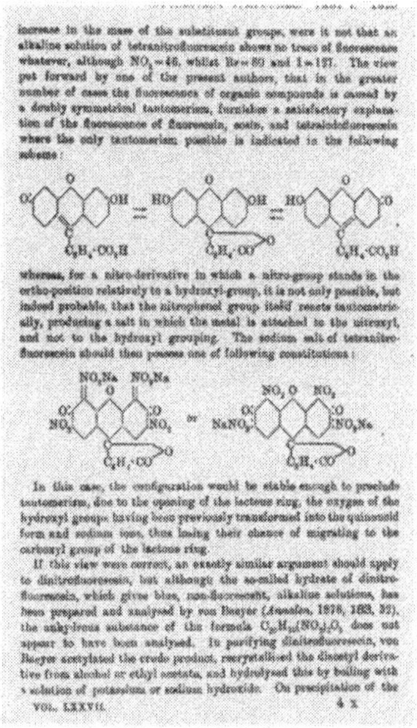

increase in the mass of the substituent groups, were it not that an alkaline solution of tetranitrofluorescein shows no trace of fluorescence whatever, although $NO_2=46$, whilst $Br=80$ and $I=127$. The view put forward by one of the present authors, that in the greater number of cases the fluorescence of organic compounds is caused by a doubly symmetrical tautomerism, furnishes a satisfactory explanation of the fluorescence of fluorescein, eosin, and tetraiodofluorescein where the only tautomerism possible is indicated in the following scheme:

Fig. 20. - Reversible reactions and agreed formulae, 1900.

between chemistry and physics was closing, and chemistry's century of independence was coming to an end.

We can end with a page from a chemical paper of 1900[55] showing reversible transformations of aromatic compounds of complex formulae,[56] involving the sort of reasoning Laurent had suggested half a century earlier. Without agreed symbols, such reasoning would have been impossible: non-verbal communication was therefore flourishing in chemistry. But it was inaccessible. We

[55] WILLIAM RAMSAY, *The Gases of the Atmosphere*, London, 1896, p. 214.

[56] J. T. HEWITT & B. W. PERKINS, in *Journal of the Chemical Society, Transactions*, LXXVII, 1900, 1325.

still find some portraits of chemists,[57] and views of laboratories; but the diagram and the symbol have almost completely displaced the picture. Initiates can understand them, and they now have clear and distinct meanings; and because chemistry is everybody's essential service science, there are plenty of initiates. But the visual language of chemistry, while still essential within the science, has ceased to be accessible to outsiders, or to resonate; and this may be partly why in the twentieth century chemistry has come to be seen as narrowly technical rather than as fundamental and exciting.[58]

Accessibility only to an *avant-garde* was a characteristic of modern painting as well as modern chemistry; but it was perhaps also a feature of verbal communication in science. F. L. Holmes[59] sees the "dense narrative" of seventeenth-century virtuosi give way to a schematic account of what had happened: "the 'story' that modern research papers tell is not expected to be a narrative such as we try to tell as historians. It is a synopsis: a story reduced to the elements deemed essential to its outcome, pared not only of contingent circumstances encountered along the way, but of all details of procedure and background that readers sharing the author's professional expertise will be able to supply from their own experience". He sees this happening in France by the early eighteenth century with chemical texts; it certainly happened with chemical illustration during the nineteenth century. Synoptic diagrams and symbols, rare at the beginning, had become ubiquitous by the end of the century; the visual language of chemistry had become much more powerful, and yet also impoverished.

---

[57] NORMAN ROBINSON & ERIC FORBES, *The Royal Society Catalogue of Portraits*, London, 1980.

[58] PRIMO LEVI, *The Periodic Table*, tr. Raymond Rosenthal, London, 1985.

[59] FREDERIC HOLMES, "Argument and Narrative in Scientific Writing", in PETER DEAR (ed.), *Literary Structure*, pp. 165-181; 180f.

XIX

ACCOMPLISHMENT OR DOGMA: CHEMISTRY IN THE INTRODUCTORY WORKS OF JANE MARCET AND SAMUEL PARKES

THE last years of the eighteenth century are well known as a time when large crowds flocked to hear lectures on chemistry; and indeed when chemistry seemed to be a science more fundamental than physics, penetrating below the shell and surface of things. When in the 1790s the young Humphry Davy taught himself the science, as an ambitious apothecary's apprentice, he used the textbooks of Lavoisier and of William Nicholson: but these were fairly stiff, and in 1806 there appeared two works which were directed specifically at beginners.[1] These were Jane Marcet's anonymous *Conversations on Chemistry*, and Samuel Parkes' *Chemical Catechism*. They were enormously successful, with editions going into double figures; and both their authors found their way into the *Dictionary of National Biography*. Parkes later wrote more advanced chemical works, but Mrs Marcet turned her hands to other fields, writing *Conversations* on Political Economy and on Natural Philosophy: her most eminent readers were the young Faraday and the young J. S. Mill, and her work was imitated in the Fittons' *Conversations on Botany*.[2]

The *Conversations on Chemistry* is aimed at women; there is a preface justifying teaching chemistry to them, and the characters in the book are Mrs B. (the tutor), Caroline and Emily. The girls are exceedingly sharp, and would be very welcome in any tutorial group; Mrs Marcet points out in her preface that this is necessary to keep the book moving. Caroline is slightly more impetuous, but the characters are not strongly sketched: they are sophisticated townees, but Caroline's father has a lead mine in Yorkshire. There is little to make it a girls' book: cookery or needlework are not prominent, but we do learn that Emily and Caroline were wearing muslin dresses; and Mrs B. will not let them breathe nitrous oxide or expose themselves to the dew.

The Fittons quoted Maria Edgeworth on Chemistry:[3]

> It is not a science of parade, it affords occupation and infinite variety, it demands no bodily strength, it can be pursued in retirement; there is no danger of its inflaming the imagination, because the mind is intent upon realities. The knowledge that is acquired is exact; and the pleasure of the pursuit is a sufficient reward for the labour.

It was therefore eminently suitable for women as well as men; but it is perhaps curious that we find no suggestion of the dangers of the laboratory, as we do for example in Davy's *Consolations in Travel* of 1830, where he remarks that a good eye and a steady hand are useful auxiliaries, but unlikely to be long preserved in a chemist.[4]

What is curious is that Parkes' *Chemical Catechism* was also in the first instance written for a girl—his daughter. But his preface is clearly written with the parents of sons in mind. If they have land, their son should know chemistry to exploit it; if he is going into medicine, he must know chemistry; and if he is going into any kind of industry then chemical knowledge will be extremely valuable to him. Here Parkes is clearly an apostle of applied science, looking forward towards the world of the steam intellect societies;[5] but his epigraphs from Tilloch

262

and from Fourcroy are concerned with the place of chemistry in a liberal education. Mrs Marcet was not expecting to generate professional chemists—indeed "professional" in contemporary contexts implied something like "talking shop"—but Parkes wrote from Haggerstone Chemical Works, London, and gives a fair amount of detail of chemical processes of industrial importance.

The written dialogue from the days of Plato was an important way of conveying knowledge, used by Galileo, Boyle, Berkeley and Hume; and conversation, especially if it involved women, was about 1800 recognised as the civilised way to education. Mrs Marcet wrote in a clear and attractive style, and her book is good to read: she began with simple substances, and then went on to treat of compounds. There are illustrations interspersed with the text, one of them showing Mrs B.'s hands and chin as she operates a blowpipe. Experiments are described: as Davy later pointed out, one of the great changes in his lifetime was the change from large-scale experiments involving furnaces, to small-scale ones with spirit lamps which could be done on the table.[6]

The catechism has a rather different history. It was a device for inculcating dogma—and as Kuhn tells us this has a major role in science.[7] The idea was that children should learn by heart a series of questions and answers about their faith; and there are great numbers of catechisms to go with different dogmatic traditions. Parkes' book is not a particularly easy read, and learning it would have been a painful business, requiring a real determination for self-improvement. He did not expect that the details of processes must be committed to the memory, and so each page has footnotes which take up more room than the text. The idea was that after going through the questions and answers, the reader would then come back to the notes: which as well as chemical information, contain poetic effusions especially from Erasmus Darwin, and copious references to the Creator. Parkes was a keen Unitarian, and an admirer of Priestley; and while natural theology is present in Mrs Marcet's book, it plays a much more prominent part in Parkes' volume.

Both books are deeply loaded with theory. Indeed they are an excellent guide to the state of chemical theory in 1806, just on the verge of Davy's electrochemical discoveries. They present the new French nomenclature,[8] and with it the caloric theory in which gases are compounds of bases with caloric, and exothermic reactions ones in which caloric is liberated, perhaps along with other elements. Light is also a "simple body"; and oxygen is the basis of acidity. As well as being a uniquely useful science, chemistry at this date had just acquired theoretical coherence; it was this which meant that it could be presented as dogma, and that knowledge of it could be an accomplishment. The books are full of taxonomies, sometimes revealing as where Parkes puts chromium among the brittle rather than the malleable metals; which tells us something about contemporary standards of purity. Both books, but especially Parkes', display an international perspective: Parkes in his notes has copious references to French publications. German or Scandinavian work on the other hand was little known, and probably less available. The French wars cut British scientists off from their allies more than from their enemies. Both authors have little diagrams illustrating double-decomposition through some mysteriously quantified measures of affinity; and Parkes in an appendix refers to the work of Richter. He also has a glossary of chemical terms, and lists over 250 experiments, some suggested by Davy.

Parkes, with his separation of text and footnotes, has affinities with modern authors who separate simple material in larger type from more detailed or advanced information which not all readers will want. But the whole idea of the catechism is that there is a body of doctrine to be learned which will not change radically with time. Indeed Parkes was readily able to take account of Davy's work on the alkali metals of 1807; it had been suggested by Lavoisier and others that the alkalies were compound, and the proof that they were so did not lead to any loss of coherence. Davy's idea that chemical affinity and electricity were manifestations of one power was also not completely new; what he provided was experimental evidence for it. This too could be incorporated into the chemical dogma of 1806. But more radical changes proved difficult to fit into this framework, altough the tenth edition of 1822 claims on its title-page to be "carefully corrected, and adapted to the present state of chemical science".

Mrs. Marcet's book is much more of a work of art. This means that it is much better to read; but also that it is even harder to keep up to date. By the eleventh edition of 1828 high-waisted muslin dresses had given way to full skirts and large sleeves, and the chemistry of 1806 was equally out of fashion. In the tenth edition a new chapter, or dialogue, on the steam engine had been added; in the eleventh, eighteen years after Davy's papers on the subject had begun to come out, the term "oxymuriatic acid" was replaced by "chlorine", except in a few headlines where the printer and proof-reader had missed it. But the problem was that this change was not one readily assimilable to the chemistry of the opening years of the century. Oxygen had been given its very name as the generator of acids; if there was none in muriatic acid then this was a serious matter, and oxygen could no longer be seen as the centre of chemistry.

The author herself pointed out that terms like "muriate of soda" were false under the new interpretation, for common salt contained neither muriatic acid nor soda—it was $NaCl$, not XO,Na_2O, where X stands for the assumed basis of the acid, and XO_2 would be oxymuriatic acid. The whole idea of property-bearing elements was called in question, and the nature of acidity and of exothermic reactions was again an open question. The later parts of the book are not reconcilable with the earlier, and all coherence is gone. There are also less-serious problems, such as the statement that there are fifty-six elements, while the list that follows contains fifty-eight, still including caloric and light; all authors of textbooks must be familiar with these minor problems of updating, and sometimes have gloomily to wonder whether the whole book should not be recast.

The theory of matter in the book is still a Newtonian corpuscularian one, with the various attractions of affinity, aggregation and gravitation described, along with repulsive powers.[9] Although Dalton had in 1826 been awarded a Royal Medal of the Royal Society for his laws of chemical combination, commonly called the atomic theory, they or it are not dealt with in the 1828 edition. There was more to know of chemistry than there had been two decades earlier, and perhaps those exciting days had given way to a period of consolidation; one in which the amateur would find it harder to keep up, to see what was at issue, and to applaud at the proper time. The science was thus perhaps less attractive as an accomplishment than it had been: and perhaps it had moved down-market to the mechanics' institutes, where utility could properly be the major concern of its students; or further into the world of medical education, where the factual lectures of Brande were a better bet than the eloquence of Davy.

The emphasis on Providence in Parkes, notably in his remarks on the oxides of nitrogen where, from only two elements, air, the delightful nitrous oxide, and the acid fumes of the higher oxides were all derived, was a feature of the time and had not dated by the 1820s; indeed it was in the following decade that the Bridgewater Treatises appeared. The chemical one, by William Prout, indicates the strongly medical and teleological framework within which chemists worked in these decades.[10] For Mrs. Marcet, pharmacy was a professional matter to be omitted from liberal *Conversations*;[11] but in both her book and Parkes' we find the kind of vitalism characteristic of the day.[12] Life is lent to us; during life rather different chemical reactions go on from those which happen after death, and which lead to putrefaction and fermentation. The taint of materialism was something which both authors would wish to avoid; the Baconian commonplace that "every acquisition of knowledge will prove a lesson of piety and virtue" was prominent in both, as indeed in most authors of the time—especially when they were writing elementary works.

These books then were not exactly textbooks, in the way that Thomas Thomson's was and that those of J. F. Daniell and Edward Turner were to be in the 1830s: they were not devised to go with a particular course of instruction.[13] They belong to a time in which elementary instruction in science was hard to come by; and in which scientific lectures were extremely popular. Mrs. Marcet herself alludes to finding it hard to follow such lectures at the Royal Institution until after she had gone through some elementary chemistry with a friend, and done some experiments herself. Then

> Every fact or experiment attracted her attention, and served to explain some theory to which she was not a total stranger; and she had the gratification to find that the numerous and elegant illustrations, for which that school is so much distinguished, seldom failed to produce on her mind the effect for which they were intended.

Lectures like those of Davy had a syllabus, but they were not intended to be a formal course leading to anything like a qualification, or to systematic knowledge. All lectures ought to be entertaining, but those of this time were genuinely theatrical occasions; and the Royal Institution was simply the most elegant and fashionable centre for such rational entertainment.[14]

The union of dogma and elegance is thus a feature of the period, in which general interest and utility were not seen as opposed; the contrast being between science and practice, or knowledge and mere rule of thumb. The books survived in their later later editions into a different world from that into which they had been born; but there were a succession of little easy-to-read introductions to chemistry even later in the century. They are the kind of work from which it is possible to recognise a norm at a given date, and in their way illuminate their period just as much as more original productions; and to be supercilious about them is to deny oneself a route into the science of the past.

REFERENCES

1. [J. Marcet], *Conversations on chemistry*, 11th ed., London, 1828; S. Parkes, *The Chemical Catechism*, 4th ed., London, 1810; this was published by the author, but printed by Richard Taylor, on whom see W. H. Brock and A. J. Meadows, *The Lamp of Learning*, London, 1985. These books replaced James Parkinson's *Chemical Pocket-book*, 2nd ed., London, 1801, perhaps as Parkinson moved into palaeontology; see my paper "Chemistry in Palaeontology", *Ambix*, 21 (1974), 78–85.

2. [E. and S. M. Fitton], *Conversations on Botany*, 6th ed., London, 1828; the *Conversations* were all published by Longman.

3. Fitton, note 2, p.vi; on the Beddoes and Edgeworth circle, see T. H. Levere, "Dr Thomas Beddoes" *BJHS*, 17 (1984), 187–204, and D. A. Stansfield, *Thomas Beddoes*, Dordrecht, 1984.

4. H. Davy, *Collected Works*, ed. J. Davy, London, 1839–40, vol 9, p. 365. On Davy, see S. Forgan (ed.) *Science and the Sons of Genius*, London, 1980.

5. I. Inkster (ed.), *The Steam Intellect Societies*, London, 1985.

6. Davy, note 4, *ibid*. G. L'E. Turner, *Nineteenth-century Scientific Instruments*, London, 1983.

7. T. S. Kuhn, in A. C. Crombie (ed.), *Scientific Change*, London, 1963, pp. 346–69.

8. W. C. Anderson, *Between the Library and the Laboratory*, Baltimore, 1984; M. P. Crosland, *Historical Studies in the Language of Chemistry*, 2nd ed., New York, 1978; and on chlorine, his *Gay-Lussac*, Cambridge, 1978, and J. H. Brooke in Forgan, note 4, pp. 121–75.

9. A. J. Rocke, *Chemical Atomism in the Nineteenth Century*, Columbus, 1984; and my *Transcendental Part of Chemistry*, Folkestone, 1978.

10. W. H. Brock, *From Protyle to Proton*, Bristol, 1985; see R. Yeo's "The Principle of Plenitude", forthcoming in *BJHS*; and my "William Swainson: Types, Circles, and Affinities", in J. D. North and J. J. Roche (ed.), *The Light of Nature*, Dordrecht, 1985, pp. 83–94.

11. Marcet, note 1, vol. 1, p. 4, later quotations are from p. vi; and from vol. 2, p. 338.

12. See my paper "The Vital Flame", *Ambix*, **23** (1976), 5–15.

13. A. J. Meadows (ed.), *The Development of Science Publishing in Europe*, Amsterdam, 1980; a similar work to those under discussion but a generation later is J. Scoffern, *Chemistry no Mystery*, 2nd ed., London, 1848.

14. S. Forgan, "Context, Image and Function", forthcoming in *BJHS*, and see my *The Age of Science*, Oxford 1986, Ch. 8.

XX

Lavoisier; Discovery, Interpretation and Revolution (**)

In August 1812, the young Michael Faraday wrote in great exitement[1] that «I would wish you not to be surprised if the old theory of Phlogiston should be again adopted as the true one tho I do not think it will entirely set aside Lavoisiers». It did not: there was no chemical counter-revolution. Although imperfections in Lavoisier's theory had become evident, he remained on his pedestal as the founder of modern chemistry. If in the late twentieth century we accept that all knowledge is somewhat unstable, we may wonder at this longevity.

Preamble

Lavoisier[2] was indeed extraordinarily successful. He got his interpretation of acidity and combustion accepted, and with it his list of simple substances or elements, so completely that we are apt to see his work simply as discovery in the literal sense. Like his contemporary Captain James Cook actually landing on what had been «Terra Australis Incognita», Lavoisier seems to have taken the lid off the phenomena of chemistry, seeing what was really there for the first time. Australia, and its plants and kangaroos, had always been there: so, we accept, had oxygen and hydrogen, but not phlogiston. When Lavoisier died, two hundred years ago, many chemists were converts to his views; but not all were. We wonder why; and tend to compare Joseph Priestley and others with the reactionary professors who refused to look through Galileo's telescope and see what the Moon was really like. We look back at the chemistry of the late eighteenth

---

(**) Comunicazione presentata il 20 ottobre 1994 al Seminario internazionale per il bicentenario della scomparsa di Antoine Laurent Lavoisier (1743-1794).

[1] F.A.J.L. JAMES (ed.), *The Correspondence of Michael Faraday*, vol. 1, London, 1991, p. 17.

[2] B. BENSAUDE-VINCENT, *Lavoisier*, Paris, 1993; A. DONOVAN, *Antoine Lavoisier: Science, Administration, and Revolution*, Oxford, 1993.

century through Lavoisier's eyes; and see the work of chemical, electrical and pneumatic philosophers as leading inevitably to his labours and interpretation.

Science may be a more-or-less steady advance in knowledge, a progressive business in which each generation sees more than its predecessor; but even then, we would have to admit that in the excitement of new discovery older truths have sometimes been forgotten or neglected. If, on the other hand, science is the imposition of paradigms upon booming, buzzing confusion, then Lavoisier's new order will have prevailed because it seemed to those active in the science to be the best available: we should have to see how he and his associates propagated it, rather as a political party promulgates its views, or a church wins converts. Either way, then, it is worth exploring what other interpretations of nature were available, which seemed to some able contemporaries to be both convincing, and fertile in suggesting experiment.

This is more difficult because Lavoisier self-consciously promoted a scientific revolution. He lived through, supported, and was in the end the victim of, the French Revolution. The English Revolution of 1688 had seemed a return to the Good Old Days before the Norman Yoke had been imposed in 1066; but the French looked forward rather than back. With his fashionable interest in language, Lavoisier (as Linnaeus had done) changed the terms used in his science so that they became more definite. Older chemists had used a language rich in overtones and suggestions, where the names of discoverers, the appearance of substances, or their geographical whereabouts determined how things were referred to; and where terms came from a variety of European and exotic tongues. The new language was systematic;[3] and Lavoisier and his associates launched a new journal[4] in which it was used exclusively. Once it became accepted, with some modifications in different countries,[5] it was hard to understand what users of the old nomenclature were talking about. This is another reason why they seem obscurantists. George Orwell in his novel *Nineteen Eighty Four* imagined a language, Newspeak, in which it was not possible to think old thoughts; Lavoisier had already achieved it.

In 1814 the Bourbons were restored to power in France; and much earlier than that Napoleon had declared that the Revolution was over. Political revolutions can be reversed, the wheel of fortune can revolve, though of course one can never go back to the past and start again. Even the language of revolutions,

[3] GUYTON DE MORVEAU et al., *Methode de Nomenclature chimique [1787]*, intr. A.M. Nunes dos Santos, Lisbon, 1992; D. KNIGHT, *Chemistry and Metaphors*, «Chemistry and Industry», 24, 996-9 (1993).

[4] M.P. CROSLAND, *In the Shadow of Lavoisier: the Annales de Chimie and the Establishment of a New Science*, Faringdon, 1994.

[5] A conference on the new language was held in Paris in May 1994, organized by Dr Bernadette Bensaude-Vincent; it will be published.

as we saw when Leningrad reverted to being St Petersburg, can be dropped in favour of the old nomenclature. What is striking about Lavoisier's intellectual revolution is that there was no serious counter-revolution; despite the work of the next generation demonstrating that some of his interpretations were false, it is upon Lavoisier's labours that modern chemistry has been built. We do not have to go all the way with Adolphe Wurtz and his «Chemistry is a French science: it was founded by Lavoisier of immortal fame»,[6] but he had a point. Let us now look at what might have been a counter-revolutionary tradition, with its origins in Britain and in Italy.

Chemical Philosophy

Lavoisier quantified chemistry in terms of weights. Chemists in the narrow sense, concerned with pharmacy or metallurgy and involved in doing analyses for the most part, may have found this a straightforward way to proceed: but it was not the most obvious, or what contemporaries would have called the most philosophical, route for all those involved in what came to be called chemical philosophy.[7] Joseph Priestley wrote about the impact of electricity and optics, his own route into science:[8]

Hitherto philosophy has been chiefly conversant about the more sensible properties of bodies; electricity, together with chymistry, and the doctrine of light and colours, seems to be giving us an inlet into their internal structure, on which all their sensible properties depend. By pursuing this new light, therefore, the bounds of natural science may possibly be extended, beyond what we can now form an idea of. New worlds may open to our view, and the glory of the great Sir Isaac Newton himself, and all his contemporaries, be eclipsed, by a new set of philosophers, in quite a new field of speculation.

Priestley like Lavoisier drew upon what came to be called experimental physics; but unlike Lavoisier, who believed that theories of matter must be metaphysical and indefinite,[9] Priestley hoped for real understanding of its internal structure, with point atoms that were centres of force.[10] In the Newtonian

[6] W.H. BROCK, *The Fontana History of Chemistry*, London, 1992, p. 87.
[7] M.J. NYE, *From Chemical Philosophy to Theoretical Chemistry: Dynamics of Matter and Dynamics of Disciplines, 1800-1950*, Berkeley, 1993, ch. 3.
[8] J. PRIESTLEY, *The History and Present State of Electricity [1775]*, intr. R. Schofield, New York, 1966, I, pp. xiv-xv.
[9] A.L. LAVOISIER, *Elements of Chemistry [1790]*, trans. R. Kerr, intr. D. McKie, New York, 1965, p. xxiv.
[10] J. PRIESTLEY, *Disquisitions Relating to Matter and Spirit*, 2nd ed., London, 1782, vol. 1, pp. 34 ff. See also F. JAMES, *Reality or Rhetoric? Boscovichianism in Britain: the cases of Davy, Herschel and Faraday*, in P. Bursill-Hall (ed.), *R.J. Boscovich: Vita e Attività scientifica*, Rome, 1993, pp. 577-85.

tradition, a quantified chemistry would for Priestley have been based upon particles and forces. Not with ordinary inanimate brute matter was the natural philosopher primarily concerned, but rather with forces and powers; and in particular with the «imponderables», light, heat and electricity. The weights of things were by contrast banal; the balance was not the only or obvious route into chemical understanding. We shall take these three imponderables as our guide.

Colour, or colour-blindness, was one of John Dalton's ways into science; and the young Humphry Davy's[11] first chemical speculations had to do with the role of light, via the supposed compound *phosoxygen*. Although *light* came at the top of Lavoisier's list of elements or simple substances,[12] Davy believed that he had failed to appreciate the part combined light plays in chemistry: Davy connected light closely with electricity, as others had who noticed that transparent bodies, fullest of light, were «electrics», capable of accepting charge. Thomas Young, the Newtonian heretic who proposed a wave theory of light, was also one of the last to publish a numerical table of elective affinities.[13] William Hyde Wollaston, the chemical analyst called by his friends «The Pope» because he was infallible, invented the optical «reflective goniometer» for measuring the angles of crystals;[14] but spectroscopy lay far in the future, so that investigating light and colours did not, for Priestley's and the next generation, much illuminate chemistry.

On the other hand, the study of electricity, and of Priestley's later favourite, gases (those compounds of heat), proved extremely fruitful; opening doors into new territories, and leading to a vision qualitative as well as partly quantitative which was rather different from that of Lavoisier. For Davy at the end of his life in 1829:[15]

Chemistry relates to those operations by which the intimate nature of bodies is changed, or by which they acquire new properties. This definition will not only apply to the effects of mixture, but to the phenomena of electricity, and in short to all the changes which do not merely depend upon the motion or division of masses of matter.

Chemistry was thus the fundamental science, while Mechanics (which Romantics always despised) was of minor significance: we might note that in Britain, Davy's protégé Michael Faraday counted as a chemist. The study of both gases and electricity have the further advantage for us that they had a strong input from both Italy and England. The serious physical study of the atmosphere had begun with Evangelista Torricelli, and had been carried on in

---

[11] D.M. KNIGHT, *Humphry Davy: Science and Power*, Oxford, 1992, p. 23.

[12] LAVOISIER, *Elements of Chemistry*, p. 175.

[13] «Philosophical Transactions», 99, 148-160 (1809).

[14] «Philosophical Transactions», 99, 253-8 (1809).

[15] H. DAVY, *Consolations in Travel; or, The Last Days of a Philosopher*, 5th ed., London, 1851, p. 262, (italics original).

the laboratory by the Accademia del Cimento in the 1650s;[16] and their work was then taken up by Robert Boyle. In the eighteenth century came the realization that the air involved or evolved in chemical changes was not just good or bad, but belonged to distinct kinds: fixed air, then vital air and inflammable air and other sorts were collected and identified, notably by Priestley,[17] another great admirer of the French Revolution.

Just two hundred years ago, in the Dissenting Academy at Hackney (near London), whither he had gone after a mob had sacked his house in Birmingham, Priestley delivered a course of lectures on chemistry.[18] They were full of phlogiston: and indeed Priestley invoked Newton in his support against the latest French ideas:

It is one of the principal rules of philosophizing to admit no more causes than are necessary to account for the effects. Thus, if the power of gravity, by which heavy bodies fall to the earth, be sufficient to retain the planets in their orbits, we are authorized to reject the *Cartesian Vortices*. In other words, we must make no more general propositions than are necessary to comprehend all the particulars contained in them. Thus, after having observed that iron consists of a particular kind of earth united to phlogiston, and that it is soluble in acids; and that the same is true of all other metallic substances, we say, universally, that all metals consist of a peculiar earth and phlogiston, and that they are all soluble in some acid.

Later, he reported that «alkaline air», or ammonia, «consists chiefly of phlogiston»; and discussed at some length the composition of water, where his views conflicted with those of the recently-executed «Mr. Lavoisier and most of the French chemists». For Priestley, the new theory was both unnecessary, and inconsistent with facts. We are apt to see Lavoisier's quantitative argument, that phlogiston would have to have negative weight, as crucial; but Priestley did not — his chemistry was qualitative. His suggestion was that:

it seems probable, that water united to the principle of heat, constitutes atmospherical air; and if so, it must consist of the elements of both dephlogisticated and phlogisticated air; which is a supposition very different from that of the French chemists.

It is indeed; and Priestley's lectures also included a remark about a «wise provision in nature»; though unorthodox, he belonged in the tradition of natural theology. His interpretations both of chemical changes, and of the world generally, thus differed from Lavoisier's — and indeed most of ours.

Nevertheless, Priestley's attempts to save the interpretation he had grown

---

[16] *Essayes of Natural Experiments [1684]*, tr. R. Waller, intr. A.R. Hall, New York, 1964.

[17] J. PRIESTLEY, *Experiments and Observations on Different Kinds of Air [1790]*, 3 vols., reprinted New York, 1970.

[18] J. PRIESTLEY, *Heads of Lectures on a Course of Experimental Philosophy, particularly including Chemistry [1794]*, reprinted New York, 1970; quotations from pp. 3f, 38, 128, 132, 134 and 142.

272

up with was unsuccessful. Henry Cavendish, whose experiments on inflammable air had led Lavoisier to the view that water was a compound, gave up chemistry. In Priestley's circle, the Lunar Society of Birmingham,[19] Josiah Wedgwood provided financial support for a Pneumatic Institution in Bristol where Thomas Beddoes with James Watt treated the sick with factitious airs such as oxygen — for they and their young protégé, Davy, used the new terms. There, Davy took up Priestley's work, his first major publication[20] (in 1800) being on «nitrous oxide», its chemistry and its effects as laughing gas; and this led to his appointment at the newly-founded Royal Institution in London.

Priestley's friend Benjamin Franklin had shown the electrical character of thunder and lightening; and Priestley was a great user of electrical discharges to set off chemical reactions in airs — especially to test for the «goodness» or respirability of samples of air, in a eudiometer.[21] By 1800 he had lost his battle to keep phlogiston at the centre of chemistry in Britain;[22] the new language had been generally adopted (despite quibbles about details), and although authors of textbooks tried to keep theory and facts apart, language and theory went hand in hand. But we might note that when Davy was ill in the 1820s, he was given antiphlogistic remedies (to reduce fever) — and phlogiston in medicine had quite a long run after that time. But Priestley's advocacy of electricity did bear fruit in a dynamical chemistry.

Tiberius Cavallo had been another of those working where chemistry and electricity met; but it was Luigi Galvani and his adversary Alessandro Volta[23] who made electricity central to chemical philosophy. Volta's «pile» of metallic discs in water was indeed as Davy put it, an alarm-bell to the experimenters of Europe. Chemical affinity had been a mystery; and the science awaited its Newton who would explain the phenomena in terms of forces. A polar force was required, unlike gravity which is always attractive; and electricity looked as if it might be the answer. Lavoisier had been reluctant to enter into questions of particles and forces, believing that these led only to metaphysics and would set chemistry back; but a disciple of Priestley's, sharing his Newtonian dream and feeling for natural theology, was well-placed to set chemistry in a new direction. This was Davy.

[19] R. SCHOFIELD, The Lunar Society of Birmingham, Oxford, 1963, pt. 5.

[20] H. DAVY, Researches, Chemical and Philosophical, chiefly concerning Nitrous Oxide or dephlogisticated Nitrous Air and its Respiration [1800], reprinted, London [1972].

[21] J. GOLINSKI, Science as Public Culture: Chemistry and Enlightenment in Britain, 1760-1820, Cambridge, 1992, chapter 4.

[22] D.M. KNIGHT, Ideas in Chemistry: a History of the Science, London, 1992, chapter 6.

[23] M. PERA, Radical Theory Change and Empirical Equivalence: the Galvani-Volta Controversy, in W. Shea (ed.), Revolutions in Science: their Meaning and Relevance, Canton, Mass., 1988.

Counter-revolution or Synthesis?

Using a giant Voltaic battery, Davy[24] isolated the new and anomalous metals sodium and potassium; and went on to infer that the «oxymuriatic acid» of Lavoisier and C.L. Berthollet was in fact an element, which he called *chlorine*. Davy thus illuminated the faults in Lavoisier's theory of acidity, demonstrating that the caustic alkalis soda and potash contain large amounts of oxygen, the acid-maker; whereas the acid from sea-salt contains none. In his papers about the new metals, there is a footnote indicating that the discoveries might be explained using the phlogiston theory.[25] Faraday, who as a bookbinder's apprentice attended Davy's later triumphant lectures on chlorine, took this occasion to write[26] to his friend Benjamin Abbott that Lavoisier's chemistry might be set aside and phlogiston might come back again. But although Davy delighted in having proved that chemistry was not a French science, and rebuked Berthollet and his associates for their dogmatism, and the baseless fabric of a vision[27] that they had erected, he did not achieve, or even seriously attempt a counter-revolution.

Like Naploeon dismissing the Holy Roman Empire, Davy remarked of chlorine that[28] «to call a body which is not known to contain oxygen, and which cannot contain muriatic acid, oxymuriatic acid, is contrary to the principles of that nomenclature in which it is adopted». He added his conviction that names «should be made independant of all speculative views, and that new names will be derived from some simple and invariable property». One of the terms which Lavoisier did not replace was *acid*: but (like *mass* in physics) it changed its meaning, from a sour substance, to a basis or radical combined with oxygen; and then through Davy's puzzlement, to Auguste Laurent's idea[29] of a compound in which hydrogen is replaceable by a metal, to G.N. Lewis' *proton donor or electron acceptor*; the reference is much the same, but some substances are acids according to one account, but not another. Davy knew[30] that «hydrogene is disengaged from its oxymuriatic combination, by a metal, in the same manner as one metal is disengaged by another» but he had not got a theory to replace Lavoisier's. He seems to have felt that acidity was the outcome of a particular balance of forces or powers.

[24] D. KNIGHT, *Humphry Davy: Science and Power*, Oxford, 1992, chap. 5 & 6.

[25] H. DAVY, *Collected Works*, ed. J. Davy, vol. 5, London, 1840, p. 89n.

[26] See above; F.A.J.L. JAMES (ed.), *The Correspondence of Michael Faraday*, vol. 1, London, 1991, p . 17 — see also following pages.

[27] This is a quotation from Shakespeare's *Tempest*.

[28] H. DAVY, *Experiments ... on Oxymuriatic Gas*, «Philosophical Transactions», *101*, 32, 35 (1811).

[29] A. LAURENT, *Chemical method*, tr. W. Odling, London, 1855.

[30] H. DAVY, *Muriatic Acid in its different States*, «Philosophical Transactions», *100*, 240 (1810).

274

Oxygen, which had occupied a privileged position in Lavoisier's chemistry, had to share its throne with chlorine; and acidity, which had seemed to be explained, became once again problematic. Sulphuric acid, supposed by Lavoisier and by Davy and his contemporaries to be composed of sulphur and oxygen only, and the acid made of hydrogen and chlorine only, had no element in common: and for Davy this vindicated the belief that forces rather than material components were crucial in chemistry — a belief that went back to his work on the oxides of nitrogen, which had very different properties, though composed of the same two elements. Davy had also long rejected the idea that heat was a substance — like light, it came as «caloric» on Lavoisier's list of Simple Bodies — because with Count Rumford he believed that it was the motion of particles. During the 1820s belief in the substance of heat waned generally in the scientific community; but this again led to modification of Lavoisier's schema, and not to its abandonment.

Dalton, Davy's contemporary, hit upon his chemical atomic theory when thinking about the composition of the atmosphere in the light of caloric theory — questioning why it was uniform, and not a sandwich with the densest gases at the bottom. He thus adopted Lavoisier's view of heat, but believed that atoms were a part of science rather than metaphysics. His atomism and Davy's more Romantic electrochemistry were synthesized by J.J. Berzelius[31] in a way that consolidated Lavoisier's revolution, and also gave us in time our modern chemical notation and equations.

Dalton's beliefs about atoms have almost all been falsified, and yet our chemical atomism is the direct descendant of his, rather than that of Lucretius, Galileo, Gassendi or Boyle which was indeed not testable chemically. In the same way, although many of Lavoisier's crucial ideas have been proved wrong, the foundation for the science which he laid have proved capable of bearing the load of later discoveries and interpretations. Physical chemistry does have its debts to the dynamical tradition associated with Priestley, Volta and Davy; and perhaps even with phlogiston;[32] but it developed within Lavoisier's structure, or perhaps we should say «paradigm». Modern chemistry incorporates a number of traditions, and yet we can see that it was with Lavoisier that the science took perhaps its most important change of direction; and the critical feature was probably the new language. Language and leadership in chemistry have gone together throughout its modern history.[33] Those who described chemistry in

[31] E.M. MELHADO and T. FRÄNGSMYR (ed.), *Enlightenment Science in the Romantic Era: the Chemistry of Berzelius and its Cultural Setting*, Cambridge, 1992, chaps 3 and 4 (by G. Eriksson and A. Lundgren).

[32] W. ODLING, *The Revived Theory of Phlogiston*, «Proceedings of the Royal Institution», 6, 315-25 (1870-2).

[33] M.J. NYE, *From Chemical Philosophy to Theoretical Chemistry*, Berkeley, 1993, p. 270.

Lavoisier's terminology had to see nature as he did. When willy-nilly Beddoes and Davy adopted Lavoisier's language, they had to translate into it the discoveries of Priestley and others; and the Italian proverb tells us that translation is treason. The new language was not just a matter of new names for old things, like rechristening Van Diemen's Land as Tasmania; it involved interpretations, which like discoveries were all bound up in Lavoisier's revolutionary insight.

XXI

Words that make Worlds

If there really are two cultures, then we might expect that science and poetry would be polar opposites. Music has intimate links with mathematics and with physics, the great name of Hermann Helmholtz being prominent amongst its theorists; while Alexander Borodin was both chemist and composer. Painting has similar connections: perspective is a matter of geometry, colour engaged the attention of Isaac Newton, James Clerk Maxwell and Wilhelm Ostwald, and illustration is an important feature of many sciences and of technology. Visual language is indeed inescapable, in pictures, diagrams and symbols; some in science have considerable aesthetic merit, while others may tell lies or even jokes. Ordinary scientific language, on the other hand, seems inescapably prosy. Journalists might perhaps be hard pressed without catalysts, chain reactions and quantum leaps; but they would probably get along with sea changes and other poetic imagery.

Rhetoric is important in science, because anyone who has found out something new needs to get the attention of the scientific community: but the most successful rhetoric is very plain, and has been since the Royal Society was founded in the 1660s. Recent studies indicate that Robert Boyle's wordy writings were deliberately written that way, tedious though they seem to us; and our children are coached by their science teachers into describing experiments in the passive voice, with abstract nouns where possible. Yet S.T.Coleridge, the poet and philosopher, said that he went to lectures on chemistry by his friend Humphry Davy to improve his stock of metaphors.

Alchemists had often written in verse, because their texts had at least a double meaning, as in a poem of 1633 containing the stanzas:

Through want of Skill and Reasons light
Men stumble at Noone day,
While busily our Stone they seeke,
That lyeth in the way.

The Eagle that aloft doth fly
See that thou bring to ground;
And give unto the Snake some wings,
Which in the Earth is found.

Probably there is a definite chemical meaning to this second verse, but it is apparent only to the adept, who has spent years of recondite study with a master. The language of modern chemistry is very different. It was devised by Lavoisier (executed as a taxman 200 years ago this May) and his associates. It is said that languages begin as poetry and end as algebra; the new language for chemistry was certainly to be a kind of algebra, and with the symbols adopted twenty years later even looked like it - though the early HO for water gave way to our H_2O. Names like potassium nitrate and calcium carbonate were clear and systematic, but lacked resonance or overtones which might tempt the poet. Other sciences soon followed chemistry into exact, defined and rigorous linguistic forms. Scientific papers and monographs are still referred to as "the literature", but they do not make a good read: they convey information concisely to the specialist.

Two hundred years ago Erasmus Darwin was as famous as his grandson Charles was to be, with his poems starting with *The Botanic Garden* of 1789-91. The second part of this, *The Loves of the Plants*, actually appeared first; and it was a very entertaining and effective way of getting across the Linnean System of classification, based as that was on the sexual parts of plants. Like Lucretius, he used verse to communicate science. The poems were accompanied with long and learned footnotes, which today might be a death sentence; but the mixture of imagery, from the classics and from biology, in the verses made them delightful until the new century brought a new generation of poets, Coleridge and Wordsworth among them, with a different ideal of language.

It is not easy to see in Coleridge's poems any precipitate from Davy's lectures; but Davy himself wrote poetry, and had even corrected the proofs of Wordsworth's *Lyrical Ballads*, putting in appropriate punctuation. He was also the friend of Walter Scott and of Lord Byron. His descriptions of chemical reactions, even for very formal journals like the Royal Society's Philosophical Transactions, were lively. Chemistry after all is the science of the secondary qualities, with interesting colours, tastes and smells, and even noises: and Davy's account of the isolation of potassium on passing an electric current through fused potash conveys the excitement that he showed at the time by dancing round the laboratory in delight. In "a vivid action" the potash began to fuse at both points of electrization. There was violent effervescence at the upper surface; at the lower ... small globules having a high metallic lustre ... appeared, some of which burnt with explosion and bright flame.

His poetry is generally less exciting, but here is a rhapsody:

> Oh, most magnificent and noble nature!
> Have I not worshipped thee with such a love
> As never mortal man before displayed?
> Adored thee in thy majesty of visible creation,
> And searched into thy hidden and mysterious ways
> As Poet, as Philosopher, as Sage?

Poetry, scientific knowledge (natural philosophy) and wisdom all went together in his self image. He was preoccupied with death; and one of his stanzas engages with his science:

> If matter cannot be destroy'd,
> The living mind can never die;
> If e'en creative when alloy'd,
> How sure its immortality!

If Davy had not been a great scientist, we would not read his poetry; but it does tell us about him, and it has some good lines. The winner of the pentathlon rarely holds the record in any one event: Davy had, in effect, had to choose between a career in literature or in science; the time and thought put into chemistry paid off, but meant that poetry was a hobby for him.

John Herschel was an eminent early Victorian physicist and astronomer, for whom again science on its own was incomplete:

> To thee, fair Science, long and dearly loved,
>   Hath been of old my open homage paid;
> Nor false. nor recreant have I ever proved,
>   Nor grudged the gift upon thy altar laid.
>   And if from thy clear path my foot have strayed,
> Truant awhile, -'twas but to turn, with warm
>   And cheerful haste; whilst thou didst not upbraid,
> Nor change thy guise, nor veil thy beauteous form,
>   But welcomedst back my heart with every wonted charm.

But often scientists' verse has been lighter than this, exploiting the possibilities of using scientific terms to comic effect; like Maxwell's "Tyndallic Ode" of 1874, mocking John Tyndall:

> I come from empyrean fires,
>   From microscopic spaces,
> Where molecules with fierce desires
>   Shiver in hot embraces.
> The atoms clash, the spectra flash,
>   Projected on a screen
> The double D, magnesian b
>   And Thallium's living green.

He disliked the agnostic "scientific naturalism" of Tyndall and T.H.Huxley; so his comic verse had a serious purpose.

Davy, Herschel and Maxwell had not had the benefit of a modern scientific education; but in our day too there are eminent scientists who write poetry. Any good poetry has to express the human condition, and poets must be concerned with the words that make worlds: occasionally, these aims may come together as they did for Erasmus Darwin in promoting science; but not very often. Davy's concern with death, and Herschel's with beauty are still with us; but in our day sexiness can be more directly expressed, and it is. So is play with words. Thus Desmond King-Hele reflects on the advantages of Latin:

>   Six over infinity
> is mathematically zero.
>   Sex super omnia
> is a motto fit for a hero.

Carl Djerassi, discoverer of the pill, has a poem "You wash this shirt like a chemist", asking

> Do I touch you like a chemist?
> Grip your wrist
> The way I grip the necks of Erlenmeyer flasks?

He comes close to Erasmus Darwin in describing the amours of the spiders; as does Roald Hofmann telling about the western pine beetle "which has an aggregation pheronome/ calling all comers (of that species)." But he is also not far from Maxwell with:

> In its crystalline beginning
> there was order, there was a lattice.
> And the atoms - cerium, lanthanum,
> thorium, yttrium, phosphate - danced
> round their predestined sites,
> tethered by the massless springs
> of electrostatics
> and by their neighbours bulk.

As with Davy's, these poems let us see the person behind the scientist. But they have a value way beyond that. Newton may have believed that poetry was a kind of nonsense; some is, like that of Lewis Carroll the mathematician and Edward Lear the zoological illustrator - but theirs is serious nonsense, with resonances for us all. Scientists' poetry reminds us that the verifiable kind of knowledge expressible in exact prose is not the only kind, and that the well-rounded and well-educated person responds to all kinds of intellectual excitement in the effort to make sense of experience and enjoy life.

Coleridge spoke of increasing his stock of metaphors; and yet the ideal of a clear and distinct language, like Lavoisier's, is to stamp out metaphors. Our ideas of scientific method are often prosy. Many of those who feel that science has nothing for them have met it in flat and dogmatic form; while others, looking at those practitioners who no longer try to be natural philosophers in search of wisdom, see in it a moral blindness. The scientist poet draws attention, as Tyndall did at the British Association in 1870, to the role of the imagination in science. Making intuitive leaps, taking metaphor seriously, spotting analogies, seeing significance in experience that to others seems ordinary (as our alchemist poet urged us to do); these are an essential part of the scientific process. In recognizing that, we shall cease to worry so much about those two cultures.

Reading:
C.Djerassi, *The Clock runs Backward*, Brownsville, OR: Story Press, 1991.
J.Heath-Stubbs and P.Salman (ed.), *Poems of Science*, Harmondsworth: Penguin, 1984.
R.Hofmann, *The Metamict State, and Gaps and Verges*, Gainsville, Florida, 1987 and 1990.
D.King-Hele, *Animal Spirits*, Farnham: THE X PRESS, 1983.
D.M.Knight, *Humphry Davy: Science and Power*, Oxford: Blackwell, 1992.
K.Moore, "Space and Time Forgot: John Herschel as a Poet", in D.King-Hele (ed.), *John Herschel 1792-1871*, London: Royal Society, 1992.

XXII

From science to wisdom: Humphry Davy's life

Creative science is a game for the young. Those excel in it who retain a child-like curiosity about the world down to an age when most of their contemporaries have got interested in other things like sex, power and money. While politicians, historians and playwrights (whose jobs depend upon understanding people) improve like claret with age, scientists may go off. Those engaged in scientific biography, therefore, face in particularly acute form the problem of dealing with a drama which comes to a climax early on, and then tails off. This makes for a poor read.

Davy's work on laughing gas was done when he was twenty-one; his electrochemical researches led to his discovery of potassium when he was twenty-nine; by his middle thirties he had elucidated the nature of chlorine, and invented the safety lamp for coal miners. If we concentrate upon his life in science as a matter of making discoveries which are still of importance in our own day, then his later life will have little interest for us. This is the approach in Harold Hartley's biography,[1] which is excellent when dealing with the scientific discoveries but where Davy's later years are briefly dismissed. At forty-one, in 1820, he was elected President of the Royal Society; in early retirement from 1827 he wrote dialogues about fishing and then about life in general; and he died abroad, at Geneva, in 1829 after travelling in fruitless search of health. A biography in which all this is an anticlimax is somehow defective as a work of art: we ought to impose order on life, whether it is our own or someone else's, so that it makes sense as a whole; and its interest must not therefore die before its subject does.

Davy died relatively young, at fifty; and yet in his case the problem is compounded by the change in his reputation among his contemporaries over his

[1] Hartley (1966), ch. 10.

last decade. Richard Holmes reflects of Coleridge[2] how differently we would have seen him had he died in 1804 when he set out hopelessly for Malta (but cheered on by a splendid valediction from Davy). Had Davy died in 1819 at forty soon after being awarded a baronetcy for his safety lamp, there would have been tremendous sadness at his early demise. He would have seemed a genius cut off at the height of his powers. He had enemies, among those who smarted at his reformation of chemical theory, and who supported the claims of George Stephenson[3] to have invented the safety lamp, seeing Davy as an arrogant metropolitan; but they were few and unimportant. His reputation as the Newton of Chemistry and the apostle of applied science stood extremely high even if (or perhaps because) he had made chemistry seem something depending on genius, superb apparatus and technique rather than accessible to all[4] – unlike engineering, in which Samuel Smiles seems to suggest that anyone could emulate his heroes through strength of character and organised common sense.

Then in 1820 Sir Joseph Banks died.[5] He had been President of the Royal Society since just before Davy was born; from being an immensely attractive young man who had sailed with Cook to Tahiti and botanised at Botany Bay, he had become an unpopular autocrat. Being President of the Royal Society gave wonderful opportunities for influencing people, but not for making friends: though 1993, being Banks' 250th anniversary, has led us to a new and clearer view of his achievements, with a number of conferences and publications. Davy speedily declared his candidature and as the inventor of the safety lamp he was unstoppable; Banks had written him a magnificent letter in 1815:

> Much as, by the more brilliant discoveries you have made, the repu-
> tation of the Royal Society has been exalted in the scientific world, I
> am of the opinion that the solid and effective reputation of that
> body will be more advanced among our cotemporaries of all ranks
> by your present discovery, than it has been by all the rest. To have
> come forward when called upon, because no one else could discover
> means of defending society from a tremendous scourge of
> humanity, and to have, by the application of enlightened philos-
> ophy, found the means of providing a certain precautionary meas-
> ure effectual to guard mankind for the future against this alarming

[2] Holmes (1989), 362–4.
[3] Smiles (1975), ch. 6.
[4] Golinski (1992), ch. 5.
[5] Carter (1988), parts 2 and 3; Banks (1994).

and increasing evil, cannot fail to recommend the discoverer to
much public gratitude, and to place the Royal Society in a more
popular point of view than all the abstruse discoveries beyond the
understanding of unlearned people.[6]

Great but impossible things were hoped for from Davy's Presidency, for
he brought enormous scientific distinction to the post, and was not hostile
as Banks had been to specialised societies: he saw them as complementary
rather than as threatening to the Royal Society. But Banks was a landed
gentleman and Oxford graduate; and Davy was not. His father had been a
woodcarver in Penzance, and his mother had for a time kept a shop. He had
been an apothecary's apprentice, and had dropped out to work with Thomas
Beddoes in a medically dubious institution where gases were administered
to the sick. It was possible in Regency England to rise dramatically, and social
mobility is the key to Davy's life. Sir Thomas Lawrence, who painted Davy's
portrait (complete with lamp) similarly rose from plebeian origins to be President of the Royal Academy; but social mobility has its price, and it was difficult for Davy to maintain his authority. It had not been easy for Banks, who
had faced down a major revolt in 1783–4;[7] had Davy survived longer he
might like Banks have brought the Society into line, achieved his programme
of cautious reform, and manoeuvred between the hostile camps[8] which beset
him: but in the autumn of 1826 his health failed. He seems to have had a
stroke; and travel abroad did not work a cure. In 1827 he resigned.

To himself and to others his Presidency had been a disappointment.
Although he had been involved with Stamford Raffles in setting up the
London Zoo, had established good relations with specialised societies, had
arranged through Robert Peel for annual Royal Medals to be awarded for
distinguished science, and had begun the transformation of the Society from
a club into an academy, his bad temper and what were seen as attempts to
domineer were notorious. In particular, he was taken aback in his attempt to
prevent corrosion of the copper bottoms of warships by attaching lumps of
a more reactive metal to them. He had been one of those responsible for
establishing 'applied science' rather than trial and error as the best route to
technical progress. In this case, the principle of what is now called cathodic
protection was sound, and in the laboratory the scheme worked very well;
but in practice weeds and marine organisms adhered so strongly to the protected copper that the ships' sailing was adversely affected. He had taken

[6] Davy, J. (1858), 208.
[7] Carter (1988), ch. 9.
[8] Miller (1983), 1–48.

the whole investigation upon himself, rather than refer it to a committee (though he was assisted by Faraday,[9] who was thus introduced to electro-chemistry). To see the great Sir Humphry's theorising go adrift gave a lot of pleasure to plain men; he had expected plaudits, and he over-reacted to jokes and criticisms. His reign was neither happy nor glorious.

Historians of science now have much less trouble than they used to do in coping with those who take up institutional responsibilities though their research falls away thereby. In our day, eminent scientists may do their research largely by deputy – through research students and assistants in their laboratory – and thus maintain their profile; but this was not possible in Davy's world. He had no research school like the French were developing. Davy was keen to accept responsibility, or ambitious for power; he realised what the implications were, and about 1821 he wrote a poem about eagles teaching their young to fly up towards the Sun, the important lines of which go:

> Their memory left a type, and a desire;
> So should I wish towards the light to rise,
> Instructing younger spirits to aspire
> Where I could never reach amidst the skies,
> And joy below to see them lifted higher,
> Seeking the light of purest glory's prize.[10]

The poem, like much of Davy's heartfelt writing, is in part a lament for lost youth; and the wish expressed here was not easy for him to fulfil, as the unhappy turn of his relationship with Faraday was to show. He was happiest when he could have undisturbed at least his intermittent bursts of research activity; and conscious that in middle age his shaping spirit of imagination was no longer as potent as it had been: that there would be younger spirits going higher.

Davy was not perhaps really prepared for the problems and frustrations (as well as opportunities) involved in running science rather than doing it; and as President he was not much missed. Indeed for 1826–7 he was unable to fulfil his duties at the Royal Society, and his colleagues must have got used to his absence. It is probably significant that his Presidential Chair was filled by Davies Gilbert (formerly Giddy), a Cornish MP and Davy's first patron; a public figure perhaps but with no significant scientific research to his credit.

[9] Faraday (1991–), 330–6.
[10] Davy J. (1836), ii, 157.

287

The 1820s were a difficult decade for reformers in science as in politics; and indeed after Gilbert the Royal Society elected the Duke of Sussex in preference to Sir John Herschel,[11] thereby perhaps freeing him to do physics rather than administration. It is said that the best Popes have not been saints; the experiment of having a man of genius to fill Banks' place was felt to have failed, and the clock was turned back. Davy himself had hoped for Robert Peel as his successor: 'He has wealth and influence, and has no scientific glory to awaken jealousy, and may be useful by his parliamentary talents to men of science';[12] indicating that he shared the view of his colleagues about his own presidency.

Within science itself there had, however, been dramatic change with Oersted's discovery of electromagnetism; and although in 1826 Davy gave the Bakerian Lecture to the Royal Society, placing his own past and recent electrical work in context, when he died in 1829 his researches were all assimilated or superseded and he did not leave a great gap. Within chemistry, analysis seemed the most important field, rather to the disgust of Faraday and J.B. Daniell who were interested like Davy in the powers that modify matter, and in explaining chemical affinity, and not just in accurate recipes. Faraday's reputation was not yet sufficient to be comparable to Davy's, so he was in no way eclipsed by his former pupil; but he must have seemed like a survivor from an earlier epoch rather than a man dying before his time. A Romantic genius is anyway not a very useful role model. Analytical chemistry was by contrast relatively straightforward, a normal science in which a career was a reasonable aspiration for a talented person: one puzzle would lead readily on to another in a life of steady usefulness.

Davy had been one of the first men of science who could be described as a professional; earning his living by research and lecturing at the Royal Institution, and giving up the prospect of a life in medical practice. By 1829 a career in science in Britain had become much more of a possibility; particularly with the rise of formal medical education, which required lecturers who would make their reputation by research and move up a ladder of promotion; but Davy did not belong to this pattern. Indeed to him as to his generation, a career in science seemed a bit puzzling. It was not obvious that a lifetime spent adding to knowledge of Nature was well spent. That was why Davy was so delighted to have proved Bacon right, in that experiments of light done with flames did indeed lead to experiments of fruit with the safety

[11] Hall (1984).
[12] Davy, J. (1858), 288.

lamp: one ought not to be reclusive, but rather seek to be useful. In a sense, one should outgrow that childish curiosity that impels scientific research: and applied science made recondite theorising and experimenting respectable. But Davy also (like most of his contemporaries) saw the world in terms of natural theology; science properly understood led to wisdom and not only to knowledge. He wrote:

> Oh most magnificent and noble Nature!
> Have I not worshipped thee with such a love
> As never mortal man before displayed?
> Adored thee in thy majesty of visible creation,
> And searched into thy hidden and mysterious ways
> As Poet, as Philosopher, as Sage?

During his working life he had been a philosopher (chemical or natural), and a spare-time poet; his illness gave him time before he died to be a sage.

If the biographer has a problem with a life that falls away from a peak, his subject will have had it first and more seriously. This makes it worth our wrestling with the question of how Davy saw his life himself: and while we do not have to agree with his estimate, as far as we can reconstruct it, it is very important for our understanding. We find that whereas Davy was thrown into rage and despair by the frustrations of office and then by the illness that removed him from it, he recovered on his travels although it was clear that his sickness was mortal. He was pleased to leave his *Consolations in Travel* as a legacy to the world; and he did not in the end see his last decade as a time of failure and decline. As Georges Cuvier put it in his obituary,[13] he returned to the sweet dreams and sublime thoughts which had enchanted him in youth; and he was a dying Plato, ending an examined life in the expectation of a better and more intellectual one to come. He managed to make sense of his life, making it come full circle; and it is worth following his lead, and seeing what it meant for Davy to try to become a sage.

On 27 September 1827 he wrote gloomily in his journal:

> As I have so often alluded to the possibility of my dying suddenly, I
> think it right to mention that I am too intense a believer in the
> Supreme Intelligence, and have too strong a faith in the optimism
> of the universe, ever to accelerate my dissolution . . . I have been,
> and am taking a care of my health which I fear it is not worth; but

---

[13] Cuvier (nd), 354.

which, hoping it may please Providence to preserve me for wise purposes, I think my *duty* – G.O.O.O.[14]

The G.O.O.O. was a form of pious ejaculation often used by Davy in this
journal. Had he died at this point, his life would have been incomplete; as it
was, he had time to complete his odyssey, writing dialogues which reveal his
thoughts and feelings.

The first, *Salmonia*, was modelled on Isaak Walton's *Compleat Angler* but
dealt with the more gentlemanly sport of fly fishing, of which Davy was
extremely fond. As well as discussion of different kinds of fish, and flies with
which to tempt them on to the hook, there are topographical descriptions
giving local colour; and the dialogue form inherited from Walton allowed for
reflections on general topics and some (but not very much) variation in tone.
The book was flatteringly reviewed by Walter Scott, and Davy set about
revising it for a second edition in which he expanded the passages dealing
with life in general. In particular, he expanded a purple passage which is
clearly autobiographical:

> Ah! could I recover any thing like that freshness of mind, which I
> possessed at twenty-five, and which like the dew of the dawning
> morning, covered all objects and nourished all things that grew,
> and in which they were more beautiful even than in mid-day sun-
> shine, – what would I not give! – All that I have gained in an active
> and not unprofitable life. How well I remember that delightful
> season, when, full of power, I sought for power in others; and
> power was sympathy, and sympathy power; – when the dead and
> the unknown, the great of other ages and distant places, were
> made, by the force of the imagination, my companions and
> friends; – when every voice seemed one of praise and love; when
> every flower had the bloom and odour of the rose; and every spray
> or plant seemed either the poet's laurel, or the civic oak – which
> appeared to offer themselves as wreaths to adorn my throbbing
> brow.[15]

He also began on a further series of dialogues, beginning with a vision in the
Colosseum, which would deal with life and death, time, progress and science. These were dictated to John James Tobin, a medical student (and son

[14] Davy, J. (1836), ii, 281.
[15] Davy (1832), 325.

of an old friend) who was his companion on his last journey (his wife Jane was suffering ill health in London). They spent the summer months in Austria and what is now Slovenia, and the winter in Italy.

On 23 February 1829 Davy in Rome had another stroke. He dictated a letter to his brother:

> I am dying from a severe attack of palsy, which has seized the whole of the body, with the exception of the intellectual organ . . . I bless God that I have been able to finish my intellectual labours. I have composed six dialogues, and yesterday finished the last of them. There is one copy in five small volumes complete, and Mr Tobin is now making another copy, in case of accident to that. I hope you will have the goodness to see these works published.[16]

To his wife, he also sent the message that:

> I should not take so much interest in these works, did I not believe that they contain certain truths which cannot be recovered if they are lost, and which I am convinced will be extremely useful both to the moral and intellectual world. I may be mistaken in this point; yet it is the conviction of a man perfectly sane in all the intellectual faculties, and looking to futurity with the prophetic aspirations belonging to the last moments of existence.[17]

His brother and wife rushed to his bedside; he rallied, and they were bringing him home to England when he died in Geneva. The dialogues were indeed published in 1830,[18] in an elegant little volume, as *Consolations in Travel*; later editions were embellished with engravings, some by Lady Murchison, wife of the geologist and friend of Davy. The book went on selling past the middle of the nineteenth century, and was translated into Spanish, French, German and Swedish; there were also American editions. It must be called a success although there is little evidence of its being actually read very much.

The book is interesting for its references to Papina,[19] the blue-eyed and pink-cheeked daughter of an innkeeper in Ljubljana; but while perhaps his conversation with her was intellectual and refined, and she 'made some days

---

[16] Davy, J. (1836), ii, 346.
[17] Ibid., 384.
[18] Fullmer (1969), 98–9.
[19] Knight (1992a), 169–71.

of my life more agreeable than I had any right to hope' (as he told his wife), it seems to have chiefly been scenery and philosophy rather than company which consoled him. He enjoyed following rivers up to their sources, even when in poor health; with fishing and shooting, at which he seems to have been very skilled. In poetry[20] he compared human life to the course of a river.

When it reaches the sea, the water of a river is mingled with it and its individuality is lost. Davy did not believe this of human life; despite a period of enthusiastic materialism when with Beddoes at Clifton, he soon reverted to a belief in the personal immortality of the better part of man. It was this idea which in the end was to provide him with the consolations of philosophy. The central message of the first dialogue, from which the book grew, is that when our machinery is worn out, and our work here done, we die and migrate to another planet where we shall inhabit a more ethereal body, and live a higher and more intellectual life. Tobin recorded that it was Davy's

> pleasure and delight during his mornings at Ischl, and when he was not engaged in his favourite pursuit of fishing, to work upon this foundation, and to build up a tale, alike redundant with highly beautiful imagery, fine thoughts, and philosophical ideas.[21]

On his fiftieth birthday, Davy himself wrote to his wife describing what he was about in composing the dialogues:

> I lead the life of a solitary. I go into the Campagna to look for game, and work at home at my dialogues on alternate days. I hope I shall finish something worth publishing before the winter is over. This day, my birth-day, I finish my half century. Whether the work I am now employed on will be my last, I know not; but I am sure, in one respect it will be *my best*; for its object is to display and vindicate *the instinct or feeling of religion*. No philosopher, I am sure, *will* quarrel with it; and no Christian *ought* to quarrel with it.[22]

The first dialogue is not, however, very close to orthodox Christianity. One of the lessons of the book is that death is necessary for birth or rebirth; another that the laws of chemistry and physics do not alone govern life. The progress described in the vision is attributed to men of genius, 'a few

[20] Ibid., 178.
[21] Tobin (1832), 120.
[22] Davy, J. (1858), 307.

superior minds'; it is not usually among the upper classes that these 'benefactors of mankind' are to be found, and they received little reward for their activities:

> The works of the most illustrious names were little valued at the
> times when they were produced, and their authors either despised
> or neglected; and great, indeed, must have been the pure and
> abstract pleasure resulting from the exertion of intellectual superior-
> ity and the discovery of truth and the bestowing benefits and bless-
> ings upon society, which induced men to sacrifice all their common
> enjoyments and all their privileges as citizens, to these exertions.[23]

Davy thus saw himself as a martyr to science, rather than as just unpopular; and in a great tradition of benefactors. We may find it implausible that one so famous should consider himself to have given up all his enjoyments, but for Davy it must have been comforting, and made sense of the unpleasant features of his presidency.

The characters in the dialogue then met upon Vesuvius to discuss the vision: all three, Philalethes (lover of truth) Ambrosio (immortal), a liberal Roman Catholic, and the sceptical Onuphrio (a hermit, patron saint of a church in Rome; whose name is a version of Humphry) must in different degrees represent Davy himself. They discuss the relations between religious and scientific belief, with particular reference to immortality (always a very important matter for Davy), and to the age of the Earth. Davy's religion was strongly personal, and he belonged to no particular church; he was surpris-ingly sympathetic to Roman Catholics, and rejoiced when Catholic Emanci-pation was passed. He had little patience with William Paley's clockwork universe, though he shared his love of fishing: he saw God as inscrutable, and submission and trust as essential for the freshness of mind essential to guide the wave-tossed mariner to his home. He refused to construct a system of geology upon Genesis, but urged a progressive development of the Earth over a long period; Charles Lyell was duly provoked to rebuke the dead Davy for this in his *Principles of Geology*.[24]

Davy had referred to a time when 'the dead and the unknown, the great of other ages and distant places, were made, by the force of the imagination, my companions and friends';[25] and indeed the participants in the dialogue

---

[23] Davy, (1830), 20–1, 30, 35 (quotation), 57, 228.
[24] Lyell (1830–3), i, 144–5.
[25] Davy (1832), 325.

are joined by a mysterious figure called the Unknown. He seems to be another manifestation of Davy. Here and in a later dialogue on chemistry, he denounces materialism, and urges that chemistry is progressive and useful, and is moreover the fundamental science, wrestling with the fundamental nature of matter: it is also rather dangerous. What may have begun as a recruiting document for chemists was reworked for *Consolations* into an *apologia pro vita sua*: the chemist's life is not merely worthy, but adventurous; and scientific ambition is the highest kind, unlike that of the lawyer or the politician. The chemist both comes to understand God's world, but is also enabled to benefit his fellow men.

Davy had thus come to terms with his coming death, and with his worldly impotence. At the end, he found that his real pleasures were simple, though he loved receiving letters from the mighty, and was indignant that he had never been made a Privy Councillor as Banks was. Unlike Faraday, he had pushed social mobility to its limits; Faraday knew when to stop, remaining one of Nature's gentlemen. Faraday's religion[26] was very different from Davy's, for he was rooted in a small sect: this gave him an intense social life, but led to his eschewing worldly position and thus responsibilities. His life was almost monastic compared with Davy's, and he did not have to find out about worldly glory the hard way. Contemporaries could revere him as a kind of scientific saint, inimitable by ordinary folk; a different sort of genius from Davy's.

In the West as in China, a certain longevity is probably essential to the sage. At least he ought to survive long enough to enjoy this new found status; he may be benevolent or curmudgeonly, but should be surrounded in his declining years by disciples. Coleridge did not die in Malta, and indeed went on to outlive Davy; becoming (under the careful management of Dr Gilman) the Sage of Highgate. Thomas Carlyle similarly saw the publication of his *French Revolution* which brought him the status of a sage.[27] John Herschel returned from South Africa to assume a somewhat similar status, writing poetry and articles for the *Reviews*;[28] it was possible for a man of science to be a sage. Davy however died in the moment of victory, just as he had finished dictating *Consolations*. Had he been brought home and lived for a few more years, he might indeed have blossomed in his new role of wise man: not probably a happy one, for the man Tobin describes was grumpy.

---

[26] Cantor (1991), ch 4.
[27] Rosenberg (1985), part 2.
[28] King-Hele (1992), 115ff (by M.B. Hall).

114

Perhaps then Davy died at the right moment, though his reputation was therefore insecure: it may have been wise to die just at the point of acquiring wisdom, and avoid the risk of losing it; and perhaps inscrutable Providence was on Davy's side at the end.

Bibliography

Banks, R.E.R. *et al* (1994) *Sir Joseph Banks: a Global Perspective*. London: Royal Botanic Gardens, Kew.

Brock, W.H. (1992) *The Fontana History of Chemistry*. London: Fontana.

Cantor, G. (1991) *Michael Faraday: Sandemanian and Scientist*. London: Macmillan.

Carter, H.B. (1988) *Sir Joseph Banks, 1743–1820*. London: British Museum (Natural History).

Cuvier, G. (nd) *Eloges Historiques*. Paris: E. Durocq.

Davy, H. (1832) *Salmonia, or Days of Fly Fishing*, 3rd edition. London: John Murray.

Davy, H. (1830) *Consolations in Travel, or the Last Days of a Philosopher*. London: John Murray.

Davy, J. (1836) *Memoirs of the Life of Sir Humphry Davy*, 2 vols. London: Longman, Rees, Orme, Brown Green and Longman.

Davy, J. (1858) *Fragmentary Remains, Literary and Scientific, of Sir Humphry Davy*. London: John Churchill.

Faraday, M. (1991–) *Correspondence*. ed F.A.J.L. James. London: Institution of Electrical Engineers.

Fullmer, J.Z. (1969) *Sir Humphry Davy's Published Works*. Cambridge, MA: Harvard University Press.

Fullmer, J.Z. (1980) Humphry Davy, Reformer. In S. Forgan (ed.) *Science and the Sons of Genius: Studies on Humphry Davy*. London: Science Reviews, pp. 59–94.

Golinski, J. (1992) *Science as Public Culture: Chemistry and Enlightenment in Britain, 1760–1820*. Cambridge: Cambridge University Press.

Hall, M.B. (1984) *All Scientists Now*. Cambridge: Cambridge University Press.

Hartley, H. (1966) *Humphry Davy*. London: Nelson.

Holmes, R. (1989) *Coleridge: Early Visions*. London: Hodder and Stoughton.

King-Hele, D.G. (ed.) (1992) *John Herschel 1792–1871: a Bicentennial Commemoration*. London: Royal Society.

Knight, D. (1992a) *Humphry Davy: Science and Power*. Oxford: Blackwell.

Knight, D. (1992b) *Ideas in Chemistry: a History of the Science*. London: Athlone, and New Brunswick: Rutgers University Press.

Lyell, C. (1830–1833) *Principles of Geology*, 3 vols. London: John Murray.

Miller, D.P. (1983) Between hostile camps: Sir Humphry Davy's presidency of the Royal Society of London, 1820–1827. *British Journal for the History of Science*, 16, 1–48.

Paris, J.A. (1831) *The Life of Sir Humphry Davy*. London: Colburn and Bentley.

Rosenberg, J.D. (1985) *Carlyle and the Burden of History*. Cambridge, MA: Harvard University Press.

Smiles, S. (1975) [1874] *The Lives of George and Robert Stephenson*. Introduction by E. de Maré. London: Folio Society.

Tobin, J.J. (1832) *Journal of a Tour Made in the Years 1828–1829 through Styria, Carniola, and Italy, whilst Accompanying the Late Sir Humphry Davy*. London: W.S. Orr.

XXIII

Getting science across

'Read until you hear the voices'; so the maxim goes for those who would engage with the Victorians. Let us try with Thomas Henry Huxley:

> A great chapter in the history of the world is written in the chalk. Few passages in the history of man can be supported by such an overwhelming mass of direct and indirect evidence as that which testifies to the truth of the fragment of the history of the globe, which I hope to enable you to read, with your own eyes, tonight. Let me add, that few chapters of human history have a more profound significance for ourselves. I weigh my words well when I assert, that the man who should know the true history of the bit of chalk which every carpenter carries about in his breeches-pocket, though ignorant of all other history, is likely, if he will think his knowledge out to its ultimate results, to have a truer, and therefore a better, conception of this wonderful universe, and man's relation to it, than the most learned student who is deep-read in the records of humanity and ignorant of those of Nature.[1]

Even if we cannot exactly hear the voice, its message is clear: Bishop Berkeley soon got from Tar Water to the Holy Trinity;[2] here is Huxley in spate and at his best, doing much the same. We think of science as involving research, teaching and administration; Huxley was committed to these things, and cared passionately about how nature worked. But a substantial, and to us surprising, amount of his time and energy were devoted to bringing science into general culture; and his reputation was thereby enhanced.

Getting across gentlemen of science was one of Huxley's strengths, and getting science across was another; and sometimes the two went together.[3] He was an extraordinarily effective popularizer, collecting and holding large audiences both at the fashionable Royal Institution in London and in various venues for working men. His success in public lecturing was an important aspect of his career, opening posts to him and enabling him to make ends meet as a professional scientific man: and his eminence in scientific research ensured that he was not seen (as he saw the anonymous author of *Vestiges* of *the Natural History of Creation*) as a mere popularizer. Because his public lectures and lay sermons, rather than the lecture courses for his students, propagated agnosticism, scientific

---

1 T. H. Huxley, *Lay Sermons, Addresses and Reviews*, 6th edn, London, 1877, 176–7.

2 G. Berkeley, 'Siris', in *The Works*, 3 vols., London, 1820, iii, 259–418; a journey from the ridiculous to the sublime.

3 A. Desmond, *Huxley: The Devil's Disciple*, London, 1994, especially chs. 14–17. For *Vestiges*, see 193.

This article is reproduced with the permission of the Council of the British Society for the History of Science. It was first published in *BJHS* (1996) **29** 129–38.

130

naturalism and secularism they were particularly exciting; but Huxley fits readily into a long tradition in which science, its prospects and its world view had been diffused, especially, but not exclusively, within the metropolis.[4]

MAKING COMMON SENSE MOMENTOUS

It may seem obtuse to look back to Joseph Priestley, whose stammer excluded him from a career as a great preacher and perhaps lecturer. But Priestley threatened the English hierarchy with an air-pump just as Huxley was to do with skulls; and his materialism, Christian though it was, and democratic politics[5] generated alarm in the same way that, later on, scientific naturalism did. Priestley made his reputation in science with accessible works on electricity and light, and then on gases.[6] After his house in Birmingham was sacked in the 'Church and King' riot (the last so far) of 14 July 1791 he came to London, and lectured in Hackney,[7] which was then quite a salubrious suburb, housing not only a Dissenting Academy but also the headquarters of a High-Church pressure group, the Hackney Phalanx.[8] In the provinces he had been supported by the Lunar Society, but in London was cold-shouldered by Sir Joseph Banks[9] and most Fellows of the Royal Society, anxious to preserve it from radical taint and pilot it through the alarming 1790s.[10] Unable to make his way against the establishment, he emigrated to the USA and settled on the banks of the Susquehanna River.

Priestley saw his mantle falling upon the brilliant Humphry Davy, a young man from a lowly background who like Huxley sought to make a career out of science: but instead of taking on the powers that be, he accepted patronage from Banks and others as the route into success and prominence.[11] His was thus a smoother path than Huxley's was to be; though there was a price to pay, and he later noted resentfully how Banks 'required to be regarded as a patron, and readily swallowed gross flattery... A courtier in character... [he] made his house a circle too like a court'.[12] At the Royal Institution, Davy's skill in lecturing for a fashionable audience, and more formally to improving landowners,[13] brought him an

---

4 On science lecturing in the eighteenth century, see the special issue of *BJHS* (1995), **28**, pt 1.

5 Nelson to the Duke of Clarence, 10 December 1792, in *Nelson and Emma* (ed. R. Hudson), London, 1994, 74f.

6 J. Golinski, *Science as Public Culture: Chemistry and Enlightenment in Britain, 1760–1820*, Cambridge, 1992, 93ff.

7 J. Priestley, *Heads of Lectures on a Course of Experimental Philosophy, Particularly Including Chemistry...* [1794], reprint, New York, 1970.

8 E. A. Varley, *The Last of the Prince Bishops: William Van Mildert and the High Church Movement of the Early 19th Century*, Cambridge, 1992, ch. 3; and P. Corsi, *Science and Religion*, Cambridge, 1988.

9 M. Fitzpatrick, 'Priestley in caricature', in *Oxygen and the Conversion of Future Feedstocks*, Royal Society of Chemistry Special Publication no. 48, London, 1984, 347–69; H. B. Carter (ed.), *The Sheep and Wool Correspondence of Sir Joseph Banks*, London, 1979, 211.

10 D. M. Knight, 'Sir Joseph Banks: Mr Science, 1778–1820', *Interdisciplinary Science Reviews* (1995), **20**, 121–6.

11 R. E. Schofield, *A Scientific Autobiography of Joseph Priestley, 1733–1804*, Cambridge, MA, 1966, 313.

12 D. M. Knight, *Humphry Davy: Science and Power*, Oxford, 1992, 119–20.

13 M. Berman, *Social Change and Scientific Organization; the Royal Institution 1799–1844*, London, 1978, ch. 2.

income which by 1810 was about £1000 a year. He aimed in lecturing 'to excite feelings of interest' rather than 'give minute information'; which for popular audiences is clearly wise, and probably a good general principle.[14] His lectures gave him the means of doing research; which in turn gave authority to the lectures. His auditors had the excitement of hearing about new discoveries and interpretations from the person who had made them, with reflections upon science in general.

Perhaps because of the grind of all this, Davy (again, unlike Huxley) took the classic Dick Whittington step of marrying a wealthy widow; it was not exactly a loveless match, but it was not a happy one. He was subsequently in 1820 elected President of the Royal Society at the age of 41. Adrian Desmond brings to an end his gripping account of Huxley's early life at an equivalent point.[15] Were one to end Davy's biography there, on a note of triumph, one would miss his deep unpopularity but also his attempts to make sense of a life in science.[16] He pleased neither the gentry, nor the Cambridge network. Struck down by disease, in premature retirement he roamed around Italy and Slovenia.

The young radical of the 1790s had turned conservative, admiring the happy peasantry who knew their place, unlike the rowdy and deluded English victims of King Press and the March of Intellect; sympathetic too to the Irish, and much gratified by the passing of Catholic Emancipation at the end of his life. His last work, *Consolations in Travel*, is suffused with the pantheism which was to be characteristic of many men of science, children of both the Enlightenment and the Romantic Movement, associated with Victorian London rather than Oxbridge.[17] Huxley was to be the great exception, using agnosticism, as most of his predecessors had used more-or-less orthodox natural theology, to make sense of science.[18]

Davy's election to the Presidency was assured by his invention of the miners' safety lamp, a great achievement of metropolitan science; and indeed he is one of the first whose laboratory science led directly to a new device.[19] He later wrote that 'real philosophers, not labouring for profit, have done much by their own inventions for the useful arts'.[20] His lectures from the beginning reflected concerns with utility natural in his place and time: by Huxley's day, a gulf had opened between pure and applied science;[21] and chemistry was perhaps strikingly useful.[22] Davy's Introductory Lecture of 1802 ended magnificently:

14 J. Smith, *Fact and Feeling: Baconian Science and the 19th-century Literary Imagination*, Madison, 1994, 78.
15 Desmond, op. cit. (3).
16 I explore this in my essay, 'From science to wisdom', in *Telling Lives in Science* (ed. M. Shortland and R. Yeo), Cambridge, forthcoming.
17 See the Introduction to the *Quarterly Journal of Science* (1864), 1, especially 11f., 22; F. Abbri, 'Romanticism versus Enlightenment: Sir Humphry Davy's idea of chemical philosophy', in *Romanticism in Science* (ed. S. Poggi and M. Bossi), Dordrecht, 1994, 31–45; Smith, op. cit. (14), 231; B. Lightman, *The Origins of Agnosticism: Victorian Unbelief and the Limits of Knowledge*, Baltimore, 1987, 148f.; and D. M. Knight, 'Science and culture in mid-Victorian Britain', *Nuncius* (1996), forthcoming.
18 J. H. Brooke, *Science and Religion: Some Historical Perspectives*, Cambridge, 1991, ch. 6.
19 D. M. Knight, 'The application of enlightened philosophy', in *Sir Joseph Banks: A Global Perspective* (ed. R. E. R. Banks *et al.*), London, 1994, 77–86.
20 H. Davy, *Consolations in Travel*, 5th edn, London, 1851, 256.
21 R. F. Bud and G. K. Roberts, *Science versus Practice: Chemistry in Victorian Britain*, Manchester, 1984.
22 D. M. Knight, *Ideas in Chemistry: A History of the Science*, 2nd edn, London, 1995, ch. 8.

298

132

In this view, we do not look to distant ages, or amuse ourselves with brilliant though delusive dreams, concerning the infinite improveability of man, the annihilation of labour, disease, and even death, but we reason by analogy from simple facts, we consider only a state of human progression arising out of its present condition, – we look for a time that we may reasonably expect – FOR A BRIGHT DAY, OF WHICH WE ALREADY BEHOLD THE DAWN.[23]

Later, in his posthumously published *Consolations*, he wrote that 'science is nothing more than the refinement of common sense making use of facts already known to acquire new facts'.[24] This phrase was echoed in Huxley's famous remark that 'Science is, I believe, nothing but *trained and organized common sense*'.[25] In popularizing science, they both sought to demystify it.

This is not a necessary characteristic of popular lecturing or publishing. Physical science has ever since Copernicus (and the inertial physics of Galileo and Newton) been paradoxical, and relied upon mathematical demonstration (as John Herschel pointed out in one of his popularizations), thus defying common sense.[26] Davy's *Consolations* was intended to draw people into science, though his earlier demonstration experiments in lectures at the Royal Institution may have had the effect of daunting and dazzling the audience – he seemed a genius with magnificent apparatus providing a new spectator-sport for leisured ladies and gentlemen.[27] Perhaps contemplating a decade of lecturing, Davy 'wondered that men of fortune and rank do not apply themselves' to science, and reflected that 'we may search in vain the aristocracy now for philosophers',[28] even though science was the most honourable of activities, and the chemist godlike in the creative energy he applied to matter.[29]

THE STRUGGLE FOR EXISTENCE

Davy, Faraday and Huxley all made their reputations as much through lecturing as research. Theirs were all stories of social mobility: not quite as startling as Napoleon's, retold as a fairy story by Richard Whately – a poor boy from Corsica becomes an Emperor and marries the Princess – but dramatic enough.[30] Napoleon triumphed (until Waterloo) over his enemies; and in the intellectual realm too, images of battle seemed appropriate. This is not surprising for Davy's generation, living through twenty years of French wars. Davy's research and lecturing have to be seen against a background of rivalry with France, the enemy recognized as the scientific great power of the day. Establishing that the acid from sea salt contained none of Lavoisier's acid-generating 'oxygen' was a great victory:

no part of modern chemistry has been considered as so firmly established, or so happily elucidated; but we shall find that it is entirely false – the baseless fabric of a vision... The confidence of the French inquirers closed for nearly a third of a century this noble path of

23 J. A. Paris, *The Life of Sir Humphry Davy*, London, 1831, 89.
24 H. Davy, *Consolations in Travel, or the Last Days of a Philosopher*, London, 1830, 234.
25 Huxley, op. cit. (1), 77 (original lecture, 1854); Huxley's adjectives are both important, as Sir Andrew Huxley remarked to me.
26 J. Herschel, *A Treatise on Astronomy*, new edn, London, 1851, 5.
27 Golinski, op. cit. (6), ch. 7.
28 Davy, op. cit. (20), 239.
29 Davy, op. cit. (20), 259f.
30 [R. Whately], *Historic Doubts relative to Napoleon Bonaparte*, [1819], 7th edn, London, 1841.

investigation, which I am convinced will lead to many results of much more importance than those which I have endeavoured to exhibit to you. Nothing is so fatal to the progress of the human mind as to suppose that our views of science are ultimate.[31]

Here the enemy of truth was C. L. Berthollet, doyen of French chemists;[32] and an enemy alien both to the general British public and to Davy's excited audience, which included the young Faraday assiduously taking the notes which he was to bind up and present to the great man. The tone was suitably belligerent; and quoting Shakespeare very proper on such patriotic occasions.[33]

Davy subsequently accumulated other, domestic, enemies. Controversy over the safety lamp revealed provincial and class hostility,[34] and then his running of the Royal Society lost him further friends. With these enemies he was in the difficult position of being at the summit of science, above fighting back: Huxley seems to have managed better in seeing himself always as one of a small and beleaguered minority, the remnant of the true Israel, specializing in moral indignation.[35] Had Davy lived, his *Consolations* might have lifted him out of this warfare and qualified him as a 'sage', as Huxley was by the end of his life: societies seeking presidents, and reporters soundbites, look after all for a 'man of restless and versatile intellect' ready with words of wisdom on all and every topic.[36]

When elected President of the Royal Society Davy, in his address, compared himself to a general in the army of science, but always happy in research to act as a private soldier.[37] Faraday seems to have had no enemies, so that this element of aggression was absent from his lectures and self-image; and while his relationship with Davy became difficult, it never degenerated to anything like Huxley's with Richard Owen, or Auguste Comte's with Saint-Simon.[38] Davy's early and continuing promotion of Faraday's career at the Royal Institution, and Faraday's Sandemanian lack of worldly ambition, meant that apart from a few priority disputes he was not involved in this struggle for existence:[39] though as John Tyndall noted in an earthy metaphor, he was volcanic underneath.[40]

Huxley, in the next generation, lived in an epoch in which Britain enjoyed profound peace abroad, interrupted only by sideshows in such places as the Crimea, Bengal and the Sudan: his enemies could not therefore be his country's, but were local – notably Owen, who like Davy had risen through patronage, but also any members of what we call the 'establishment', the powers-that-be in church and state.[41] Huxley's ferocity was notorious;

31 H. Davy, *Collected Works*, 9 vols., London, 1839–40, viii, 313f.

32 M. P. Crosland, *The Society of Arcueil: A View of French Science at the time of Napoleon I*, London, 1967, ch. 2.

33 Shakespeare, *The Tempest*, 4.1.151.

34 See D. M. Knight, 'Tyrannies of distance in British science', in *International Science and National Scientific Identity* (ed. R. W. Home and S. G. Kohlstedt), Dordrecht, 1991, 39–53.

35 1 Kings 19:10; Isaiah 10:20–3.

36 Huxley on Wilberforce, in *The Life and Letters of Charles Darwin* (ed. F. Darwin), 3 vols., London, 1887, ii, 322.

37 Davy, op. cit. (31), vii, 15.

38 M. Pickering, *Auguste Comte: An Intellectual Biography*, I, Cambridge, 1993, 245–52.

39 G. Cantor, *Michael Faraday: Sandemanian and Scientist*, London, 1991.

40 J. Tyndall, 'Faraday as a discoverer' [1868], reprinted in *The Royal Institution Library of Science, Physical Sciences*, 11 vols., London, 1970, ii, 66.

41 N. A. Rupke, *Richard Owen: Victorian Naturalist*, New Haven, 1994, 294ff.

134

in 1864 a favourable essay review of his writings on comparative anatomy, and on the skull, included the passage:

> as far as the public are concerned, they either take it as a matter of course that Professor Owen will be attacked whenever Professor Huxley speaks or writes; or they crowd to the lecture hall with the same feelings as they would go to witness a prize fight; all we can say is, that it imparts to the non-scientific world a false estimate of the spirit which exists among scientific men, a very false estimate indeed, and what chiefly concerns us as reviewers is that it does great permanent injury and reduces the value of an author's works, for it is difficult to accredit a writer with strict impartiality, who cannot exercise a little control over his feelings.[42]

Huxley was perhaps a depressive curing himself by finding an external target for his rage: but failures in decorum which offended reviewers might indeed bring in crowds sure of seeing some soldiering in the cause of science; for in popularizing, as in a law court, aggression can be rhetorically compelling and can seem (and be) uncovering or discovering truth.

A good example of Huxley making war was his Royal Institution 'Friday Evening Discourse' of 10 February 1860, on the origin of species.[43] At that time, as Fullerian Professor since 1855, he was giving a discourse each year, as were Owen, Tyndall and Faraday; others who lectured occasionally included William Thomson, Darwin's former Captain (now Admiral) Robert FitzRoy, Charles Lyell, Hermann Helmholtz and John Ruskin. Huxley's lecture was very different from the norm: half of it was taken up with Darwin's book, and how far domestic selection, among pigeons in particular, justified the idea that nature 'must tend to cherish those variations which are better fitted to work harmoniously with the conditions she offers, and to destroy the rest'.[44] Unfortunately, this proposed theory did not measure up to the stringent conditions which Huxley himself had earlier laid down, mentioning Tyndall's work on glaciers (where imagination was controlled by experiment);[45] so it was only a hypothesis. What was lacking was experimental determination of the conditions under which species were producible, and proof that such conditions were operative in nature.

Huxley therefore devoted the second half of the lecture to making a kind of smoke screen. His unease about Darwin's 'hypothesis' might to us seem like Cardinal Bellarmine's about Galileo's; which is not a comparison Huxley would have relished.[46] He set about the 'unscientific' objections to Darwin, which were foolish, frightened and illogical:

> the man of science is the sworn interpreter of nature in the high court of reason. But of what avail is his honest speech, if ignorance is the assessor of the judge, and prejudice foreman of the jury?... And there is a wonderful tenacity of life about this sort of opposition to physical science. Crushed and maimed in every battle, it yet seems never to be slain; and after a hundred defeats it is at this day as rampant, though happily not as mischievous, as in the time of Galileo.[47]

42 'Comparative anatomy and classification', *Quarterly Journal of Science* (1864), 1, 544.
43 T. H. Huxley, 'Species and races and their origin', *Proceedings of the Royal Institution* (1858–62), 3, 195–200.
44 Desmond, op. cit. (13), 267–70.
45 J. Tyndall, 'The scientific use of the imagination', in *Fragments of Science*, 2 vols., London, 1899, ii, 101–34.
46 M. Sharratt, *Galileo: Decisive Innovator*, Oxford, 1994, 118.
47 Huxley, op. cit. (43), 199.

There were 'foolish meddlers' about who thought they did 'the Almighty a service by preventing a thorough study of his works'; who were ashamed of parts of the glorious fabric of the world. All this is a tremendous rant, very offensive to the enemies of science whoever exactly they were – clearly they belonged to the establishment, and this whole performance again shows Huxley biting the hand that fed him, and being admired for it.

He then moved into a patriotic key worthy of Davy, and perhaps appropriate when the democratic United States was drifting into civil war, and when many feared that Napoleon III might try to bring off what his uncle had not achieved – an invasion of the Scepter'd Isle. There was, he said, an intellectual revolution going on:

> the part which England may play in the battle is a grand and a noble one. She may prove to the world, that for one people, at any rate, despotism and demagoguy are not the necessary alternatives of government; that freedom and order are not incompatible; that reverence is the handmaid of knowledge; that free discussion is the life of truth, and of true unity in a nation. Will England play this part? That depends upon how you, the public, deal with science. Cherish her, venerate her, follow her methods faithfully and implicitily in their application to all branches of human thought; and the future of this people will be greater than the past. Listen to those who would silence and crush her, and I fear our children will see the glory of England vanishing like Arthur in the mist; they too will cry too late the woful cry of Guinever:
>
>> 'It was my duty to have loved the highest;
>> It surely was my profit, had I known
>> It would have been my pleasure, had I seen.'[48]

This senatorial eloquence is a rather curious defence of a working hypothesis, and no wonder Darwinians were rather disappointed in their gladiator.[49] Effete, prejudiced or timid opponents there were, no doubt: but not all those who would have gone, and were to go, a long way with Darwin saw things as Huxley did. Polarizing and dividing a community is a tricky business.

The rhetoric of warfare thus has its limitations, when a civil war rather than a foreign one is in prospect; but returning to the Royal Institution twenty-one years later, Huxley in a retrospective lecture saw himself as a general enjoying a Triumph.[50] He began with what in him seems bravado: 'I think it is to the credit of our age that the war was not fiercer, and that the more bitter and unscrupulous forms of opposition died away as soon as they did'; and ended with the claim that 'evolution is no longer a speculation, but a statement of historical fact'. By sidelining natural selection, and emphasizing his own work on *Archaeopteryx* and the analogies between reptiles and birds, Huxley could get away from presenting development as hypothetical. And he concluded, as he was wont to do, with a biblical reference: Darwin like William Harvey 'has lived long enough to outlast detraction and opposition, and to see the stone that the builders rejected become the head-stone of the corner'.[51]

---

48 Huxley, op. cit. (43), 200. For the verses, see J. Pfordresher (ed.), *A Variorum Edition of Tennyson's Idylls of the King*, New York, 1973, 962.

49 F. Burkhardt, D. M. Porter, J. Browne and M. Richmond (eds.), *The Correspondence of Charles Darwin*, 9 vols., Cambridge, 1985–95, viii, 84.

50 T. H. Huxley, 'The coming of age of the "Origin of Species"', *Proceedings of the Royal Institution* (1879–81), **9**, 361–8.

51 Matthew 21:42.

136

SPREADING THE WORD

Preaching the gospel of scientific naturalism is one thing, but popular scientific lecturing requires cake as well as icing. Henrietta Huxley referred to the working men 'whose cause my husband so ardently espoused'; he was not the first to lecture on science to such audiences, but he made it an important part of his mission.[52] The March of Intellect, and the coming of cheap books and periodicals, led to a generation anxious for improvement, and practising self-help; and Huxley saw himself as a plebeian with a duty to help others from his order. In 1863 he gave a famous course of six lectures on our knowledge of the causes of the phenomena of organic nature (this audience was not frightened by a long title),[53] and in 1868 an exemplary lecture on a piece of chalk at the BAAS at Norwich.[54] For these audiences, polemic and invective were inappropriate; being earnest was important, and Huxley could do that too.[55]

The common sense which Huxley espoused, and which involved his agnosticism, put the man of science and the man of sense more or less on a level;[56] and one of Huxley's strategies in this kind of lecturing was to make his hearers realize that they did science all the time, as Molière's M. Jourdain wrote prose.[57] When the evidence was clearly presented to them, an audience of working men could like an open-minded jury come to the right conclusion. We do almost hear his voice. Huxley's lectures make such excellent reading, and must have made wonderful theatre, because one feels drawn along with him in his reasoning, and not just given information. No wonder he was 'the great and beloved teacher, the unequalled orator, the brilliant essayist, the unconquerable champion and literary swordsman'.[58]

The 'Piece of Chalk' was a single lecture, beginning with an introduction, followed by various facts, then a suggested explanation, and a rhetorical flourish as conclusion: a classic pattern, which suited Huxley's determination as far as possible to keep facts and theory apart. He could also use his own research, which always makes the best lectures, on the nature of the creatures whose remains composed the chalk; and vividly evoke past epochs.[59] We cannot recover from a written text, and the statuesque photographs of Huxley's day, what the lecture must have been like: professing is after all a performance art. But clearly he always held audiences in the palm of his hand. In his conclusion, 10 per cent of the text (just as in his classic textbook on crayfishes)[60] is devoted to an exposition of evolution, because 'the mind is so constituted that it does not willingly rest in facts and immediate causes, but seeks always after a knowledge of the remoter links in the chain of

52 T. H. Huxley, *Aphorisms and Reflections* (ed. H. Huxley), London, 1907, p. vi.

53 T. H. Huxley, *Darwiniana*, London, 1894, 303–475.

54 Huxley, op. cit. (1), 174–201.

55 Oscar Wilde's play was first performed in 1895.

56 Lightman, op. cit. (17), 14.

57 Huxley, op. cit. (53), 363ff.

58 E. R. Lankester, in the entry for Huxley in R. B. Freeman, *Charles Darwin: A Companion*, Folkestone, 1978, 170.

59 See, for example, M. J. S. Rudwick, *Scenes from Deep Time: Early Pictorial Representations of the Prehistoric World*, Chicago, 1992.

60 T. H. Huxley, *The Crayfish: An Introduction to the Study of Zoology*, London, 1880, 317–46.

causation'. This ends without the bullying to which the patrician audience at the Royal Institution had been subjected: as with his students, Huxley invited his working men to

> Choose your hypothesis; I have chosen mine. I can find no warranty for believing in the distinct creation of a score of successive species of crocodiles in the course of countless ages of time. Science gives no countenance to such a wild fancy; nor can the perverse ingenuity of a commentator pretend to discover this sense, in the simple words with which the writer of Genesis records the proceedings of the fifth and sixth days of the Creation.[61]

The operation of natural causes underlies all that we see.

His peroration was splendid, using metaphors from science:

> A small beginning has led us to a great ending. If I were to put the bit of chalk with which we started into the hot but obscure flame of burning hydrogen, it would presently shine like the sun. It seems to me that this physical metamorphosis is no false image of what has been the result of our subjecting it to a jet of fervent, though nowise brilliant, thought tonight. It has become luminous, and its clear rays, penetrating the abyss of the remote past, have brought within our ken some stages of the evolution of the earth.[62]

This audience would thus have been edified without being mystified. Students would have got something of the same from the course on the crayfish; but the objectives in lecturing to them would not have been the same. They had an examination to pass. The best of them might be recruited into science; lecturers long to replicate themselves like 'selfish' genes, and Huxley was father in science to most of the prominent physiologists and zoologists of the next generation. With working men it was different; he was casting bread upon the waters, hoping that it might return after many days – but while some of his hearers may have taken up formal science, most had perforce to go on doing it informally, as Huxley had told them they were accustomed to do. They could not be trained and organized into original science; but some science would enable them to detect false reasonings and become more responsible citizens. It was an important part of an education.

Huxley confounded someone who had admired someone else's oration, by demanding to know just what had been said; and indeed the world little notes nor long remembers such details. We have seen how a reputation could be enhanced, if not made, by popular lecturing. Inaccessible texts led to exciting talks.[63] To the next generation, making a career this way was strange; and Michael Foster, Huxley's greatest pupil, accounted for it in terms of self-sacrifice: 'to a large extent [he] deserted scientific research and forsook the joys which it might bring to himself, in order that he might secure for others that full freedom of inquiry which is the necessary condition for the advance of natural knowledge'.[64] We will be less tempted to see Huxley as a Bodhisattva; finding in his bravura performances careerism as much as scientism, but certainly helping to bring in professional science.

In popularizing, he could not draw upon utility as convincingly as Davy: but his science too was in its age of innocence. He made powerful rhetorical use of images from battle and warfare and enemies, but the science he popularized still seemed harmless – unlike politics

61 Huxley, op. cit. (1), 199, 201.
62 Huxley, op. cit. (1), 201.
63 T. H. Huxley, *The Oceanic Hydrozoa*, London, 1859.
64 M. Foster, in *Huxley Memorial Lectures* (ed. O. Lodge), Birmingham, 1914, 36.

138

and religion. And above all, he could still portray it as essentially common sense; he had begun on HMS *Rattlesnake* as an explorer, and it was as an explorer rather than a proposer and tester of recondite hypotheses that he matured, and presented himself to his publics. Common sense, anger, information both wide and deep, and great rhetorical skill made him into the consummate orator whose voice, as we read, we can almost hear:

> So word by word, and line by line,
> The dead man touch'd me from the past,
> And all at once it seem'd at last
> The living soul was flashed on mine.[65]

65 A. Tennyson, *In Memoriam* [1850] (ed. S. Shatto and M. Shaw), Oxford, 1982, 112 (from section 95). Huxley admired the poet and this poem, especially its insight into scientific method: see L. Huxley, *The Life and Letters of Thomas Henry Huxley*, 2nd edn, 3 vols., London, 1913, iii, 268ff.

XXIV

SCIENCE AND CULTURE IN MID-VICTORIAN BRITAIN:
THE REVIEWS, AND WILLIAM CROOKES'
QUARTERLY JOURNAL OF SCIENCE

In the summer of 1895, Thomas Henry Huxley[1] died. He was a noteworthy anti-clerical, coiner of the term «agnostic», and particularly famous as «Darwin's bulldog». In April 1995 there was an international conference[2] to commemorate his life, held at Imperial College in London under the auspices of the Royal Society, and in the presence of his surviving grandson, Sir Andrew Huxley, himself also a former President of the Royal Society. The first volumes of new

---

[1] Alan Barr is editing a volume of essays on Huxley for University of Georgia Press, 1996.

[2] My BSHS Presidential Address on Huxley, «Getting Science Across», was delivered there and will be published in *BJHS*.

biographies of both Huxley[3] and Charles Darwin[4] have just come out; and although the latter emphasises the strength of the alliance of moderate Anglican churchmanship with science, we might be tempted to see 1859, the year when *The Origin of Species* was published, as the triumph of irreligion and scientism.[5] Huxley's lecture at the Royal Institution in February 1860,[6] and his debate with Bishop Samuel Wilberforce in Oxford in the summer, have caught the attention of commentators. It might seem that science and religion became locked in a struggle for existence in the brave new world that science opened up. And yet we may ask how new a world it was; and what the cultural position of science in mid-Victorian Britain really amounted to. One way to do this is to look at widely-read general journals, the great Reviews; and particularly at a new one devoted to the sciences.

The nineteenth century was the great age of the Review; and just as the *Edinburgh Review* in the early years had carried reviews of papers by Thomas Young, Humphry Davy and others, so in the 1850s scientific writings were regularly noticed. We are now uneasily aware of «two cultures», humanistic and scientific; but such a division came late to Britain, whatever the religious or political viewpoints of editors. anonymity[7] was a crucial feature of the reviews, though authorship was usually an open secret; and reviewers cultivated an easy, knowing style which by the 1850s was very different from the extracts from a book with some linking commentary which in the 1790s had often constituted reviewing. Sympathetic exposition rather than tremendous judgement had also become the goal of reviewing by the early 1860s,[8] and sectarian party-spirit had cooled. Regular readers of a Review would become familiar with the latest writings across a very wide range of disciplines; indeed, that was the point.[9] And yet in

---

[3] A. DESMOND, *Huxley: the Devil's Disciple*, London, Michael Joseph, 1994.

[4] J. BROWNE, *Charles Darwin: Voyaging*, London, Cape, 1995.

[5] See my «Arthur James Balfour (1848-1930), Scientism and Scepticism», *Durham University Journal*, 87 (1995), 23-30.

[6] T. H. HUXLEY, «On Species and Races, and their Origin», *Proceedings of the Royal Institution*, 3 (1858-62), 195-200.

[7] A. DE MORGAN, *A Budget of Paradoxes*, [1872], ed. D. E. Smith, New York, 1969, vol. 2, pp. 15-20; but see P. STANSKY'S introduction to JOHN MORLEY, *Nineteenth Century Essays*, Chicago, 1970.

[8] J. WOOLFORD, «Periodicals and the practice of literary criticism, 1855-64», in J. SHATTUCK and M. WOLFF (ed.), *The Victorian Periodical Press: Samplings and Soundings*, Leicester, 1982, ch. 5.

[9] See W. HOUGHTON, «Periodical literature and the articulate classes», in SHATTUCK and WOLFF, *op. cit.*, ch. 1.

1864 the up-and-coming chemist William Crookes,[10] painfully making a career, began with the scientific journalist James Samuelson something like a quarterly Review devoted to science.

The learned journal with signed research papers had been a feature of the sciences in Britain for two hundred years,[11] but played a small part in the humanities. The historian or the critic expected to publish books, or spacious reviews; and reviewing constituted a respectable or higher kind of journalism, often in fact advancing knowledge rather than simply commenting. Eminent reviewers would publish these essays in due course in book form; they were meant to have more than immediate significance. It may not seem clear how this might fit science, especially in its more technical branches, like the chemistry of the time; but the great Cambridge polymath William Whewell (1794-1866) has been described[12] as a 'metascientific commentator' or critic, and the critic's role in calling attention to the work of Emily Brontë or George Eliot was perhaps akin to high-level popularizing, then a respected activity for men like Huxley. Certainly, the reviewer in whatever discipline expects to draw attention to excellence, and to leading practitioners. At all events, the early *Quarterly Journal of Science* seems to have had important features in common with the Reviews.

In order to place Crookes's, it will be helpful to look at other publications. Walter Bagehot the political philosopher was editor of the new *National Review*, taking a liberal stance, which by 1857 had reached volume five. It contains an essay on physiology, looking at Huxley, W. B. Carpenter and Richard Owen; another on Charles Waterton and natural history; and another on the exploration of Africa. Further articles examine the Grimms' folk tales; education in the Army, and in the University of London; and the career of Henry Brougham, the lawyer and politician, with some reference to his science. Two essays are concerned with religion, one with F. C. Baur's suggestion that St John's Gospel was essentially a work of fiction, and on «higher criticism» generally, and the other on unspiritual religion. Otherwise, we read about the Brontës' life and writings, the Manchester art exhibition, the Indian Mutiny, London architecture and other things. In my copy, a contemporary reader has written at the

---

[10] W. H. BROCK, *The Fontana History of Chemistry*, London, Fontana, 1992, pp. 454-461.

[11] A. J. MEADOWS (ed.), *The Development of Science Publishing in Europe*, Amsterdam, Elsevier, 1980.

[12] R. YEO, *Defining Science* (Cambridge, 1993), pp. 62 ff.

46

end of the «Brougham» piece: «a most flimsy article, written by some tyro, who knew as much of Brougham's powers as a mere frenchman w^d have known»; but even that one reads well after nearly a century and a half.

The *North British Review*, also a quarterly, with its Scots base and its evangelical tone [13] reads differently from the *National* but its width of cover is similar. Indeed its volume 33 for 1860 contained more on Brougham, on East Africa, and on the assault on the authority of the Bible represented by the «higher criticism» belatedly reaching Britain from Germany, in the notorious volume suggestively entitled *Essays and Reviews* to which Baden Powell was a notable contributor. [14] Another essay was concerned with modern thought, or unbelief, in the writings of Sara Hennell, who was roughly handled. Straightforward science came in articles on planetary theory, with reference to Urbain Leverrier and the claim that the planet Vulcan, within Mercury's orbit, had been spotted. There is also an essay on modern theories in meteorology (looking at the possible connection between weather and sunspots), and another on the trial of Galileo; and one on colonial constitutions. Another article examines the state of logic in Britain, with special reference to William Hamilton and J. S. Mill, but also involving Henry Mansel and Augustus de Morgan. The other essays are political, religious and literary.

Science in the Reviews was therefore present, but in an Aristotelian manner, [15] for gentlemen who wanted principles rather than detail: «an educated man should be able to form a fair off-hand judgement as to the goodness or badness of the method used by a professor in his exposition. To be educated is in fact to be able to do this». For ladies in the nineteenth century, this would be even truer: science about 1860 was a component of a liberal education. In 1862, a new scientific monthly was launched: *The Intellectual Observer*. Its theme was «natural history, microscopic research and recreative science», and it had coloured illustrations as well as woodcuts. It contained a mixture of signed articles and anonymous reviews; one of the latter in volume one is concerned with Leonard Jenyns' biography of J. S. Henslow, Darwin's pious father-in-science. One of the articles, on astronomy, is by a woman, the Hon. Mrs Ward (the reader should

---

[13] See J. SHATTUCK, «Problems of patronage: the *North British Review* and the Free Church of Scotland», in SHATTUCK and WOLFF, *op. cit.*, ch. 6.

[14] P. CORSI, *Science and Religion: Baden Powell and the Anglican Debate, 1800-1860*, Cambridge, C.U.P., 1988.

[15] ARISTOTLE, *De Partibus Animalium* (tr. W. OGLE, Oxford, 1911), 639^a.

be warned that Shirley and Grace could be male names at this date); and my copies were presented to her «dear Uncle Charles by his very grateful and affectionate niece». The edges of the volumes are gilt, and the effect rather pretty.

With its emphasis upon natural history, and its mixture of signed articles and anonymous reviews and reports of meetings, this journal was not very different from others such as the *Magazine of Natural History*[16] of a generation before; but the tone is much less quarrelsome. And by including the sublime science of astronomy, very popular always with amateurs (who as in natural history could make serious contributions to knowledge), and some chemistry, it was breaking rather different ground. There is a good deal about the Exhibition of 1862; and an éloge of the Prince Consort. Again, we have magnetic and atmospheric perturbations; we also have spectroscopy,[17] announcing Crookes' discovery of thallium, one of the great scientific events of the day especially as it involved a priority dispute with a Frenchman.

The main emphasis is upon conveying information attractively in signed articles, but the volume begins with a report on the scientific work of 1861, and there is an essay on the Domestication of Science – these are more like material in the Reviews. The latter[18] celebrates our emergence «from the profound barbarism of the Georgian era» (1715-1837), and the wide welcome that the magazine had received «in every town». The domestication, meaning reception into the home, of science and literature was a «most important incident for civilization»: its lack was the reason for the failure of Mechanics' Institutes.[19] Only at the Royal Institution[20] had scientific culture really grown, and even there the professors were paid a mere pittance; each town ought to be teaching elementary science, and Davy was quoted on behalf of science for women: «Let them make it disgraceful for men to be ignorant, and ignorance will perish; and that part of their empire founded upon mental improvement will be strengthened and exalted by time, will be untouched by age, and will be immortal in its youth». Domestic life

---

[16] See my «William Swainson: Naturalist, author and illustrator», *Archives of Natural History*, 13 (1986), 275-290, esp. 275-276.

[17] See my *Ideas in Chemistry*, London, Athlone, 2nd ed. 1995, p. 160.

[18] *The Intellectual Observer*, 1 (1862), 471-475.

[19] I. INKSTER and J. MORRELL, *Metropolis and Province: Science in British Culture, 1780-1850*, London, Hutchinson, 1983.

[20] See my *Humphry Davy: Science and Power*, Oxford, Blackwell, 1992, reprint Cambridge, C.U.P., 1996.

without ideas is worse than a bottle with no wine in it; and the happy home is the intelligent one, where nature is studied and human interests not forgotten. Such homes make the most useful of families, making all wiser, removing evil and doing good. Homes and families were thus the especial target of this journal, making it very characteristic of its period.[21]

Two years later, in 1864, the first issue of the *Quarterly Journal of Science* appeared, with a manifesto preceding the articles. The mixture of anonymity and of signed «original articles» reminds us of the *Intellectual Observer*; but there is only one coloured illustration (in a paper by Samuelson), and the effect is rather more serious although the appeal is meant to be wide. There are extensive reports and criticisms of the recent meeting of the British Association, in Bath: the narrowness of focus of the Presidential Address by Darwin's mentor Charles Lyell[22] is deplored, but his agreement on basics with Sir Roderick Murchison is stressed in case people might suppose that science is a business of personalities. Elsewhere, in one of the reviews, the bitter hostility of Huxley to Owen[23] is noted:[24] «as far as the public are concerned, they either take it as a matter of course that Professor Owen will be attacked whenever Professor Huxley speakes or writes; or they crowd to the lecture hall with the same feelings as they would go to witness a prize fight». The reviewer deplored such a spirit among scientific men, for whom impartiality was essential; for the possibility of permanent injury to science through ill-temper was real. Science was an earnest matter requiring the gentlemanly ethos. The *Quarterly Journal of Science*, like the Reviews, was clearly aimed at the upper middle class.

In the other great *Reviews* questions of religion kept on coming up. It would be difficult for us to take Victorian religion too seriously.[25] We tend to see religious language as a cloak for other more genuine (and generally base) feelings; but when our ancestors said «Come into my heart Lord Jesus» they meant it, and even when they lost their faith they could not leave the subject alone. Religion was not a matter of indifference, as it has become in contemporary Britain. In particular,

---

[21] D. CHERRY, *Painting Women: Victorian Women Artists*, London, 1993, ch. 7.

[22] *Quarterly Journal of Science*, 1 (1864), 736, 740.

[23] This is central to N. RUPKE, *Richard Owen: Victorian Naturalist*, New Haven, 1994. Owen had primed Wilberforce for his debate with Huxley.

[24] *Quarterly Journal of Science*, 1 (1864), 544.

[25] M. WOLFF, in SHATTUCK and WOLFF, *op. cit.*, p. 374.

SCIENCE AND CULTURE IN MID-VICTORIAN BRITAIN

atheism was associated with immorality; that was part of the reason why Huxley coined the word «agnostic», which also implied a lack of dogmatism appropriate to the man of science. In the *Quarterly Journal of Science* there is a hostile review of an atheistic work by the German materialist Louis Büchner.[26] The author is handled very roughly, and the review ends with Bacon's aphorism: «A little philosophy inclineth man's mind to atheism, but depth of philosophy bringeth men's minds to religion». The morality, or rather immorality, of atheists and materialists is particularly remarked upon; and the author is blamed for having done his very worst for science and for himself, as a fanatic who in another age would have been an equally unreasoning religious zealot. We might note that Huxley and his circle were the first generation to make irreligion respectable; but this had not happened by 1864. There is also a sympathetic review[27] of a pamphlet defending the idea of a «vital principle», though elsewhere there are various favourable accounts of the work of Louis Pasteur and others.

Religion, and concern about the cultural position of science, duly feature in the manifesto which forms the introduction to this first volume of the journal.[28] Here we find Science praised both as a component of a liberal education, and for its usefulness, in comparison with Art. The «devotees» of science, we are told «are rather men of thought and action than of wordy eloquence», and are often unappreciated by comparison with the talented historian going into politics, or the eloquent theologian becoming a Bishop. The «grandest revelations» of science (note the religious language) «being frequently held up to scorn and obloquy, and twisted and tortured until they were made to appear the teachings of the Evil One», makes scientific men a kind of persecuted minority. We are reminded of Huxley's remark about extinguished theologians surrounding the cradle of every science like the serpents round that of Hercules. Gentlemen saw personal religion as generally worthy of respect, but felt free to attack «theologians»: it is not quite clear who these people were, and utterances like these may perhaps best be seen as Metropolitan ripostes to comfortably-off gentlemen[29] from Oxford and Cambridge

---

[26] *Quarterly Journal of Science*, 1 (1864), 545-554.

[27] *Quarterly Journal of Science*, 1 (1864), 729-731.

[28] *Quarterly Journal of Science*, 1 (1864), 1-23.

[29] J. MORRELL and A. THACKRAY, *Gentlemen of Science*, Oxford, O.U.P., 1981; on PALEY, J. H. Brooke, *Science and Religion*, Cambridge, C.U.P., 1991, and my *The Age of Science*, Oxford, Blackwell, 2nd ed. 1988, chapter 3.

50

who like the famous William Paley saw the world as made by a Cosmic Watchmaker.

Certainly, if one were looking for great theologians, Victorian Britain would be the wrong place. Just as it was a «land without music», so it was more or less destitute of distinguished thinkers in this field. There had been S. T. Coleridge;[30] there were F. D. Maurice, Mansel, J. H. Newman and perhaps E. B. Pusey, the writers of *Essays and Reviews*, and Bishop Colenso. Several of these lost their jobs through alleged heresy: being a theologian in fact led to persecution much more regularly than being a scientist ever did. Equally, these major theologians (except perhaps Pusey[31]) were not in any way opposed to or afraid of science as such; and if we were to include amongst theologians those who thought deeply about God, freedom and immortality, then we might have to include laymen such as Alfred Tennyson, and even John Tyndall:[32] Mansel's *via negativa* was close to agnosticism.[33]

The Romantic pantheism we find for example in Davy's *Consolations in Travel*[34] (1830) was as important to men of science as anything more orthodox, and indeed in the *Quarterly Journal* we find the meteorologist and balloonist James Glaisher quoted[35] on the religious influence upon him of aerial flights. «I have experienced the sense of awe and sublimity myself ... I am an overwrought, hard-working man, used to making observations and eliminating results, in no way given to be poetical, and devoted to the immediate interest of my pursuit, and yet this feeling has overcome me in all its power. I believe it to be the intellectual yearning after the knowledge of the Creator, and an involuntary faith acknowledging the immortality of the soul». Scientific knowledge, unless misused, will never, according to the manifesto, «lower man's religious nature»; but because it drives superstition before it, it may make some enemies.

Zealots and dogmatists, rather than theologians, were thus the target for this manifesto; in which science is displayed as both intellectually exciting, and also useful. There are mentions of improvements to marine steam engines, and also of acclimatization: we

---

[30] S. T. COLERIDGE, *Aids to Reflection*, ed. J. Beer, Princeton, 1993.

[31] P. BUTLER (ed.), *Pusey Rediscovered*, London, 1983, ch. 1.

[32] W. H. MALLOCK, *The New Republic [1877]*, intr. J. LUCAS, Leicester, 1975.

[33] B. LIGHTMAN, *The Origins of agnosticism: Victorian Unbelief and the Limits of Knowledge*, Baltimore, 1987, pp. 65-67.

[34] D. KNIGHT, *Humphry Davy: Science and Power*, Oxford, 1992, ch. 12.

[35] *Quarterly Journal of Science*, 1 (1864), 11.

look with unease upon the transportation of convicts, animals and plants to Australia, for example, but to our ancestors this seemed a straighforward case of progress. The introduction ends with a purple patch, urging that science deserves a foremost place because in disciplining the mind it imparts a purer and more elevated conception of God, and prepares us for a purely spiritual existence, and comprehension of the highest truths. Not only the tone of its reviews, but also its interest in religion, make the *Quarterly Journal of Science* not unlike other Reviews; and altogether unlike Crookes' other (and more successful) journalistic venture, *Chemical News*, which was directed to working chemists, making a career in academe or in industry, and formed a model for Norman Lockyer's more general journal, *Nature*.

On the other hand, the *Quarterly Journal* also contained original signed articles like other scientific journals; and it speedily incorporated the *Edinburgh New Philosophical Journal*. The papers in the first number are more popular than those in a research journal: Crookes for example writes about the Atlantic Cable rather than describing his current experimental work; and while there is a review of progress in chemistry, that science is much less prominent than we might have expected, given Crookes' role. Whereas the ordinary scientific journal of the day aimed at communicating information, the Review is more Aristotelian in trying to get understanding across: though by this time, as John Herschel had pointed out in his *Outlines of Astronomy*[36] some years earlier, precise understanding without mathematics or other symbolic notation would not be possible. Popular science, and even *haut vulgarisation*, has its limitations; and the public for the *Quarterly Journal of Science* was unspecialized.

Other Reviews carried on as before, incuding some science in their catholicity. Thus the *Edinburgh Review* for the second half of 1866 had a paper on meteorology, and another on the exploration of Africa; something about the Indian Mutiny (not in the author's opinion a «rebellion»); about a European single currency, based on the franc; on the Prussian army; on George Eliot's *Felix Holt*; and on Lives of Jesus, with praise for John Seeley's (anonymous) *Ecce Homo* rather than for Strauss and Renan. There was also an essay on Mill and Hamilton – in short, the mixture as before: but the problem remained, of how the educated layman was to keep up with more technical science, and with rather more of it than was possible in some 10% of a general Review.

---

[36] J. F. W. HERSCHEL, *Outlines of Astronomy*, new ed., London, 1851, pp. 5-9.

52

The *Quarterly Journal of Science* had had its origin[37] in a letter from Crookes to Samuelson in 1861,[38] offering a «very perfect retrospect of Chemical Science», and adding that «I am so much in the habit of writing on *any* scientific topic which may arise, and have such excellent opportunities for getting the best scientific information, that I have no doubt I could please you on any subject connected with my department of science which you may suggest». Crookes was unsalaried, trying to live by consultancy and journalism; but following the discovery of thallium he had been elected FRS in 1863, and his position vis-a-vis Samuelson had changed by the end of the decade. Though the journal never made money, it gave Crookes an outlet and a position: but during the 1860s it went through various editorial changes, with Samuelson assuming greater prominence and other names being added to Crookes'. In 1869 Crookes wrote a «rather pungent» letter[39] to Samuelson, noting that «I am vain enough to think that I confer more than I receive» and offering to resign. The upshot was that by 1871 Crookes alone became both editor and proprietor; he was thus «in sole control».

By that year, the *Quarterly Journal* had diverged from the Review model, and become in effect a vehicle of popular science for a generally well-educated readershp. There are a few anonymous articles, notably on the patent system, but these have more the character of editorials. The book reviews, except for an essay on Darwin's *Descent of Man*, have become brief, though they are still unsigned (until 1873); there is no room for extended discussion of context, or for discursive commentry. There are still extensive reports of new developments in the sciences, which include meetings of scientific societies, but these are in smaller type. The Journal now has a new sub-title: *Annals of Mining, Metallurgy, Engineering, Industrial Arts, Manufactures and Technology*. This dauntingly «applied» emphasis seems belied by the actual contents: which include papers on the Great Pyramid by Piazzi Smyth (the eccentric Astronomer Royal for Scotland), on spectra, germ theory and astronomy, and A. W. Hofmann's reminiscences of the early days of the Royal College of Chemistry, where Crookes had studied. Volume 1 had contained papers about artillery, and gun-cotton: and these topics of «War Science», are also prominent in

[37] This context is discussed in D. KNIGHT, *The Age of Science*, Oxford, 1986, pp. 198-202.

[38] E. E. FOURNIER D'ALBE, *The Life of Sir William Crookes*, London, 1923, p. 69.

[39] FOURBER D'ALBE, *op. cit.*, pp. 186-188.

1871, when there had been a terrible accidental explosion of gun cotton, and when Britons were uneasily aware of the Franco-Prussian war and the place of science in it.

The first volume had contained one theoretical paper of some significance,[40] William Odling's «periodic table»; but neither Crookes nor Odling took that particular idea further, and the paper seems to have attracted no notice in the chemical community. In 1871, Piazzi Smyth's work on the Pyramida[41] was accompanied by some of Crookes'[42] on «psychic force», making the journal seem some way from the main stream of science. As well as popularizing pure and applied science, the *Quarterly Journal* thus provided space for papers which other editors would reject, and allowed Crookes to ride what others saw as his hobby-horse, psychical research – which only later became of major intellectual importance in Britain, as Henry Sidgwick and others especially from Cambridge sought evidence for life after death.[43]

By 1871 then the *Quarterly Journal* had lost its resemblance to a Review; but in that same year, the *North British* ceased publication, while the *National* had run only from 1855-64. The tide was moving against the quarterlies. Monthlies, and magazines with signed essays and with short stories, began to replace them. Crookes sold the *Quarterly Journal* in 1879, and it became a monthly in a third series from 1879 to its demise in 1885; but a change in frequency was not the answer. Specialization had come to Britain, and the old attempt to take all knowledge for a journal's province, or even all science, came to an end. In the 1880s, the Royal Society's journals were divided into Physical and Biological parts; only the Royal Institution's *Proceedings*, printing the Discourses held there, and the British Association's *Report*, continued to cover almost everything. Elsewhere, there was a new world of experts, and a narrower vision of what everybody ought to know a bit about. It looks again from this evidence as if specialization and the growth of professions are more significant in

---

[40] W. ODLING, «On the proportional numbers of the elements», *Quarterly Journal of Science*, 1 (1864), 642-648.

[41] C. P. SMYTH, «The Great Pyramid of Egypt», *Quarterly Journal of Science*, 8 (1871), 16-35, 177-214.

[42] W. CROOKES, «Experimental investigation of a new force», *Quarterly Journal of Science*, 8 (1871), 339-349, 471-493.

[43] See my paper, «Observation Experiment, Theory – and the Spirits», *Durham University Journal*, 83 (1991), 55-58.

54

understanding the role of people like Huxley or Crookes than any clear divide between organized religion and science.

XXV

Observation, Experiment, Theory - and the Spirits

"Let us commend to the love of God with silent prayer the soul of a sinful man who partly tried to do his duty. It is by this wish that I saw over his grave these words and no more."[1] This bleak farewell was written by Henry Sidgwick for his own funeral in 1900, when his "old hope of returning to the Church of his fathers had not been fulfilled"; but in the event the Church of England burial service "was used without question" when his body was buried in the village churchyard at Terling.

Sidgwick is a very important figure in the secularisation of intellectual life in the later nineteenth century; although he remained a theist rather than becoming an agnostic or an atheist; theism being that Victorian descendant of Socinianism which involved belief in a Creator not wholly separate from the world.[2] It was his high-minded resignation from his Cambridge Fellowship in 1869, because he no longer felt he could subscribe to all the doctrines of the Established Church, which made the imposition of religious tests at the universities of Oxford, Cambridge and Durham generally repugnant, and led to their removal; Sidgwick being eventually elected to the Chair of Moral Philosophy at Cambridge in 1883. He wrote the classic Utilitarian *Methods of Ethics* (London, 1874), divorcing morality from religion; and also taught Political Economy [3] in what became an important school, and which he hoped to put upon a scientific basis. He was one of the Cambridge Apostles[4]; he and his wife were pioneers of Higher Education for women through Newnham College, Cambridge; and he was astonishingly well-connected, one brother-

---

1 A.E.M. S[idgwick], *Henry Sidgwick: A Memoir*, London, 1906 ,pp.598ff.
2 W.E.Gladstone, *The State In Its Relations With the Church*, 3rd ed., London, 1839, pp. 23, 289, R. Flint, *Anti-Theistic Theories*, 5th ed., Edinburgh, 1894, pp. 441ff. C.B. Upton, *Hibbert Lectures*, London, 1894, pp.328ff
3 S.Collini, D.Winch & J.Burrow, *That Noble Science of Politics*, Cambridge, 1983, pp.339-63
4 P. Allen, *The Cambridge Apostles*, Cambridge, 1978

2   OBSERVATION, EXPERIMENT, THEORY - AND THE SPIRITS

in-law becoming Archbishop of Canterbury (E.W. Benson), another Prime Minister (A.J. Balfour), and another President of the Royal Society (Lord Rayleigh). As a disciple of J.S. Mill, he had a great respect for the sciences; and throughout his life he could be depended upon to be on the right side of such things as battles to reform Cambridge life and introduce new disciplines there. In 1882 it was beneath his aegis that the Society for Psychical Research (SPR) was founded, and he was the first President. Unlike Crookes in the 1870s, Sidgwick made the subject seem appropriate for careful study; and the SPR soon attracted a large membership, and a strong Council of men eminent in science and in public life in Britain and overseas.

Spiritualism is an Anglo-American topic, for it came from the revivalist and burned-over districts of the USA to Britain; where by the 1860s it had moved up market, and seances had become a craze.[5] Mediums were usually young women, and the unusual opportunities for groping in the dark meant that they were not generally respectable; some men, like the Rev. Stainton Moses, were mediums, and the celebrated question "Where was Moses when the light went out?" is supposed to have been asked about him rather than about the Patriarch. In her recent and valuable study[6] Janet Oppenheim has described the Spiritualists; but while in the introduction she remarks of psychical researchers that "their concern and aspirations placed them - far from the lunatic fringe of their society - squarely amidst the cultural, intellectual, and emotional moods of the era", she uses the unhelpful category "a pseudoscience" to describe their activities in the text. It seems better to see it as an unsuccessful science; not for want of trying, and especially not for want of discussion of methods and aims which could be taken for granted in better-established disciplines; in which many of its practitioners were also engaged, often very successfully.

The great hope was that human immortality, or at least survival of "bodily" death, might be proved; and in an era of scientism,[7] empirical science seemed the only respectable method of proof. Not only was this a consolation for the bereaved, and a way of making sense of life here, but it seemed a necessary cement for society. Without the prospect of judgement to come, it seemed to the earnest, and to the affluent in an age ever aware of the possibility of revolution, that there would be no reason for people to behave well; "let us

---

5 C.M.Davies, *Unorthodox London*, 3rd ed., London, 1875

6 Oppenheim, *The Other World*, Cambridge, 1985, W.R. Cross, *The Burned-Over District*, Ithaca, 1950. On the nineteenth-century ideas of death as family reunion, see P. Ariès, *The Hour Of Our Death*, Harmondsworth, 1981, pp.409ff

7 See my book, *The Age of Science*, Oxford, 1986; new ed. 1988, esp. chapter 11

eat, drink and be merry for tomorrow we die." In their *Unseen Universe* (1875) two Professors of Physics, Balfour Stewart (later President of the SPR) and P.G. Tait, argued for a future life using the principle of continuity, and ideas from thermodynamics; but also glumly reported on the delinquencies of the times, and more cheerfully on the prospects of deterring criminals by torturing them with electric shocks.[8] Their book was assailed by W.K. Clifford, a high-churchman turned agnostic and a spokesman for triumphalist science; but it was a powerful ally to churchmen in the intellectual uncertainties of the day, as well as a skilled popularization of up-to-date physics. The arguments were however merely analogical; no direct evidence of immortality was forthcoming through such a route. In contrast, psychical research offered the chance of an experimental test.

The problem from the start was the possibility of fraud; a difficulty in all science[9] but especially acute where phenomena were not reproducible in broad daylight and on all occasions. For science to work at all, one has to assume that people do generally tell the truth; and the inculcation of truth to fact was one of the arguments for the place of science in a liberal education. The SPR came to rely upon the testimony of respectable persons, generally from the professional classes; and on attempts by Sidgwick and others to control the circumstances, especially where "physical" manifestations took place. If a medium were caught out in fraud, then all observations in which he or she had ever been involved were deemed to be discredited. If a phenomenon could be produced by a professional conjuror (by the nineteenth century this formidable word meant entertainer) then this cast doubt upon its spiritual origins in a seance.

This corrosive scepticism meant that Sidgwick himself found no satisfactory evidence for immortality by the time of his own death; and Frank Podmore, one of the SPR's team of investigators, could in 1897[10] report only a grim story of falsifications and unreliability both of sense data and of witnesses; sometimes involving fraud, and generally the imposition of "dramatic unity" by which we make sense of experience. He could draw satisfaction only from experiments on telepathy, which could be done under much better conditions, in laboratories rather than darkened drawing-rooms,

---

8 [B. Stewart & P.G.Tait], *The Unseen Universe*, London, 1875, p. 143; W.K. Clifford's review is in his *Lectures and Essays*, London, 1879, vol. 1., pp. 228-53. See also G.C. Knott, *The Life and Scientific Work of P.G. Tait*, Cambridge, 1911
9 A.Kohn, *False Prophets*, Oxford, 1986; A. Gregory, "The Anatomy of a fraud", *Annals of Science*, 34 (1977) 449-549
10 F Podmore, *Studies in Psychical Research*, London, 1897

4   OBSERVATION, EXPERIMENT, THEORY - AND THE SPIRITS

and which might account for some phantasms and for clairvoyance; though an explanation of how telepathy might work was still a desideratum.

The SPR investigated particular mediums, finding many things more strange than true (objects flew around the room, and spirits assumed bodily forms); but their great work had been on phantasms seen when a loved one is in some great crisis, often at the moment of death. Such visitations were widely reported in stories which stood up to investigation; in them the tyranny of distance was overcome, for sons or husbands facing death in India or in Australia appeared by English bedsides. Indeed the material collected and published in 1886[11] gives a fascinating insight into social life when the Empire was at its height. The investigators, Podmore, Edmund Gurney and Frederic Myers, thought that they had established that the connections were not just a matter of chance; a census of hallucinations showed that they were much more frequent at such crises than ordinarily, and generally there was no reason for apprehension at that particular time. They went on to do experiments in telepathy, with cards held up in different rooms; and again found occasions on which some telepathic communication could not apparently be doubted.

Myers, reviewing this and later investigations, came to the conclusion that survival of death was proved; and in his classic work, *Human Personality*,[12] dedicated to two friends who happened to be dead (Sidgwick and Gurney) and published posthumously, he argued strongly for it. His first volume has recently been praised by Oliver Sacks[13] for its descriptions of hypnosis and dissociated personalities, and its remarks on our unconscious or subliminal parts; it was Myers who first introduced Freud to English readers: but his second volume is harder going, being concerned with phantasms of the dead. Some hallucinations come from our own subliminal regions; some by telepathy from other minds; and some, he believed, from the minds of the bodily dead. He tried, in the Darwinian manner, to establish a continuous series through these categories; and believed that the evidence was good enough to make his cumulative argument convincing.

Myers believed that one characteristic at least of a genius was that he or she was on good terms with his subliminal mind; but (as Oppenheim remarks[14]) since ordinary people are for Myers a hierarchy of personalities, mind may indeed not be equivalent to brain - but it is hard to pick upon one fragment of

11   E. Gurney, F. Myers and F. Podmore, *Phantasms of the Living*, London, 1886
12   F.F.H. Myers, *Human Personality and Its Survival of Bodily Death*, London, 1903
13   O. Sacks, *The Man Who Mistook His Wife for a Hat*, London, 1986, pb., pp. 193f, 232
14   J. Oppenheim, op.cit. note 5, p.260

321

OBSERVATION, EXPERIMENT, THEORY - AND THE SPIRITS   5

personality to survive death. While the whiggish historian of psychology would find much of interest, as a piece of science in its own right the book leaves much to be desired. Darwinian arguments were based upon a mechanism, natural selection, and not just on continuous series of cases; and Myers and the others could not, though in the era of cathode rays and X-rays, find a plausible mechanism for telepathy even between living people. Their studies remained a series of anecdotes for which so far no explanation was forthcoming, like accounts of miracles (an analogy Myers saw, believing the SPR could vindicate stories of miracles against Humeans); and thus not really a branch of science, where theory must order facts.

The idea that the sciences were bodies of positive knowledge which stood upon their own two feet was challenged by Frederick Temple in his Bampton Lectures of 1884. [15] Temple had been Headmaster of Rugby, Sidgwick's old school; had contributed to *Essays and Reviews* (1860), that liberal bombshell; was Bishop of Exeter, and was to become Archbishop of Canterbury. In a Kantian manner, he saw physics as imposing an order upon things; and as resting upon a "Supreme Postulate" of Uniformity of Nature; he also believed that we were conscious of our own permanence through the flux of our lives, and as a Bishop believed in eternal life.

A more radical challenge to the sciences came from Balfour, one of Sidgwick's earliest pupils and his colleague in the SPR, in his book on doubt in 1879; [16] it was to have had the title *A Defence of Philosophical Scepticism*, but Balfour then realized that this would be interpreted as religious scepticism - whereas it was to science that he directed his corrosive doubts, as befitted a Conservative statesman. Britain has never had the benefit of a philosopher-King, but in Balfour it had a philosopher Prime Minister; the experiment was not very successful, but it is curious that in Maurice Cowling's idiosyncratic investigation of secularization Sidgwick has only a walk-on part among the baddies, and Balfour does not appear.[17] Balfour's uncle and predecessor Lord Salisbury does, as a good pessimist; and for him science offered the hope of precision which he would have liked to find in politics, where history offered the best laboratory there was. But in the sphere of religion Salisbury was one of the Church of England's most devout sons, keeping the realm of faith

---

15 F. Temple, *The Relations Between Religion and Science*, London, 1885
16 A.J. Balfour, *A Defence of Philosophic Doubt,* London, 1879.
17 M. Cowling, *Religion and Public Doctrine in Modern England,* Cambridge, 1980, 1985, in progress

6   OBSERVATION, EXPERIMENT, THEORY - AND THE SPIRITS

distinct from that of empirical science. [18] Balfour, with his close connections with Sidgwick, could not occupy quite the same position.

For him, science was a matter of belief and not of reason; and it was in danger of becoming a new scholasticism: "That men of Science should exaggerate the claims of Science is natural and pardonable, but why the ordinary public, whose knowledge of Science is confined to what they can extract from fashionable lectures and popular handbooks, should do so, it is not quite easy to understand".[19] Science for Balfour rests upon assumptions, and is based on coherence; and in a later book he sought to disconnect it from agnostic empiricism, which he called naturalism,[20] arguing for theism, for the possibility of miracles, and for the view that the sufferings of Jesus should make us accept our own without questioning God's goodness, thus going beyond theism into orthodox Christianity.

If science is a fascinating business but not a royal road to truth - and it is sensible for its historians to see it that way [21] - then psychical research could not be ruled out: antecedent probability was a poor guide to what might turn up. Faraday was alleged to have said that investigators ought to approach an inquiry with "preliminary notions of the naturally possible and impossible"; [22] this aroused the fury of a psychical researcher but is clearly sensible, what scientists do faced with tall stories, and need not commit anybody to belief that all science is true.

Balfour's *Foundations of Belief* aroused sufficient interest to lead to the formation of Synthetic Society in 1896; it was "to consider existing agnostic tendencies, and to contribute towards a working philosophy of religious belief". Its membership included Myers and Sidgwick, the very elderly James Martineau and the young William Temple, Lord Rayleigh and Oliver Lodge. Papers of a fairly informal kind were circulated; Balfour and others urged that faith was the road to knowledge in all fields; the place of the expert was questioned; and Roman Catholics, Anglicans and theists discussed ecumenism. From the printed papers[23]we can get an interesting picture of

---

18 P Smith (ed), *London Salisbury on Politics*, Cambridge, 1972, pp.21, 67

19 A.J. Balfour, op.cit. note 16, p.307

20 A.J. Balfour, *The Foundations of Belief: Being Notes Introductory to the Study of Theology*, 2nd ed., London, 1895

21 D.M. Knight, "The History of Science in Britain: a Personal View", *Zeitschrift Fur Wissenschaftstheorie*, 15 (1984) 343-53

22 Anon, "Human Levitation: Illustrating certain Historical Miracles", *Quarterly Journal of Science*, 12(1875)31-61, p.33

23 *Papers Read Before the Synthetic Society, 1896-1908,* London, 1909, esp. pp.20ff, 193f, 212ff, 269

liberal intellectuals interacting over a decade. The value of the SPR's researches was urged by Myers, meeting some of the scepticism earlier directed by the orthodox (who wanted something more than theism) against natural theology in its Paleyan or Bridgewater manifestations; but by the early twentieth century, and with the deaths of its leaders, psychical research was beginning to lose some of the éclat it had had a generation earlier in the heyday of theoretical scientism. Nevertheless, C.D. Broad, a successor to Sidgwick in the Knightsbridge chair of Moral Philosophy at Cambridge and in the Presidency of the SPR, continued to urge[24] the importance of psychical research; but by now with much less support from eminent scientists or clergy.

The idea of supplying an empirical basis for religious faith, which is more akin to believing in Britain than to believing in quarks, has always been doubtful; and yet Christianity is an historical religion, and miracles have formed part of its claim. One of the troubles with psychical research was and is that the information garnered from the spirits, if one believed it, was generally banal; apart from that, there were extraordinary manifestations of doubtful significance, except perhaps to indicate how ignorant we were of the laws of nature. Evidence of human levitation would undermine classical physics; in the same way modern "creationists" hope that anomalies will weaken or refute Darwinism. But scientists prefer order to confusion, as Bacon said they would, and can live with a few anomalies, especially if the evidence for them is shaky. We have become used to living with a more provisional science than Sidgwick's generation, who put their faith in the luminiferous ether; but we do still put faith in science, as Faraday recommended.

Scientific investigations, like statistics or psychical research, may arise out of religious or political situations and convictions; but they move away from the questions of value which at first make them so exciting, especially to those on the outside. Ambiguous evidence, the use of scientific terms simply as metaphor, the tedious business of testing, all conspire to make things less interesting as a field matures. Psychical research turned out to be profitable only isnofar as it could be transformed into experimental psychology; to which its relationship is perhaps like that of alchemy to chemistry - an ancestor of which one should perhaps be ashamed, though (or perhaps even because) it involved men of restless and versatile intellect rather than sober specialists.

---

24 C.D. Broad, *Religion, Philosophy and Psychical Research*, London, 1953

324

8   OBSERVATION, EXPERIMENT, THEORY - AND THE SPIRITS

Perhaps we should end with Edward Caird, giving an address on Immortality as Master of Balliol College in October 1905[25] "Direct proof of immortality cannot be had, or not in a conclusive form, but if we believe in God, immortality seems to follow as a natural, perhaps we should say as a necessary consequence. For if we think of the world as the manifestation of a rational and moral principle - and that it must be so conceived seems to be a necessary presupposition of all our mental and moral life - we must regard it as existing for the realization of that which is best and highest; and that best and highest we can hardly conceive as anything but the training and development of immortal spirits." The philosophical reign of Mill and Spencer was over and there was a new confidence that science and naturalism were not enough: psychical research was no longer necessary as a prop to faith, and could be returned from the scientists to the enthusiasts with whom it had begun.

---

25 E. Caird, *Lay Sermons and Addresses,* Glasgow, 1907

XXVI

Arthur James Balfour (1848-1930), Scientism and Scepticism[1]

There is a small group of distinguished people who have declined an official nomination to the Presidency of the Royal Society. Two of them are Robert Boyle, author of the *Sceptical Chymist* (1660) in which his spokesman Carneades wrestles with the dogmatists of the day and argues that scepticism is necessary to the progress of inquiry in science: and A.J.Balfour. I was astonished to find that Balfour had in 1920 been invited to consider the office by J.J.Thomson on behalf of the Society's Council, because it seemed to go against all the preconceptions current amongst historians of science.

Balfour had been elected in 1888 not as a man of science but as someone prominent in public life with scientific interests; and historians of the science of the nineteenth century have studied the process of professionalization extensively, coming to believe that they were charting the end of the reign of the amateur. Indeed, in contrast perhaps to historians of philosophy, the majority of historians of science have chosen to examine the social history of their subject[2] rather than confine themselves to looking closely at the writings of distinguished practitioners; which can be reductive or rewarding depending chiefly on who is doing it. But what this invitation shows is that in 1920 Balfour had a high reputation among scientists, who would have been prepared to have him as their figurehead. He turned the offer down, when it was explained to him that work was expected of the President; and became in 1921 President of the British Academy instead. He was also Chancellor of the Universities of Edinburgh and Cambridge.

These offices came to him as, using the words of his entry in the *Dictionary of National Biography*, a "philosopher and statesman": as a statesman he is now chiefly remembered for the Balfour Declaration of 1917 which led ultimately to the State of Israel, but as a philosopher he is not much remembered at all. This is a pity, because while he was not a great original thinker he had interesting things to say, particularly about the scientism[3] of the late nineteenth century. When he was born, in 1848, there had been two major recent developments in the sciences. In 1844 the anonymous book *Vestiges of the Natural History of the Creation* had been published, synthesizing astronomical, biological and anthropological data within an evolutionary framework. This book, which turned out to have been written by the popularizer and publisher Robert Chambers of Edinburgh, sold more copies in the nineteenth century than the *Origin of Species* did; it was denounced by experts, but was an enormous success in promoting the idea of progressive evolution by natural causes alone. Then came the discovery or postulation of energy conservation, something which Thomas Kuhn has seen as a classic case[4] of simultaneous discovery; which brought together a number of previously-separate sciences (Mechanics, Electricity, Heat, Optics and others) into what became classical physics. These two syntheses in a time of increasing specialization brought a new confidence to men of science, but could bring gloom also: Nature, as in Tennyson's *In Memoriam* cared for nothing, all must go, in the pointless evolutionary whirl which would in the end come to a halt in the heat-death of universal tepidity forecast in thermodynamics[5].

Balfour was educated chiefly at home by his mother Blanche Cecil, sister of the Lord Salisbury who was to be Balfour's political patron. She was "profoundly religious and brilliantly amusing", and both these qualities seem to have been passed on to her son, who went to Eton, and then to Trinity College Cambridge where he was one of that obsolescent aristocratic and privileged group, the Fellow Commoners; there he read Moral Sciences and got to know Henry Sidgwick very well. During Balfour's childhood, Henry Mansel had in 1858 published his Bampton Lectures on *The Limits of Religious Thought* in which[6] he brought scepticism to bear upon natural theology, in what he intended to be support of genuine religion. Trying to tread the narrow line between rationalism and dogmatism, he stressed the unreliability of our senses and intellect.

In the following year, Darwin published the *Origin of Species*, making evolution by natural selection scientifically respectable[7]: Balfour met Darwin's son George at Cambridge, and visited Down; at University he became enthusiastic about science, like his uncle, and his later sceptical doubts were all intended to promote scientific inquiry and not to stifle it. Although Mansel became Dean of St Paul's, his via negativa in theology did not become popular among theologians; but was taken up by agnostics, to such an extent that in a recent history[8] he appears among them. Both Darwin and Mansel, therefore, were important in forming the world-view of religious sceptics of the 1860s and 70s.

On coming down from Cambridge Balfour did not have anything very obvious to do, and fell in with his uncle's proposal that he should stand for parliament for the safe "family" seat of Hertford. In that year two significant events were the publication of J.W.Draper's *History of the Conflict between Religion and Science*, and John Tyndall's *Belfast Address*.[9] Draper was an American, teaching in New York, and a prominent agnostic - probably for that reason excluded from the American Academy of Sciences. His speech had set off the Wilberforce-Huxley debate at Oxford in 1860[10] and his view was that extinguished theologians lay about the cradle of every science like the serpents about that of Hercules, though that phrase is T.H.Huxley's.

Tyndall was a more important man of science, indeed one of the leaders among those proud to see themselves as professional scientists[11]. He followed Faraday as Professor at the Royal Institution in London, and carried on there the tradition of popular lecturing to a select audience at a very high level. Presidents of the British Association for the Advancement of Science were expected to dilate upon the triumphs of the last year or two, and the prospects for the immediate future; and generally they also urged the need for more support and recognition from government. Tyndall chose instead to give a kind of agnostic sermon, profoundly shocking to his hearers and denounced from all pulpits in Belfast, Protestant and Roman Catholic, on the following Sunday. Tyndall's particular work had been on heat and its explanation in terms of motion of particles; and when the Chemical Society of London had in 1867 and 1869 debated the atomic theory, he had been one of those who favoured a realistic rather than instrumental attitude to it[12]. By the 1870s, the successes of structural organic chemistry (which could not be understood without atoms differently arranged in space) and the needs of an educational system now organised by the state, meant that Tyndall's belief in atoms had become general though not universal in the scientific community.

Tyndall also believed in the aether, the curious substance which carried the light waves; for the undulatory theory of light was one of the great successes of the nineteenth century, and James Clerk Maxwell's *Treatise*, linking optics to electromagnetism, had come out in 1873. Atoms, aether and evolution were the basis of Tyndall's science, and he believed that they represented *scientia*, rather than provisional generalizations. With this basis he claimed the whole realm of cosmological theory from the theologians.

His address represents a high point in the confidence of scientists of the later nineteenth century; though it would be a mistake to see it as typical of the men of science of the period, it seemed characteristic of some of the best of them. His great ally, T.H.Huxley, in a famous phrase said "science is I believe, nothing but *trained and organized common sense*"[13]; an idea Balfour, perhaps in part because of his closer connections with physical scientists concerned with more abstract questions, was to assail.

Also in 1874 William Crookes came to the end of a series of experimental seances in his investigations of spiritualism. In the presence of respectable and distinguished professional men, the most extraordinary manifestations had taken place; especially involving the medium D.D.Home, who had floated about in the air. Crookes, an eminent analyst and spectroscopist elected to the Royal Society for his discovery of the element thallium, believed that it was the duty of scientists to study psychic phenomena; and submitted his results to the Royal Society for publication. Amidst a row, the paper was turned down; and Crookes had to publish it in his own journal.

Crookes' work led to Sidgwick and his wife Nora (Balfour's sister), Balfour and others themselves working in this field; and in 1882 the Society for Psychical Research was founded[14]. Sidgwick was hoping for an empirical basis for religion, and Balfour's fiance May Lyttleton died in 1875; but their interest does not seem to have been the intense hope of the bereaved to contact the departed, but a revulsion at the dogmatism of those who shared Tyndall's materialistic world view. Open-minded scepticism was the way forward in inquiry for Balfour; dogmatism closed doors behind which important things might be found.

In 1878 Balfour accompanied Disraeli and his uncle to the Conference of Berlin; but he was at this stage little more than Lord Salisbury's sidekick. Darwin's cousin, Francis Galton, in his *Hereditary Genius* (1869),remarked how many eminent men had eminent uncles, and attributed it to heredity; but Balfour's career seemed to be under the direct control of his eminent uncle. Then in 1879 he published his first book, *A Defence of Philosophic Doubt*; it was to been Philosophic Scepticism, and the book was apparently advertised under that title; but because "scepticism" had come at that date to be applied almost exclusively to religion, Balfour changed the title at the last moment, and explained why in the preface. Being sceptical about religion in 1879 was easy and common; but being sceptical about science was more unusual, and to take on the dogmatism of Tyndall and Spencer while admiring science (being the Mansel of the "church scientific", to use another phrase of Huxley's) required some audacity.

The book is well written, like all Balfour's writings, and the examples are amusing[15]. He dismissed Kant as being "technical without being precise" and remarked of modern German philosophy that "if it gives little light it is not because it is hidden under a bushel". The point of his run through modern philosophy was to argue that there is no certainty to be found in science, or in philosophy, which for him was the analysis of beliefs. He was duly sceptical about Mill on induction, and Jevons on probability; and remarked on the enormous difference between the world views of common sense and science: of tables and chairs, and of colourless atoms and mysterious aether. We can make no inferences to things in themselves; and we have no reason to prefer the scientific world view to any other:

> Science ... fails in its premises, in its inferences, and in its conclusions. The first, so far as they are known, are unproved; the second are inconclusive; the third are incoherent. Nor am I acquainted with any kind of defect to which systems of belief are liable, under which the scientific system of belief may not properly be said to suffer.

Science had simply come to replace dogmatic theology as dominant; but like everything else, science rested upon belief. Whether we like it or not, we are all believers. Balfour's was not the kind of scepticism that kept him awake at night; he accepted and enjoyed science, religion and beauty, but believed that we must not be too worried about inconsistencies between them, a necessary feature of our finite minds.

He was particularly indignant about evolutionism, admirer of Darwin though he was, because scientific theories should not be made into world-views. In particular, he saw no reason why if our minds are simply the products of natural selection, leading to our ancestors having fed themselves and left descendants, they should be good at arriving at true beliefs. Our beliefs might be just a stage in the progress of the world. There was therefore no certainty to be found in the world of Herbert Spencer.

This seems to us perhaps a paradox, in that we might expect evolution to generate true beliefs, about tigers, sharks and toadstools for example; but Balfour was well aware of how we can be misled by senses or intellect, and was also thinking about beliefs more abstract than this, as A.R.Wallace did in his writings on natural selection as he came to diverge from Darwin[16]. Wallace could not accept that the mathematical and philosophical capacities which suddenly blossomed in ancient Greece could have been in latent form of any use to the first humans; and felt that these capacities must have been planted by God in mankind. But for Balfour all that is necessary is to accept that one cannot be certain that evolution would in all cases lead to true beliefs.

All our beliefs for Balfour are corrigible and provisional; and this view perhaps did not go down very well in 1879. Balfour had got a second-class degree, having a "mind perhaps too independent for a curriculum"[17] and his book might have been better received if there had been more history of philosophy in it, and perhaps if it had been more solemn. It was nevertheless, unlike his subsequent writings (and unlike the *Origin of Species*), a scholarly volume with footnotes. But in the 1880s Balfour became a very prominent public man. In 1885 the motion that brought down Gladstone's government and divided his party after the failure of the Home Rule bill for Ireland was drafted in Balfour's house; and in 1887 he was put in charge of Irish affairs in his uncle's government. His task was to impose law and order upon Ireland, and amid general surprise he proved astonishingly effective. The languid aesthete coerced the Irish more effectively than toughs like Oliver Cromwell and William of Orange; and when in 1891 W.H.Smith died, Balfour was the obvious candidate to be Leader of the Conservatives in the House of Commons. The government of Britain until 1902 when Balfour became Prime Minister was (with a brief break) in the hands of a lugubrious aristocrat in the House of Lords and his witty nephew.

As Leader, Balfour introduced reforms of procedure which curbed the possibilities of delay of which the Irish members had become masters; and also changed the half-day from Wednesday to Friday, affecting the pattern of social life among the governing classes. It has been suggested that all Balfour's thought is rooted in the class struggle of his day; this seems unnecessarily reductive[18], but his conservatism and his philosophy went very easily together, and his political importance made his philosophy interesting to contemporaries.

In 1894 he published his second book, *The Foundations of Belief: being Notes introductory to the Study of Theology*; and this one aroused great curiosity and sold very well. The subtitle of his first book had been *Foundations of Belief*, so Balfour was evidently not expecting to say anything very different. What are striking here are the even more explicit parallels between science and religion, as systems resting upon belief - the uniformity of nature, and the existence of God. He sought to disentangle true science from "Naturalism", a monistic and agnostic empiricism which he found inadequate and incoherent[19]. Our various beliefs, about God, Nature and Beauty are not coherent, but we should not worry about that: God reconciles our beliefs and their origins, and anyway all our beliefs are approximations at best. We must not impose a dogmatic pattern upon the world. In his portrayal of the gloomy and selfish world associated with naturalism, Balfour sometimes rose to heights of eloquence:

Man will go down into the pit, and all his thoughts will perish. The uneasy consciousness, which in this obscure corner has for a brief space broken the contented silence of the universe, will be at rest. Matter will know itself no longer. "Imperishable monuments" and "immortal deeds", death itself, and love stronger than death, will be as though they had never been. Nor will anything that is be better or be worse for all that the labour, genius, devotion, and suffering of man have striven through countless ages to effect.

While adopting sceptical doubts to weaken naturalism, he believed that naturalism fully embraced could not but lead to complete scepticism. Here we find touched upon lightly that long standing fear that such scepticism will lead to social chaos, the jungle of Hobbes, found for example in the *Unseen Universe*[20] written by the prominent physicists Balfour Stewart and P.G.Tait: but Balfour can manage humour as well as seriousness. His horror of all-embracing world views is perhaps typical of the conservative tradition since Burke.

One effect of Balfour's book was the foundation of the Synthetic Society[21], devoted to exploring the issues he had raised and modelled on the Metaphysical Society which had brought together earnest doubters and religious thinkers a generation earlier. We duly find among the papers of that society Balfour urging that "All human beliefs are due to non-rational causes". Henry Sidgwick's last evening[22] before the operation for cancer from which he never really recovered was spent with Balfour at 10, Downing Street; and in 1902 Balfour became Prime Minister in succession to Lord Salisbury.

Another of Balfour's sisters had married the physicist Lord Rayleigh, successor of Maxwell at the Cavendish Laboratory in Cambridge, and predecessor of J.J.Thomson; one of the greatest of classical physicists. Balfour could therefore through another brother in law pick up the latest ideas in the sciences; and in the 1890s physics was undergoing a scientific revolution[23], especially with the work of Crookes and then Thomson on the cathode rays, and the discovery of X-rays and of radioactivity. The electron, both the first sub-atomic particle and the unit of electrical charge, was identified as composing the cathode rays following Thomson's work at the Cavendish in 1897[24]. The old certainties of classical physics were collapsing, and a sceptical attitude in science seemed called for. Physics was becoming even more remote from common sense.

In 1904 the British Association met in Cambridge, and Balfour was President: the thought of a Prime Minister with a high enough reputation in the intellectual world to hold such office is astonishing to us today. Balfour chose as the title of his address, "Reflections suggested by the New Theory of Matter"; it was only moderately new, for 1904 was Einstein's *annus mirabilis* and quantum theory and relativity were not yet known to Balfour[25]. Nevertheless, he had plenty of grist for his mill. A brisk run through some history of science with a particular focus on Cambridge men led him to the reflection that our understanding of the material world was very new. The wave theory of light, and thus the aether, went back only a century; and the history of the serious study of electricity was not much longer - in 1700 electricity seemed no more than a toy. Now our understanding of the universe depends on these things, and the idea of "mass" must be revised and explained in electrical terms, following Thomson's work on the electron. Matter occupies space only metaphorically, as soldiers occupy territory.

We had been primarily concerned with the feebler forces of nature:

> but if the dust beneath our feet be indeed composed of innumerable systems, whose elements are ever in the most rapid motion, yet retain through uncounted ages their equilibrium unshaken, we can hardly deny that the marvels we directly see are more worthy of admiration than those which recent discoveries have enabled us dimly to surmise.

There were no promises of utilitarian value in this new physics, which thus restored to us the sense of wonder; and we could not tell whether it would survive or be replaced by some new drawing on the scientific palimpsest. Scientists should not wait patiently upon experience:

> the plain message is disbelieved, and the investigating judge does not pause until a confession in harmony with his preconceived ideas has, if possible, been wrung from the reluctant evidence.

Science is the imposing of patterns, leading in this case to matter being analysed in such a manner that it is explained away. Science is not concerned with phenomena, but with the inner character of physical reality; natural selection has produced our senses for utility, but not for theoretical science. This realism about physical theory was a part of the Cambridge ethos[26] of the nineteenth century; but for Balfour it was not the whole story, or the final one: there was no reason to believe that the world was simple, and every reason to believe that incoherence was inevitable.

If the new physics was true, then down to about five years before we had lived in a world of illusions about what we saw and handled. Perhaps falsehood had been more useful than truth: anyway, the senses and the intellect cannot be trusted; science is a matter of extracting from experience beliefs which experience contradicts. He concluded that the new science, which could rouse absorbing interest among those who like him were no specialists, fortified his "own personal opinion, that as Natural Science grows it leans more, not less, upon a teleological interpretation of the universe." Balfour's address could hardly be more different from that of Tyndall, reflecting not only his own preconceptions but also the change through which physicists had come; and it was a good piece of haute vulgarisation.

Balfour lost the election of 1906, and was out of office until the Great War, when he entered the Cabinet first at the Admiralty following Winston Churchill, and then at the Foreign Office, where he promulgated the Balfour Declaration. His published writings are reticent about the details of his religion, but there is an interesting suggestion[27] that British policy had an apolyptic dimension, so that restoring the Jews to the Holy Land might hasten on the Second Coming of Jesus and the Millenium. If Balfour believed that, he did not put it into his philosophical writings; and it seems the rather enthusiastic sort of belief that he would have tended to be sceptical about. By this time he was seen as an elder statesman, happy to serve his country in its hour of need but somehow above the pettinesses of party; and through most of the 1920s he continued in office in Baldwin's governments, where he took charge of the administration of government science and the promotion of scientific and industrial research, including Kapitza's magnetic studies at the Cavendish[28].

Just before the War, he had been invited to deliver a course of Gifford Lectures, and these were completed in 1922-3 in Glasgow, and published[29] under the titles *Theism and Humanism*, and *Theism and Thought: a Study in Familiar Beliefs*, in 1914 and 1923. Balfour had by 1923 been interested in philosophy for more than half a century, but he was at pains to stress his own consistency, and we do not see much sign of development from the basic ideas he had published in 1879. The use of methodological doubt, and the conclusion that we all live by faith, are familiar; but in these lectures, delivered extempore and then written down, there is more analysis of common-sense beliefs. Compared to the British Association *Address*, with its emphasis upon the scientist's wrestling with reality, we find less confidence: science is again said to be based upon falsity, or at least upon uncertain assumptions and untenable empiricism. The problem as before is that the worlds of common sense and experience, and of science, are so different.

We also find causes and reasons for belief being carefully distinguished, the former being evolutionary and the latter logical: there being no grounds for being sure that evolution will lead to truth. Most interesting is his engagement with Bertrand Russell, a philosopher from a similar social background but associated with another party. Balfour was delighted at Russell's use of sceptical arguments, but saw this new logic as leading to genuine scepticism: whereas for him, though "neither the authority of science, nor the consentient belief of all mankind, nor the individual opinions of eminent philosophers, are permitted to confer immunity or paralyse criticism", this criticism was the

work of reason. "If reason be on trial, it also presides over the court".
Balfour's was therefore no "barren scepticism": "though the method I follow
be often critical in form, it leads, I think, to conclusions which, however
tentative and provisional, are none the less constructive in substance and
rational in method". Using doubt constructively was what he had always been
trying to do.

By 1923 not only had the kind of atomic theory espoused by Tyndall been
overtaken by events, but the evolutionary synthesis which had also come in in
Balfour's childhood seemed less certain: in particular, Darwinism was in
eclipse, because natural selection did not seem to proponents of the new
genetics[30], like William Bateson, sufficient to produce major evolutionary
change. A jumpy progress through relatively gross mutations was the received
opinion of the day: but such eclipse of a theory did not worry Balfour, for
whom science might genuinely progress but never actually arrive at indubitable
truth. As an appendix to the 1923 volume, one of the 1914 Gifford Lectures
was reprinted; it is of some interest in that it deals with probability theory, and
Balfour distinguishes qualitative and quantitative probability. Admiration for
Berkeley comes out throughout his writings, and there is also some for Butler,
so generally respected throughout the nineteenth century in Britain.

During this period, Butler's idea of probability as the guide to life seemed
generally convincing; and the rise of statistics and of inductive and deductive
probability theories brought mathematical certainties into the realm of chance.
Maxwell's theory of gases of 1859 depended quantitatively upon probability,
while Darwin's theory published in the same year rested on a more qualitative
version. Balfour believed that not all qualitative probabilities could in principle
be quantified. Just as in the *Foundations of Belief* we had an interesting
discussion of authority, with the feeling that there is a good deal to be said for
it; so here we find some remarks about intellectual fashion: which is certainly
something with which the historian of science has to be concerned, and cannot
be ignored in the history of philosophy either.

Balfour's writings are not a bit dry and technical, and he raises all sorts of
questions attractively. He told a fellow M.P. in 1916 that if philosophy was
worth anything there was no reason why it should be made dull[31]. Professional
philosophers, a grouping that perhaps began in England with Sidgwick, the
first person not ordained as a clergyman to hold the Chair of Moral Philosophy
at Cambridge and who taught an ethics quite separate from religion, probably
did not take Balfour's philosophical writings as seriously as they might have
done; he was perhaps a layman's philosopher rather than a philosopher's
philosopher. It would be hard also to say that he was a philosopher of immense
originality and profundity; but he was a public man who took his

philosophizing seriously, and he did engage with some of the great issues of his time. In particular, he took on scientific naturalism at a time when it was carrying all before it; and lived to see himself apparently justified in the revolution in physics beginning in the 1890s.

His obituarist in the *Dictionary of National Biography* was struck with his opposition to nationalism in Ireland, and his promotion of it in Israel, and compared him[32] to "Halifax the trimmer" of the Restoration period; but saw in him a political genius which was "transitional, evolutionary and in that sense creative". His own nephew, the second Lord Rayleigh, wrote his obituary for the Royal Society, remarking that Balfour had often expressed the wish that he had been a scientific man, but that he had not enough patience with drudgery, and was uninterested in precision of measurement (the first Lord Rayleigh's forte).

He drew attention to Balfour's dislike of following prescribed courses at university, and his uninterest in the history and the literature of philosophy, and also to his indifference to newspapers and to praise or criticism; which one may suspect was a mandarin pose. He reports Balfour's sympathy with psychical research, and his acceptance of the reality of telepathy; and tells us how Balfour made Rayleigh explain Maxwell's work to him, being good at seizing upon essentials though unsystematic. Most interestingly, he remarks of Balfour's sceptical attitude that "it must be admitted, however, that an attitude very like this has been taken up in modern science", referring to wave-particle dualism.

Balfour would have liked to hear that; and indeed on July 16th 1928, when he was nearly eighty, he addressed[33] the British Institute of Philosophical Studies, of which he was President, at their third annual meeting. He told them that we "lived in a world of illusions"; suggested that "those made no effort to get beyond the teaching of common sense should do so in no boastful or self-confident spirit"; and added that "whenever they found common sense opposed to science, let them throw in their lot every time with science". But he reminded scientists that science was always incomplete and not always true, though on the way to truth: "its ultimate basis, deeper even than experiment and observation, was faith". Could there be, he asked, a better justification for a philosophical institution? On that note, as apt for us as for them, the meeting ended with cheers.

12   J. A. BALFOUR - SCIENTISM AND SCEPTICISM

Notes

1. This paper was like Balfour's Gifford Lectures delivered extempore and written up afterwards, and I would like to thank those who asked such interesting questions, some of which I hope I have been able to answer. This practice was I found in defiance of the advice given by Henry Sidgwick, in his "Lecture against Lecturing" in his *Miscellaneous Lectures and Addresses*, London, 1904, pp.340-52 where he urged that things should be written down.

2. J.Morrell and A.Thackray, *Gentlemen of Science*, Oxford, 1981; M.B.Hall, *All Scientists Now: the Royal Society in the 19th Century*, Cambridge, 1984; M.Rudwick, *The Great Devonian Controversy*, Chicago, 1985, and "A year in the life of Adam Sedgwick and company, geologists", *Archives of Natural History*, 15 (1988) 243-68; J.Golinski, *Science as Public Culture: Chemistry and Enlightenment in Britain, 1760-1820*, Cambridge, 1992; and my "The History of Science in Britain: a Personal View", *Zeitschrift fur allgemeine Wissenschaftestheorie*, 15 (1984) 343-53.

3. See *D.N.B.*, "A.J.Balfour". *O.E.D.* gives the first use of "scientism" in 1877, and "scientistic" as a depreciative adjective in 1878. On scientism, see my *The Age of Science*, Oxford, new ed. 1988.

4. The paper is reprinted in T.S.Kuhn, *The Essential Tension*, Chicago, 1977.

5. J.A.V.Chapple, *Science and Literature in the Nineteenth Century*, London, 1986, ch. 1 & 2.

6. H.L.Mansel, *The Limits of Religious Thought examined in Eight Lectures*, 4th ed., London, 1959; this has an interesting preface. The reference to Balfour's mother is from D.N.B., "Balfour".

7. On the reception of the *Origin*, see D.L.Hull, *Darwin and his Critics*, Cambridge, Mass., 1973. See also P.J.Bowler, *Charles Darwin: the Man and his Influence*, Oxford, 1990; and A.Desmond and J.Moore, *Darwin*, London, 1991.

8. N.Smart, J.Clayton, P.Sherry and S.T.Katz (ed.), *Nineteenth Century Religious Thought in the West*, 3 vols., Cambridge, 1985, vol.2, ch. 7 and vol.3, ch.1. They ordered these things differently in Germany; see F.Gregory, *Nature Lost? Natural Science and the German Theological Traditions of the Nineteenth Century*, Cambridge, Mass., 1992.

9. The copy of Draper's book that I have used has Herbert Spencer's name stamped in it , but he has not annotated it; it is the London, 1875 edition. Tyndall's "Belfast Address" was reprinted in his *Fragments of Science*, 10th imp., London, 1899, vol. 2, pp. 135-201; and in part in T. Cosslett, *Science and Religion in the Nineteenth Century*, Cambridge, 1984.

10. J.Vernon Jensen, "Return to the Wilberforce-Huxley Debate", *BJHS*, 21 (1988) 861-79; J.H.Brooke, *Science and Religion: some Historical Perspectives*, Cambridge, 1991, esp. chapter 7.

11. On such terms, see my *Companion to the Physical Sciences*, London, 1989.

12. W.H.Brock (ed), *The Atomic Debates*, Leicester, 1967; A.J.Rocke, *Chemical Atomism in the Nineteenth Century*, Columbus, Ohio, 1984; and my *Ideas in Chemistry; a History of the Science*, London & New Brunswick, NJ, 1992, esp. chapters 9 and 11.

13. T.H.Huxley, *Lay Sermons, Addresses and Reviews*, 6th ed., London, 1877, p.77. Davy had said much the same; see my *Humphry Davy: a Study in Science and Power*, Oxford, 1992, chapter 4. F.W.H.Myers, *Human Personality and its Survival of Bodily Death*, 1919 ed. reprinted Norwich, 1992, p. 6, saw 1873 as the crest "of perhaps the highest wave of materialism which has ever swept over these shores", and psychical research as a reaction to it.

14 A.Gauld, *The Founders of Psychical Research*, London, 1968; R.Haynes, *The SPR: a History 1882 to 1982*, London, 1982; H.Oppenheim, *The Other World: Spiritualism and Psychical Research in England, 1850-1914*, Cambridge, 1985; C.D.Broad, *Religion, Philosophy, and Psychical Research*, London, 1953; M.Wheeler, *Death and the Future Life in Victorian Literature and Theology*, Cambridge, 1990; and my paper, "Observation, Experiment, Theory - and the Spirits", *Durham University Journal*, 83 (1991) 55-8. Crookes' first papers appeared in *Quarterly Journal of Science*, 7 (1870) 316-21, 8 (1871) 339-49, 471-93.

15. A.J.Balfour, *A Defence of Philosophic Doubt: being an Essay on the Foundations of Belief*, London, 1879. Quotations are from pp. 108, 86, 293. This was a presentation copy to H.H.Howarth.

16. A.R.Wallace, *Darwinism*, 2nd ed., London, 1890, ch.1.: *Balfour, Philosophic Doubt*, ch.13.

17. *D.N.B.*, "Balfour".

18. L.S.Jacyna, "Science and Social Order in the Thought of A.J.Balfour", *Isis*, 71 (1980) 11-34. H.W.Lucy, *Memories of Eight Parliaments*, London, 1908, has a chapter on Balfour.

19. *O.E.D.* gives the first use of "naturalism" in this sense from Wallace's translation of Hegel's Logic, 1874 : "Materialism or Naturalism, therefore, is the only consistent and thorough-going system of Empiricism". A.J.Balfour, *The Foundations of Belief: being Notes introductory to the Study of Theology*, 2nd ed., London, 1895; the quotation is from p.31. On the general background, see O.Chadwick, *The Secularization of the European Mind in the 19th century*, Cambridge, 1975; E.Jay, *Faith and Doubt in Victorian Britain*, London, 1986; P.Corsi, *Science and Religion: Baden Powell and the Anglican Debate, 1800-1860*, Cambridge, 1988; and for the special case of Faraday, G.Cantor, *Michael Faraday: Sandemanian and Scientist*, London, 1991.

20. [B.Stewart and P.G.Tait], *The Unseen Universe*, London, 1875.

21 *Papers read before the Synthetic Society, 1896-1908, and Written Comments thereon circulated among the Members of the Society*, London, 1909, "for private circulation"; see pp.20ff.

22. A. and E.M.S[idgwick], *Henry Sidgwick: a Memoir*, London, 1906, pp. 589f.

23. W.Shea (ed.), *Revolutions in Science: their Meaning and Relevance*, Canton, Mass., 1988.

24. I.Falconer, "Corpuscles, Electrons and Cathode Rays: J.J.Thomson and the Discovery of the Electron", *BJHS*, 20 (1987), 241-76.

25. A.J.Balfour, "Presidential Address: Reflections suggested by the New Theory of Matter", *Report of the British Association*, Cambridge Meeting 1904, pp.3-14. Quotations are from pp.8, 9 & 14.

26. G.G.Stokes, *Mathematical and Physical Papers*, vol. 2, Cambridge, 1883, p.97 : "A well-established theory is not a mere aid to the memory, but it professes to make us acquainted with the real processes of nature in producing observed phenomena".

27. I owe this idea to Richard Popkin. O.Chadwick, *The Victorian Church*, vol. 2, London, 1970, describes the problems faced by Balfour in framing his Education Act of 1902, which made funds available to Church schools, and the setting up of a Royal Commission on Ecclesiastical Discipline, 1903-6 (Balfour having been opposed to a bill to discipline high-churchmen in 1899); and remarks that Balfour, as a Scottish Presbyterian, in fact raised no

problems over Church Establishment, believed in autonomy for the Church of England, and publicly respected Convocation, unlike his predecessors (pp. 306, 324, 357, 341, 365). He also describes him as a Christian philosopher of the first rank, p.424. In F.L.Cross (ed.), *Oxford Dictionary of the Christian Church*, it is stated that Balfour was a Communicant in the Churches of bothScotland and England; entry under "Balfour".

28. Lord Rayleigh, *Lord Balfour in his relation to Science*, Cambridge, 1930, pp.30ff.

29. A.J.Balfour, *Theism and Thought: a Study in Familiar Beliefs*, London, 1923. The copy I have used was R.B.Braithwaite's review copy for *The Nation and the Athenaeum*, and he has noted the remarks on philosophy and science on p.52, and on the specious present on p.168. On Balfour's consistency, see eg. p.236 note; amd the quotation is from p.213f.

30. R.Olby, "William Bateson's introduction of Mendelism to England: a Reappraisal", *BJHS*, 20 (1987) 399-420.

31. From an MS note in the copy of *A.J.Balfour, Theism and Humanism*, 2nd imp., London, 1915, in the hand of E.M.Pollock (later Lord Hanworth) to whom the book previously belonged; on whom see *D.N.B.*

32 *D.N.B.*, "Balfour"; Lord Rayleigh, op. cit. note 28, p.19.

33. This comes from a newspaper cutting inserted in my copy of Rayleigh's biography of his uncle.

Index

INDEX

INDEX

INDEX

INDEX

349

INDEX

Owen, R. 152, 160n, 204n, 299-300, 307, 310
Oxford, University 3, 5, 147-9, 233, 285, 311, 317, 327

painting 16, 56
palaeontology 103, 110
Paley, W. 24, 48, 56, 82, 90, 101, 107, 114, 168, 292, 312
pantheism 87-90, 98n, 101n, 106-7, 145
Paracelsus 15
paradigm 2, 18, 177, 209, 268
paradigm shifts 16
Paris, J. A. 239n, 294n, 298n
parish registers 9
Parisian Institut 2
Parker R. D. 223-4n
Parkes, S. 140, 261-5
Parkinson, J. 110, 117n, 135n, 264n
Parry, W. E. 126-30, 134-5n
Partington, C. 242n, 244
Pasteur, L. 311
Peacock 110
Pearce, W. 24
Pearson 17
Peel, R. 285, 287
Pennant 128
perception 51
Petry, M. J. 15n, 17n, 92-5, 96n, 99n, 104n
pharmacy 48
Philalethes 27-32, 33, 39, 292
Philip, Gov. 124
philosophy 2, 4, 8, 10, 16-7, 36-7, 41-2, 52, 62, 65, 76-9, 88, 91, 96, 99n, 117n, 163-72, 203, 216, 234, 238-9, 255, 269, 288, 291, 298, 307, 311, 317, 321-30, 333-5
philosophy, dynamic 80, 91, 101, 104 *see also* science, dynamic; chemistry, dynamic
philosophy, of science 4-5, 10, 19, 70, 84-5, 88, 91-3, 107, 225
phlogiston 80, 111, 113, 251-2, 267, 271-3
photography 242-3, 249
phrenology 47, 177
physical science *passim* 78-101, 103, 156, 163-72, 203, 238, 298, 315
physics *passim* 7-8, 28, 34, 38, 51, 60-2, 66, 70, 76-8, 81-5, 92-4, 98n, 102, 104, 116, 121, 145n, 156, 161n, 164, 201, 211, 213, 232, 258-9, 261, 273, 287, 291, 298, 319, 326, 331-3
physiognomy 177
physiology 34, 53-4, 83, 89-90, 101-7, 142-3, 149, 166, 178, 228, 235, 303, 307

Piper, H. W. 57n, 88-9, 98n, 101
plants 21, 50, 67, 69, 94, 103, 110, 123-4, 129, 133n, 137, 140-5, 156, 169, 174-90, 200, 220, 222, 225, 228-31, 235, 267, 278
Plato 50, 57n, 62, 78, 81, 96n, 106, 262, 288
Platonism 81-2, 88, 97n, 101-2, 113, 234
Playfair, J. 33, 39n
plays 19
Plot, R. 140
Plotinus 49, 81, 97n
pneumatic chemistry 34, 141
Pneumatic Institute 54, 141
Pneumatic Institution 272
pneumatics 102
Podmore, F. 319-20
poetry 16, 19, 48-9, 52, 61, 64, 71n, 82, 88-9, 102, 105, 118n, 177, 226-7, 233n, 262, 277-81, 286, 288, 291, 304
Poggendorff, J. 254, 255n
Polanyi, M. 83, 97n
politics; political science 88, 104, 131
Pollock, M. 2, 20n, 178n
Pond 110
Poole, T. 27
Porphyry 81
Porter, R. 119n, 160n, 232n
portraits 239-43, 260, 285
positivism 3, 42, 59, 61, 70, 76
Powell, B. 308
Prichard, J. C. 145n, 205
Priestley, J. 18, 52, 62-4, 88, 90, 102, 112, 118n, 138, 164-5, 170-1n, 198, 223n, 238-9, 251, 262, 267-75, 296
printing 28, 154, 173, 211, 216
Proclus 81
progress 27, 37, 44
Prout, W. 35n, 87, 114, 264
psychiatry 226
psychology 3, 11, 90, 97n, 177, 226, 234-5, 320-3
Ptolemy 15
public lectures 10, 261, 264, 295
public records 9
Pugh, F. 20n, 238n
Pusey, E. B. 312

radioactivity 60-1, 62n
Raffles, S. 124, 152, 285
Rafinesque-Schmaltz 197, 210
Raine, K. 81, 96n
Rambottom, J. 210
Ramsay, W. 59n, 150, 254n, 258, 259n
rationality, of science 6, 94
Ray, J. 101, 181n, 203, 212, 227
Rayleigh, Lord 318, 322, 331, 335, 339n
Reason; Rules of Reason 41-6, 88, 102

INDEX

INDEX

T - #0010 - 160425 - C0 - 234/156/20 [22] - CB - 9781138644441 - Gloss Lamination